Environmental Public Interest Litigation in China

Xi Wang • Xiaobo Zhao • Noeleen McNamara
Editors

Environmental Public Interest Litigation in China

Springer

Editors
Xi Wang
Law School of Kunming University of Science and Technology
Kunming, China

Xiaobo Zhao
School of Law & Justice
University of Southern Queensland
Ipswich, QLD, Australia

Noeleen McNamara
School of Law and Justice
University of Southern Queensland
Ipswich, QLD, Australia

ISBN 978-3-031-26525-9 ISBN 978-3-031-26526-6 (eBook)
https://doi.org/10.1007/978-3-031-26526-6

This Springer imprint is published by the registered company Springer Nature Switzerland AG
The registered company address is: Gewerbestrasse 11, 6330 Cham, Switzerland

Acknowledgements

The idea to produce this book was first conceived in August 2021 as a result of fruitful discussions between the editorial members. This book would not have been possible without a dedicated group of contributors. The editors are grateful to all the contributors for their contributions. They have overcome various inconveniences caused by the COVID-19 pandemic and submitted such interesting and thought-provoking chapters just in time.

Our special thanks go to Dr. Brigitte Reschke and Laura Hofmann from Springer Publishing for their invaluable assistance in bringing this book to life. We would like to thank Pradeep Kuttysankaran for editorial assistance with compiling the manuscript and ensuring that the formatting is consistent and complete. Our thanks also go to various reviewers for their contributions during the review process.

Dr Xiaobo Zhao would like to thank Professor Tianbao Qin and Professor Qun Du for their advice and support. The support of Professor Xiaoqin Zhu is also gratefully acknowledged. Dr. Xiaobo Zhao would also like to thank Professor Reid Mortensen and Dr. Lingling He for their endless support.

Kunming, China — Xi Wang
Ipswich, QLD, Australia — Xiaobo Zhao
Ipswich, QLD, Australia — Noeleen McNamara

Contents

Notes on Contributors

Xiaobo Bo is an associate professor in environmental law at the Law School of Jiangnan University. She received her PhD in law degree from China University of Political Science and Law in Beijing, China. She is teaching a range of courses in relation to environmental law, civil law, and the law of torts and has published several papers, hosted and participated in a number of scientific research projects in these areas.

Juan Chu is a lecturer in environmental law at the School of Economic Law, East China University of Political Science and Law. Dr Chu obtained her PhD in Law from the University of Washington School of Law. She is interested in studying environmental regulation and environmental litigation in a comparative setting.

Qi Gao is an associate professor in environmental law at KoGuan Law School, Shanghai Jiao Tong University, China. She has been a council member of the Chinese Society of Environmental and Resources Law since 2017. She teaches courses on Chinese Environmental Law, International Environmental Law, Economic Analysis of Environmental Law, and Environmental Law Case Studies. She received a BA in Law from Wuhan University in 2007, Master of Laws from the Research Institute of Environmental Law (RIEL), Wuhan University in 2009 (ML), and a PhD in 2013 from the University of Western Sydney. Her research focuses on judicial access to environmental justice, public participation and access to information, transboundary water management, and market-based environmental policy instruments.

Gu Gong is an associate professor with tenure in environmental law in the Law School of Peking University. He was the Deputy Director of the Institute of Economic Law, Executive Director of the Environmental Resources and Energy Law Research Centre of the Guanghua Law School of Zhejiang University. He is the 'Zhongying Scholar' of Zhejiang University, and the deputy of Secretary-General of the Environmental Resources Law Research Association, Chinese Law Society; he is also the Deputy Chair of the Environmental Law Branch of the Chinese Society of Environmental Sciences. His research interests include environmental resource law, ecological civilization theory, and public interest litigation.

Jinye Han is a Master candidate, Shanghai University of Political Science and Law, Shanghai, China.

Zhiyu Huang has been an associate professor of Environmental and Natural Resource Law at the Faculty of Law at Nanchang University in Jiangxi Province of China since 14 December 2018. She defended her PhD on the Legal issues of Ecosystem-based Disaster management in 2016. Since 2017, she has gained practical experience as an environmental lawyer at Zhong Yin Law Firm. Dr. Huang focuses on different topics of environmental law, with an emphasis on legal aspects of ecological protection and restoration.

Quan Ke is a PhD Candidate of School of Law, South China University of Technology.

Wenjun Luo is an associate professor in environmental law at the Law School, Hubei University of Economics, China. She received her LLM degree from Wuhan University in 2004 and a PhD in law (Environmental and Resource Law) from the Shanghai Jiaotong University. She is a visiting scholar at Pace University, USA and the Southern Institute of Technology, New Zealand.

Yanni Luo is a lecturer in environmental law at the Law School of Beijing University of Chemical Technology. Dr Luo received her PhD degree from the University of Frankfurt (Germany) in 2021. Her research focuses on the mechanisms of biosecurity risk decision-making and climate change litigation.

Mengxing Lv holds the position of Lecturer of Environmental Law at Civil, Commercial and Economic Law School, China University of Political Science and Law in Beijing, China, since January 2018. She defended her PhD on corporate environmental responsibility in 2017 at Maastricht University in the Netherlands. Dr. Lv focuses on different topics of environmental law, with an emphasis on legal aspects of environmental regulation and governance. She presented her research at various national and international conferences and published several journal articles and books. She is also a visiting scholar at Asian Law Institute at National University of Singapore, the Polish Research Centre for Law and Economy of China at University of Warsaw, and Faculty of Law at St. Petersburg University.

Yun Ma holds the position of associate professor of administrative law and vice director of the Administrative Law Department at the Law School of China University of Political Science and Law since August 2015. She defended her PhD at Erasmus School of Law & Erasmus China Law Centre at Erasmus University Rotterdam of the Netherlands in 2015. She was a visiting scholar at Oxford

University, Indiana University Bloomington, Utah University, Zagreb University, etc. She focuses on different topics of both administrative law and environmental law, with an emphasis on protected areas law, EIA, and environmental regulation.

Noeleen McNamara is a professor in law at the School of Law and Justice, University of Southern Queensland, Australia. Professor Noeleen McNamara has taught and written in the area of environmental law, especially mining and natural resources law, for over 20 years. Her new book *Law and Ethics for Australian Teachers* is published by Cambridge University Press in 2021.

Wenzhen Shi is a PhD candidate, Law School of China University of Political Science and Law, Beijing, China.

Huanhuan Wang is an associate professor in environmental law at School of Law of East China Normal University, China. Her research interests lie in land contamination law, energy law, and comparative environmental law. She received her PhD in law degree from Wuhan University in 2010.

Xi Wang is a professor of law at the Law School of Kunming University of Science and Technology (KUST). He is the Director of the Platform for Research on Ecological Civilization and Environmental Rule of Law (PRECERL) of the KUST. He worked and served as a professor and the Director of the Environmental and Resources Law Institute of the Law School, Shanghai Jiao Tong University, from 2002 to 2019 and served as a professor and a Vice Director of the Environmental Law Institute, Wuhan University, from 1987 to 2002.

Kaijie Wu is an assistant professor of Law at the Law School of Peking University, China. Dr. Wu has held the position of postdoctoral researcher at the Peking University Law School since July 2019. He earned his PhD in environmental law from the Renmin University of China in 2019. His PhD dissertation focused on the harmonization and codification of Chinese environmental law. He is also a Michigan Grotius Fellow and an S.J.D. candidate at the University of Michigan Law School.

Shuai Xin is a lecturer in environmental law at the Law School of Qingdao University of Science and Technology. He received his Master of Laws and PhD in law degree from the Ocean University of China. His research interests include environmental and resource protection law and marine environmental protection law.

Jianwei Zhang is a professor in environmental law at the Law School of Beijing University of Chemical Technology. He received his PhD in law degree from Wuhan University in 2007. He has been a visiting scholar at Sydney University and Ghent University. Professor Zhang focuses on different topics of environmental law, with an emphasis on legal aspects of climate change and governmental environmental responsibilities.

Shijun Zhang has been a professor in environmental law at the Law School of Shandong University of China since July 2005. He obtained his PhD Degree in Environmental and Resources Law at the Law School of Wuhan University in 2005. He worked as a postdoctoral researcher at the Environmental School of Nanjing University from July 2007 to July 2009. From September 2009 to September 2010, sponsored by CSC, he worked as a visiting scholar at Law School of Maryland University of America. Shijun Zhang focuses on different topics of environmental law, especially on environmental public interest litigation, environmental law enforcement, marine environmental law, and climate change. He published his monograph '*Study on the Standing in Environmental Public Interest Litigation*' in 2011.

Zhenglian Zhang is a PhD candidate of the School of Law of East China Normal University, Shanghai, China.

Jun Zhao is an associate professor in environmental law at the Shanghai University of Political Science and Law. She is the director of the Environmental Law Teaching and Research Section of the Law School. She is also the Deputy Director of the Environmental Law Research Centre. Dr Zhao received her LLB and Master of Laws degree from the Northwest University of Political Science and Law and her PhD degree from Wuhan University. She has been engaged in teaching and research on environmental law at universities since 1996. She is currently a director of the Environmental Resources Law Branch of the Chinese Law Society, a member of the Shanghai Law Society.

Xiaobo Zhao is a senior lecturer in law at the School of Law and Justice, the University of Southern Queensland (UniSQ). He received Master of Laws degree from the Research Institute of Environmental Law (RIEL), Wuhan University in 2008 and PhD degree in law from Western Sydney University (WSU) in 2012. Prior to joining UniSQ, he worked as a lecturer in law at the Shanghai University of Finance and Economics. He has been teaching across a range of law subjects including environmental law, Australian Tort Law, Water Resource Law, and Research Projects on sustainable development. He is specialized in contaminated land law, comparative environmental law, and torts.

List of Figures

List of Tables

Part I
EPIL in China: Historical Development and Theoretical Foundations

Chapter 1
Environmental Public Interest Litigation in China: An Overview

Xiaobo Zhao and Noeleen McNamara

Abstract The rapid development of EPIL and its profound impacts on China's environmental governance are unprecedented. Even though EPIL is still in its early stage of development in China, both the legal frameworks and mechanisms continue to evolve. Many important issues in relation to EPIL in China, such as the definition, purpose, permissible scope, cause of action and forms of remedies, remain under development. The current absence of a text that addresses these issues presents a major obstacle to academics and practitioners. It is therefore an important task of this book to meet the urgent need to develop a reliable resource and present readers with the most current research and authoritative information in the field.

Keywords Environmental public interest litigation (EPIL) · China · Environmental governance

1.1 EPIL in the Chinese Context

Environmental public interest litigation (EPIL) is a rather recent development in environmental governance at a national level in China. In the Chinese legal context, EPIL can be defined broadly to include a judicial process, including court-based mediation and adjudication, in which the qualified plaintiffs, including individual persons, legal persons or social groups can bring a lawsuit against other individual person(s), legal person(s) or social group(s), whose conduct has harmed or has the potential to harm the public interests in relation to the environment, due to the nonfeasance or misfeasance of the defendants. The main purpose of EPIL is to prevent, mitigate, remediate or compensate for harm to the environment. As a novel legal proceeding in China, EPIL enables a broad range of stakeholders, like the

X. Zhao (✉) · N. McNamara
School of Law and Justice, University of Southern Queensland, Ipswich, QLD, Australia
e-mail: bob.zhao@usq.edu.au; Noeleen.McNamara@usq.edu.au

X. Wang et al. (eds.), *Environmental Public Interest Litigation in China*,
https://doi.org/10.1007/978-3-031-26526-6_1

judges, procurators,[1] lawyers and eligible environmental groups to participate in the judicial process.

The primary impetus behind the emergence of EPIL in China has been the growing environmental awareness of its citizens and the Chinese Communist Party's (CCP) determination to pursue Ecological Civilization—the core of which is to create a society that is more in harmony with the natural environment. As a result of rapid industrialisation and urbanisation, China was experiencing numerous serious environmental problems, including rampant air pollution, soil contamination, and unsafe drinking water at the beginning of this century, prompting widespread public concern over environmental pollution and ecological deterioration. The Chinese government has therefore made it a political priority since the mid-2000s to explore EPIL as one of the viable alternatives for handling explosive legal disputes resulting from massive environmental incidents.

EPIL formally began in China around 2012, as a result of the amendments to the *Civil Procedure Law* and the *Environmental Protection Law*. The *Civil Procedure Law*, which was amended in 2012, for the first time allows an eligible authority or relevant organisation to institute a civil action against conduct that results in environmental pollution. This means that the law does not require a litigant to suffer 'direct harm' to justify its standing to sue in public interest litigation. This provision, as a general rule, sets the legal basis for the civil EPIL in China. Subsequently, China's basic environmental law—the *Environmental Protection Law* (EPL) confirmed that environmental pollution and ecological damage can be causes of action for civil public interest litigation, and that government agencies and NGOs have the standing to bring lawsuits against environmental polluting activities that harm the public interest.

From 2015 onwards, the Supreme People's Court (SPC) of China has issued a series of judicial interpretations to clarify relevant procedural and substantive issues in relation to EPIL. Thus far, three forms of EPIL have been recognised under the Chinese legal system: (1) EPIL lawsuits filed by social organisations; (2) EPIL lawsuits initiated by procuratorial authorities; and (3) ecological and environmental damage compensation lawsuits filed by provincial and municipal governments, including their designated departments and institutions provided by the law. While the first form of EPIL is dominated by environmental NGOs, the other two forms are dominated by public sectors. It is also worth noting that since 2014, China has set up several hundred specialised environmental tribunals within the SPC, the provincial

[1] According to the Law on the Organization of People's Procuratorates, procuratorates in China are the State's organs for legal supervision that exercise the power of prosecution. They are elected by and report to the People's Congress at the same level. The main tasks of the procuratorates include initiating public prosecution and supporting public prosecution for criminal cases; supervising the legality of trials conducted by courts, as well as supervising civil and administrative trials of courts. The Supreme People's Procuratorate (SPP) is the highest prosecutorial power in the People's Republic of China. It conducts reviews of rulings and investigations performed by local and special procuratorates. It also issues judicial interpretations, which are official and legally binding interpretations of the law.

high courts, and the local intermediate level people's courts to handle environmental cases, and one important mission of the environmental courts is to handle EPILs.

Along with environmental legislative developments and institutional construction, the emerging role of EPIL in China's environmental governance is more evident than ever. The number of lawsuits seeking damages and remediation for air, water, and land pollution through EPIL has skyrocketed over the past 5 years. In 2019, courts across the country accepted a total of 179 EPIL lawsuits filed by social organisations, which was about 1.75 times the number in the previous year. Procuratorial authorities initiated 2309 EPIL lawsuits in 2019, among which 312 were civil EPILs, 1642 were EPIL lawsuits collateral to criminal proceedings and 355 were administrative EPIL lawsuits.[2] In addition, 49 ecological environmental damage claims were recorded in 2019, which meant a notable increase rate of 145% on a year-on-year basis, and these numbers continue to grow in 2020.[3]

Another notable development in this area is that the law recognises the standing of prosecutors in EPIL. In a pilot program started in July 2015, the Chinese procuratorial authorities in 13 provincial areas were authorised to initiate public interest litigation, including EPIL. Since then, an increasing number of environmental public interest lawsuits have been filed by prosecutors. In 2018, Chinese prosecutors have handled 118,012 EPIL lawsuits, accounting for 54.96% of the total litigations initiated by procuratorial authorities. During the first 10 months of 2020, approximately 68,000 EPIL lawsuits were handled by procuratorial authorities, among which 4214 EPIL cases were directed to the people's courts. The impacts were enormous—over 13,000 km^2 of damaged lands, forests and water sources were restored, with over 10,000 polluted facilities ordered to be rectified or closed.[4]

Today, it has been widely acknowledged that the introduction of EPIL is a milestone of China's judicial reform and an important step towards law-based environmental governance in China. By far, three forms of EPILs—civil EPIL, administrative EPIL and procuratorial EPIL, together with the litigation for ecological and environmental damage compensation, and environmental torts litigation, comprise the environmental litigation system in China. Some critical issues, such as the future direction of EPIL and the legislative order of the EPIL system, have been broadly discussed by leading Chinese environmental scholars.[5] EPIL has, due to its

[2] As we will discuss in the subsequent chapters, by far, three are three forms of environmental public interest litigations in China, namely civil EPIL, administrative EPIL and procuratorial EPIL. There is no clear criterion to differentiate these forms of public interest litigation. Civil EPIL and administrative EPIL are differentiated by the objects of a litigation. Generally, civil EPIL is against a company or individual and administrative EPIL is against a governmental agency. Individuals, eligible social organisations and the procuratorate can initiate civil EPILs. However, only the procuratorate can file administrative EPILs. See Article 21, *Interpretation of the Supreme People's Court and the Supreme People's Procuratorate on several Issues Concerning the Application of Law to Procuratorial Public Interest Litigation Cases* (2020) (Fa Shi (2018) No. 6).

[3] SPC (2020, 2021).

[4] Ibid.

[5] See, e.g., Wang (2016a), pp. 49–68; Wang (2016b), pp. 101–114.

special importance, been the focal point of many environmental law colloquiums, research symposiums and discourses among environmental law researchers, judicial organs, and environmental regulators in recent years.

1.2 Purposes in Writing This Book

Our purposes in writing this book are as follows.

1. The primary purpose of writing this is to provide an accessible and broad-ranging guide to the major themes and research strategies about EPIL that have burgeoned in China over the past decade. The rapid development of EPIL and its profound impacts on China's environmental governance are unprecedented. Even though EPIL is still at its early stage of development in China, both the legal frameworks and mechanisms continue to evolve. Many important issues in relation to EPIL in China, such as the definition, purpose, permissible scope, cause of action and the forms of remedies, remain under development. The current absence of a text that addresses these issues presents a major obstacle to academics and practitioners. It is therefore an important task of this book to meet the urgent need to develop a reliable resource and present readers with the most current research and authoritative information in the field.
2. A second purpose of this book is to identify the changing roles of major stakeholders in EPIL and the interactions between them. Courts are at the forefront of handling EPIL lawsuits—China is no exception. Early EPIL cases witnessed the innovative practice of local courts with respect to the tests for standing to sue, for differentiating public and private interests, onus of proof, remedies and enforcement, the management of fines and fees for ecological reparation and the like. As EPIL lawsuits skyrocketed in recent years, the magnitude of the pressure on the courts becomes obvious. The SPC, local courts and special environmental courts and tribunals are confronted with a whole range of challenges that will affect the coherence between different forms of EPIL. Thus, in this book, special attention has been given to the changing role of courts, especially that of the environmental courts, in EPIL. This will help the judiciary identify the factors that may impact their role in EPIL proceedings and orient and design appropriate reforms in terms of institutional organisation, court efficiency, judiciary effectiveness and procedural transparency.

   Another outstanding development in this area of law is the introduction of Chinese procuratorial authorities in EPIL. Procuratorial public interest litigation has been broadly used to protect clean air, water, forest, wetlands and wildlife, and has recently been expanded to protect cultural relics and cultural heritage, which are subjects of environment conservation in a broad sense. While the investigative power of the prosecutors was thought to be one great advantage in procuratorial EPILs, there is still a lack of adequate regulations on issues, such as definitions of administrative omissions and preliminary evidence, limitations of evidence, trial procedures of administrative public interest litigation and the status of procuratorial authorities' lawsuits.[6] These issues ultimately affect the 'sustainability' of the procuratorial authorities' role in EPILs.

The adoption of EPIL has significantly changed the balance of power between governments, courts and procuratorial authorities. Traditionally, Chinese governments at all levels, mainly through their environmental regulators, play a significant role in environmental law enforcement, ranging from imposing administrative fines to monitoring the implementation of administrative orders to ensure compliance. The involvement of procuratorial authorities in EPIL complicated the traditional role of government in resolving environmental problems. Governments are now playing multiple roles in the context of EPIL—they can be a legally mandated plaintiff to bring civil public interest litigation against environmental polluters; supporters of qualified public interest litigation plaintiffs; defendants of administrative public interest litigation brought by procuratorial authorities; and plaintiffs in ecological compensation litigation in the name of state-owned natural resources owners.[7] Some of these roles are contradictory, and courts are taking the lead in initiating and supervising collaboration among various stakeholders when delivering judgements in EPIL. It is therefore an important task of this book to analyse the interactions between public sectors in the context of EPIL.

Finally, although EPIL in China has received sound development in the past decade, it is still rather new to courts, procuratorates and administrative departments. It has been observed that in many cases, Chinese procuratorial authorities are not ready to file lawsuits against peer government agencies, and the same concern applies to local courts that decide cases. Moreover, although environmental NGOs are playing an increasingly important role in EPIL in China, the political pendulum remains swung in favour of public litigants, especially procuratorial authorities. A primary aim of this book is to shed light on these phenomena.

3. A third purpose of the book is to systematically analyse the changes in theory and in legal practice that are influencing the growth and expansion of EPIL in China. To this end, we have identified various issues, at the level of both theory and practice, that have affected and are continually affecting the growth and expansion of EPIL in China. The analysis of these issues, through a historic and comparative perspective, reveals the strengths and weaknesses of the current legal regime and serves as a basis for recommendations focused on bringing about more effective EPIL in China. This book expands the literature in this area as several of these issues have not been the subject of detailed research.
4. Fourth, this book is written to remedy the gap between Chinese and English literature in this area of law. We noticed that by March 2021, more than 2800 journal articles with the keyword 'EPIL' had been indexed by the *China National*

---

[6] See, e.g., Huang (2018) and Li and Wu (2016).

[7] Ma (2019), pp. 269–291.

Knowledge Infrastructure (CNKI), the largest and most-used legal database in China. Moreover, EPIL has been the research subject of 11 doctoral theses and 716 master's dissertations over the past decade. Other forms of publications, including 214 conference papers, 439 newspaper articles and 42 yearbook publications are found relevant to EPIL. Surprisingly, English literature on the same theme is extremely limited. At the time this book proposal was made, no more than 30 journal articles/book chapters in leading English legal databases, such as the HeinOnline's Law Journal Library and Westlaw, were directly related to EPIL in China. We consider it is time for a book that can fill this gap.

1.3 The Major Themes and Research Methods of the Book

For the above purposes, we identified various issues, at the level of both theory and practice, that have affected and are continually affecting the growth and expansion of EPIL in China. The analysis of these issues, through a historic and comparative perspective, reveals the strengths and weaknesses of the current legal regime and serves as a basis for recommendations to bring about more effective EPIL in China. The book also looks closely at some of the topics that were not dealt with in the existing literature. The major themes and research questions of the book are as follows.

The Definition and Contested Nature of EPIL Although the expression 'the public interest' is widely used in Chinese legislation, no clear definition of what it means exists in the legislation. This makes the concepts of 'public interest' and 'environmental public interest' the subject of considerable debate. What emerged as an important issue is the norms to differentiate the public interest and private interest when dealing with EPIL. EPIL can be, as suggested by the SPC, distinguished from the environmental private litigation in many aspects. Factors have been identified include: (1) the purpose of the litigation; (2) the standing to sue; (3) the status of the litigants and the defendants; (4) the form of remedies; and (5) the role of judges. These criteria are quite general and pose considerable challenges for courts when assessing whether there is 'public interest' in a particular case. Therefore, a thorough understanding of the legislative history and the statutory bases, including laws, regulations and administrative orders that may be made in EPIL, is essential for resolving this contested issue.

Obstacles and Challenges for Developing EPIL in China It is an important objective of this book to identify the factors that are restraining the development of EPIL. While EPIL is strongly supported by scholars and NGOs, EPIL was not legalised in China until 2012, which is approximately 10 years after the notion of EPIL first appeared in academic discussion in China. Many researchers have attributed the slow development of EPIL to the lack of appropriate environmental legislation and sufficiently broad standing provisions.[8] These findings, although helpful for readers to understand the emergence and development of EPIL in China, have generally failed to understand the fundamental issues. The main constraints we currently observe are the weak capacity building of major participants, the enforcement of law and a general lack of financial resources. Keeping this in mind, this book seeks to situate the debate in a richer theoretical context and focuses on fundamental issues: Why should EPIL receive judicial recognition in China? Why does EPIL warrant special treatment as to access? What is the value to Chinese jurisprudence and society for the judicial recognition and nurturing of EPIL? Reflective assessments of these issues comprise the main body of our discussion.

Standing in EPIL At the core of the argument surrounding the establishment of EPIL is the question about who is entitled to commence proceedings in EPIL, or, who has the standing to sue in EPIL. In the traditional civil actions, only parties whose rights or interests are damaged are qualified to bring lawsuits against the defendants. In EPILs, the law removed such restrictions and broadly recognised the standing of NGOs and procuratorial authorities. However, various restrictions remain on the scope of eligible litigants. To have standing to sue in EPILs, NGOs should be registered with a prescribed civil affairs authority, and they must have been engaged in environmental protection activities in the public interest for at least 5 consecutive years without a record of violating laws. In addition, NGOs are not permitted to file an EPIL against public authorities, or to file administrative EPIL lawsuits. Due to such restrictions, the vast majority of environmental NGOs, including all foreign NGOs, are excluded from the 'eligible organisations' that are able to file an EPIL. Key issues to be argued in relation to the law of standing include the following: Why should the standing of NGOs in EPILs be restricted? Whether the current restrictions on standing should be removed in favour of open standing? Should a simple standing test be introduced? Whether new rules aimed at reducing cost, delay and abuse of process should be introduced to enhance institutional capacity and legal frameworks, and ultimately to facilitate the implementation of EPIL in China?

Compared to NGOs, including those sponsored by the government, procuratorates have obvious advantages in initiating EPILs. These include unrestricted legal standing, which is achieved through amendments to the *People's Procuratorates Organic Law* in 2018 and *Public Procurators Law* in 2019, and possession of overwhelming authority to conduct investigations and collect evidence. Defendants, especially government agencies, are more likely to be responsive to procuratorates' pre-litigation demands and threats of litigation because of procuratorates' unique constitutional mandate to supervise the overall legal compliance in China. Nevertheless, procuratorates' standing in EPILs seems to contradict their traditional role of legal supervision. Other debates concerning procuratorial EPILs in terms of burden of proof, court fees and ability to assess environmental

[8] Xie and Xu (2021), p. 449.

professional issues are missing under the current legal system.[9] Moreover, some scholars also argue that standing should be expanded to environmental administrative departments due to the *de facto* standing being recognised in early EPIL cases.[10] In such a context, substantive arguments to justify procuratorates' standing in EPILs are particularly essential. This question has been intensively addressed by contributors in Chaps. 5 and 8.

Remedies and Enforcement Under the current legal framework, remedies available in EPILs include cessation of infringement, removal of obstruction, elimination of danger, restoration to the pristine state, compensation for losses, and formal apologies. These civil remedies were provided by the former *General Principles of the Civil Law* and the *Tort Liability Law* and then incorporated into China's *Civil Code* which came into force on January 1, 2021. In environmental litigation, it has been commonly accepted that civil remedies can be used to redress personal injury and property loss that has arisen from environmental pollution or ecological damage.[11] However, for most EPIL cases, there is no obvious infringement on the personal or property interest of a particular person. This gives rise to an important question about whether traditional civil remedies can be expanded to compensate for 'pure ecology damage'—that is, damage to the natural environment or eco-system itself.[12] Apart from the theoretical challenges, courts are also confronted with technical challenges in terms of investigation, baseline determination, causation and environmental value assessment when deciding ecological damages. Key issues to be argued include: How could the interim loss and resulting restoration funds be calculated? How should the restoration funds be disbursed? How should clean-up and restoration be monitored and by whom? Should punitive damages be awarded in EPILs? Some of these issues have been addressed by Chinese environmental authorities through technical guidelines, but more work needs to be done in this area.

Meanwhile, we also observed that innovative and flexible use of market instruments in judicial enforcement is one notable feature of EPIL in China. The SPC has incorporated certain innovative remedies that have been awarded by lower courts. It provides that a monetary alternative, including restoration costs, costs for restoration planning and compliance monitoring, can be awarded if the complete restoration of the environment to its original status is not feasible. Reflecting on these innovations and exploring a range of options will help improve the effectiveness and efficiency of judicial enforcement in EPIL.

The Case Guidance System Due to the novel nature of EPIL, some typical EPIL cases were released as guiding cases under the case guidance system. The case guidance system is a ground-breaking system established by the SPC (*Notice of the Supreme People's Court on Issuing the Provisions on Case Guidance* No. 51

---

[9] See e.g., Lv (2016), p. 250.

[10] See, e.g., Ma (2019), p. 269; Qin (2019), p. 91; Liu (2017), p. 3, Liu (2021), p. 33.

[11] See, e.g., Gong (2019), pp. 127–147.

[12] Lv (2017).

[2010]) in November 2010 and was eventually confirmed by Article 18 of the *Organic Law of the People's Courts of the People's Republic of China* (2018 Revision). Under this system, Chinese court judgments are selected and circulated as *de facto* binding Guiding Cases. Courts are expected to follow the decisions when dealing with similar cases in subsequent hearings to ensure the uniform application of law. The SPC and the Supreme People's Procuratorate (SPP) have released guiding cases in relation to EPIL to help lower courts and procuratorial authorities deal with comparable cases in court trials. These cases have been broadly discussed by our contributors, which will help us understand the application of law and the rationale behind court decisions.

Expert Witnesses and Involvement of Professionals Due to the extreme complexity of assessing damages in environmental pollution cases, courts often refuse to hear cases to avoid the difficulties and risks of having to carry out the assessment themselves. The involvement of experts in EPIL has therefore been encouraged by the SPC. Experts can be appointed, depending on their expertise, to review the evidence, undertake scientific investigation, and/or assess environmental damage. The court can assign a third party to perform clean-up and restoration activities, the cost of which would be covered by the monetary compensation awarded by the courts. Experience shows that expert witnesses and the involvement of professionals are important elements for a successful EPIL. It is therefore worthy to investigate the procedural and substantive laws to secure the appropriate running of these mechanisms.

In terms of research methodology, case studies and comparative studies are the two major research methods we have undertaken in developing this book. From the outset of the program, we attached special importance to case studies. In each of these case studies, various viewpoints on the subject matter have been explored. Detailed analyses of key cases in relation to the above themes were made in the form of "case studies". Another important approach that has been used is comparative study. Public interest litigation has been utilised in many jurisdictions as a vehicle for social, political and legal change, and environmental pollution has been an important cause of action. The introduction of EPIL in China was inspired by the experience of the US, India, Germany, France and Australia, which allow class actions in public environmental lawsuits and grant standing to various groups (or even open standing). Thus, a comparative legal approach is beneficial to allow for the comparison of different solutions dealing with EPIL from a range of legal systems.

1.4 Structure of the Book

In organising the contents of this book, the contributor of each chapter shares insights on a subject within their specific sphere of expertise that is closely aligned with the themes stated above. Apart from this Introduction, this book consists of

3 parts and 16 chapters, each covering a salient topic. These chapters are briefly explored below.

Part I **EPIL in China: Historical Development and Theoretical Foundations**

Part I is an introductory section that describes the evolution of EPIL as a part of Chinese law, discussing its nature, purpose, function, scope and sources of the law. This introduction is then followed by detailed sections addressing EPIL under the civil, environmental and administrative legislation, as well as various legal mechanisms under regulations and judicial interpretations.

Chapter 1 is an overview of the whole book. It covers the background information for developing this book, the objectives of the book and research methodologies that have been used by the authors.

In Chap. 2, 'Historical Development of EPIL in China', Professor Shijun Zhang examines the social and political background under which EPIL, as a new form of judicial instrument for environmental governance, emerged in China. An initial step is to review the social and environmental context under which EPIL is developed, to evaluate the integrity of the existing legal frameworks, and to discuss the theoretical frameworks that are necessary to analyse the purposes and functions of EPIL. Zhang divides the development of EPIL into three major phases. At the early stage, relevant environmental judicial cases and local practices before 2007 provided abundant practical experience for the establishment of the EPIL system. In the second phase, which is between 2007 and 2014, environmental courts were created and developed across the country, and some local courts started to promulgate normative documents to guide EPIL at the provincial and municipal levels. The third phase, from 2015 to the present, witnessed the formal establishment and quick development of EPIL at the central level. A significant development of the latter phase was the introduction of EPIL initiated by the procuratorial authorities. In this chapter, Zhang also considers the unique role and far-reaching implications of EPIL in China's progress towards the rule of law in environmental governance and the establishment of environmental civilisation.

In Chap. 3, 'Defining Public Interest Under the Environmental Public Interest Litigation System', Professor Shijun Zhang clarifies the nature of EPIL by examining the way that the 'public interest' and 'environmental public interest' are defined in the Chinese social and legal contexts. He notes that traditionally, legal interests can be divided into individual interests, group interests and social interests. According to a number of modern legal philosophers, the term "social interest" generally means "social public interest" in China, because the state sectors are entrusted to protect it. Zhang reviews the traditional way to classify legal interests from a historical viewpoint. He argues that EPIL is essentially a reflexive benefit that not only brings benefits to some individuals, but also safeguards and promotes the public interest. He also argues that in situations where the public and private interests overlap, a focus on the public interest is recommended for environmental protection. EPIL involves multiple, as well as competing, interests. A clarification on the nature of the interests protected by EPIL directly relates to the court's legal value judgments. It also relates to the unique function of the EPIL system in terms of public interest protection. EPIL can not only remedy environmental damage as the traditional litigation systems can do, but also can effectively prevent environmental damage. Dr Kaijie Wu shares the same point of view, and the preventive function of EPIL has been further elaborated by Wu in Chap. 4.

In Chap. 4, 'From Compensation to Prevention: Expanding the Function of EPIL in China', Dr. Kaijie Wu focuses on two types of EPIL in China: compensatory EPIL and preventive EPIL. While the compensatory EPIL aims to provide civil remedies, mainly in the form of monetary compensation for environmental pollution and ecosystem damage, preventive EPIL aims to prevent significant environmental risks before environmentally damaging activities take place. As observed by Wu, social organizations have filed a few preventive civil litigation actions, but they have encountered difficulties. The existing preventive civil EPIL cases are essentially tort litigation and face limitations in terms of scope, adjudication basis and litigation effect in preventing environmental damage. He argues that preventive EPIL in China should be enforcement litigation, rather than tort litigation. This is because the purpose of the preventive EPIL is to achieve a preventive function by supervising and assisting administrative authorities in implementing preventive regulations. In order to strengthen its preventive function, the EPIL system should expand the scope of cases in which such action could be brought, utilize independent pre-litigation procedures, evaluate the performance of duties with behavioral standards, and strengthen the support of environmental professional capacity. These improvements could, as recommended by Wu, be realized by revising judicial interpretations in the short term and formulating special legislation in the longer term.

In Chap. 5, 'The Standing to Sue of NGOs in Environmental Public Interest Litigation in China: A Doctrinal Analysis of Laws and Cases', Dr Qi Gao, Quan Ke and Dr Xiaobo Zhao examine one of the most important issues in EPIL—the standing to sue. Gao *et al.* trace the evolution of the law on legal standing and the resultant developments of EPIL in this chapter. Since 2012, China has expanded standing to sue in EPIL to administrative authorities, procuratorates, and non-governmental organizations (NGOs). Significantly, the law has put restrictions on allowing NGOs' standing in EPIL actions. As a result of such restrictions, only a small number of government-sponsored social organisations have filed EPIL lawsuits. Gao observes that the status quo of practice does not match expectations—NGOs are increasingly marginalized in this court-centred environmental movement. In an attempt to reveal the reasons behind this phenomenon, Gao *et al.* explores how the judiciary interprets relevant legislation in normative documents and individual cases. Representative cases are selected to analyse Article 58 of the *Environmental Protection Law* from the aspects of activity performance, activity area and activity period. Gao states that the movement to grant administrative authorities and procuratorates standing in EPILs has significantly restrained NGO involvement in EPILs in China. Such a move departs from the aim of the 'objective legality' model. She further argues that NGOs' willingness and ability to file an EPIL could be improved by removing restrictions on their standing and by enhancing their technical, financial, human resource and managerial capacity to handle EPIL cases.

In Chap. 6, 'Environmental Public Interest Litigation and the Application of Civil Code', Professor Gu Gong examines the role of China's Civil Code (enacted in

2020) and its impacts on EPIL. The Civil Code added two ecological damage liability provisions that aim at "restoration" and "compensation". These clauses together constitute a complete ecological environment damage compensation system and provide a clear substantive law basis and specific rule guidance for EPIL in China. However, as Gong points out, these provisions are still in principle and general and give there are still uncertainties in terms of legal attribute, the application of punitive damages, the definition of the concept of ecological environmental damage, the proof of causality and other specific aspects alike. Gong suggests that special legislation should be made to address these issues.

Part II **Substantive and Procedural Issues About EPIL in China**

Part II deals with substantive and procedural issues in relation to EPIL in China. It examines various mechanisms and their implementation within the EPIL system, including the roles of a range of stakeholders in EPIL, liabilities regimes, remedies and enforcement. In this section, special attention is given to procuratorial EPIL and litigation for ecological and environmental damage compensation in China—which forms the major research topic for Dr. Yun Ma and Dr. Wenjun Luo respectively. In the meantime, we have three contributors, Dr. Huanhuan Wang, Zhenglian Zhang, and Dr. Zhiyu Huang, who broadly examine the EPIL dealing with soil environment pollution and the judicial application of ecological remediation liability.

In Chap. 7, 'Procuratorates at the Crossroad: Performance, Controversies and Prospects of Procuratorial EPIL in China', Dr. Yun Ma and Wenzhen Shi discuss the procuratorial EPIL in China by examining the origin, developments and challenges and future trends of this unique legal process. While EPIL plays an important role in improving the accountability of Chinese public sectors involved in environmental governance, the introduction of procuratorial EPIL complicated the traditional role of government in resolving environmental problems. It should be noted that procuratorates still lack experience with bringing EPIL cases, and in many instances, the procuratorial authorities are not willing to file lawsuits against peer government agencies. They argue that procuratorial EPIL needs to be founded on a rational design of the procuratorates' role in the overall EPIL system. They argue that connection mechanisms among different types of PIL should be established; legal empowerment and top-down design should be adopted to reduce discrepancies and increase cohesion.

In Chap. 8, 'Litigation for Ecological and Environmental Damage Compensation in China: EPIL with Typical Chinese Characteristics', Dr. Wenjun Luo examines ecological and environmental damage compensation litigation (EEDCL), which is thought to be an important achievement of China's environmental justice reform in recent years. Luo defines the EEDCL as a special civil environmental public interest litigation system with unique operating mechanisms. It is different from the traditional environmental lawsuits in many aspects, e.g., the starting conditions of litigation, the scope of compensable damages, burden of proof, and the liability mode. Luo observes that the development of the EEDCL system is deeply influenced and shaped by the comprehensive deepening reform which has established the environmental rule of law in contemporary China. This is symptomatic of the general progress made in environmental legislation and justice in China in recent years. She believes this system still needs to be tested and improved in practice.

In Chap. 9, 'The Role of NGOs in China's Environmental Public Interest Litigation', Dr. Mengxing Lv explores the significant role of NGOs in China's EPIL system. The law has recognised the standing of NGOs since 2012. Despite this, there are substantial restrictions on the scope of eligible NGO litigants, and NGOs are excluded from filing an EPIL against public authorities and administrative EPIL lawsuits. Dr. Lv observes a decline of NGOs in PIL and attributes this to a number of reasons, which include, among others, the lack of legitimacy of NGOs in China, the lack of financial independence, the restrictions imposed on NGOs in challenging administrative authorities, and the ascendancy of administrative authorities in bringing ecological compensation suits. Some scholars described PIL in China as 'castles made of sand'. She further argues that China should consider removing the current restrictions on standing and take measures to enhance the capacity of NGOs to maximise their role in EPIL.

In Chap. 10, 'Regulating Through Litigation: Regulatory Functions of NGO-Led Civil Environmental Public Interest Litigation in China', Dr Juan Chu examines how the broad scope and flexible remedies of civil EPIL have allowed environmental NGOs to use it as a powerful regulatory tool. She highlights that NGO-led civil EPILs perform valuable regulatory functions by providing an alternative enforcement mechanism when government enforcement fails, filling various regulatory gaps when a statutory remedy is unavailable or weak, challenging the legitimacy of government decisions, and setting the agenda in the regulatory process. She also identifies several pitfalls of utilizing civil EPIL as a regulatory tool, which include threatening the primacy of government enforcement, displacing statutory remedies, unreasonable calculation of environmental damages, and the misuse of civil EPIL. She concludes that environmental NGOs should avoid the inappropriate use of civil EPIL to increase its legitimacy and regulatory significance in China's environmental governance system.

In Chap. 11, 'From Advocators to Rule Makers: Exploring the Role of Chinese Lawyers in Environmental Law Making and Public Interest Litigation', Professor Jun Zhao and Jinye Han explore the changing role of lawyers in China's environmental law and policy making, including their unique contribution to the laws and regulations in relation to the newly emerged EPIL. She argues that knowledgeable, experienced and willing lawyers are key elements for a successful EPIL system. She observes that lawyers are now able to represent clients for compensation where ecological damage has occurred and to assist NGOs and government agencies in their environmental litigation activities. With the introduction of new environmental policies and the gradual benchmarking of China's environmental standards with international standards, the need for environmental lawyers to interpret policies and promote environmental protection concepts is increasing. The author argues that lawyers can improve the prospects of EPIL provided that attention is given to innovation, value pursuit, rights protection and system innovation in the legal industry.

In Chap. 12, 'Three-Dimensional Model of Expertise-Input for EPIL in China', Professor Jianwei Zhang and Dr. Yanni Luo examine the roles of key third parties in EPIL, which include the jury, lay witnesses and expert witnesses. Due to the extreme complexity and high cost involved in assessing damages, the involvement of experts in EPIL at various stages has been encouraged by the SPC. The court can invite experts to review the evidence, conduct scientific investigation, or assess environmental damages in EPIL. The court can also assign a third party to perform clean-up and restoration, with the cost to be covered by the monetary compensation ordered by the courts. Expert witnesses and the involvement of professionals are important elements for a successful EPIL. Zhang presents a typology of how the "three-dimensional model" of expertise-input is running in China and analyses in detail the role and problems of the three types of professionals with specialized knowledge in EPIL. Zhang argues that, to address challenges confronting courts, further steps should be undertaken by the SPC to promote expert involvement in the litigation process.

In Chap. 13, 'Judicial Application of Ecological Remediation Liability in EPIL: A Commentary to the Guiding Cases, Gazettes Cases and Model Cases of the SPC in China', Dr. Zhiyu Huang begins by explaining the Guiding Cases, Gazettes Cases and Model Cases issued by the SPC on EPIL in China. She investigates the innovative approaches of the judiciary in considering ecological remediation liability and points out the problems of the current approach. These include poor implementation of the principle of the priority of ecological remediation, insufficient supervision of the assessment and management of ecological restoration, compensation for damages, and serious divergences in some innovative practices. She calls for an urgent amendment to the relevant rules, e.g., to establish operating guidelines that clearly define the responsible parties for restoration programs and parties that are to supervise such programs. The ultimate purpose of this change is to ensure that the decision is fair, legitimate and democratic in terms of public participation.

In Chap. 14, 'Looking Backwards, Looking Forwards: Environmental Public Interest Litigation in Soil Pollution Law', Dr. Huanhuan Wang and Zhenglian Zhang review the historical developments of EPIL within the context of soil pollution in China to identify what role it has played in soil pollution prevention and remediation. They find that, due to the complexity of land remediation, there is a heavy reliance on scientific evidence. A deficiency of EPIL exists especially in confirming compensation amounts, setting clean-up criteria and monitoring remediation processes. With the commencement of the Law on Soil Pollution Prevention and Control in 2019, liability for land contamination is clearly stipulated as public liability, directly imposed by environmental agencies. This liability may include undertaking a site investigation, site risk assessment, risk control, remediation, evaluation of site risk control and remediation, and long-term stewardship. They argue that, in view of the dominance of administrative power in soil pollution prevention and control, EPIL should be reshaped by focusing on the preventive function, supplementing the responsibility of soil remediation with civil public interest litigation, and strengthening the supervision of administrative power with administrative public interest litigation.

Part III **Reform the EPIL System in China: Problems and Solutions**

Part III mainly focuses on reflections on the EPIL system as it currently operates in China, and our contributors offer potential solutions to increase its effectiveness. The contributions from Dr. Shuai Xin, Dr. Xiaobo Bo and Professor Xi Wang follow these themes, with the three chapters all critical of the nature of EPIL and the interrelationship of the existing models or forms of EPIL. All the contributors put forward valuable suggestions for improvement.

In Chap. 15, 'The Erroneous Setting of the Essential Attribute of Environmental Public Interest Liability in China', Dr. Shuai Xin notes that both the legal academy and the legislators have mistaken the essential attribute of environmental public interest liability as civil liability. This has led to a series of jurisprudential and legal institutional consequences. These include (1) the conflicts of two types of environmental public interest liabilities with different attributes, (2) the conflicts between administrative enforcement procedures and civil procedures of realizing environmental public interest liability, and (3) the conflicts between social organizations' public authority and administrative agencies' public authority. Xin believes that the many factors working together has led to the erroneous setting, which includes the impact of the theory of civil remedies for environmental public interest damage, the theory of crystallization of an environmental right as a civil right, and conceptual misunderstandings in the transplantation of environmental legal regimes from the common law system. He suggests that amendments in legislation are needed to rectify the erroneous setting of the essential attribute of environmental public interest liability.

In Chap. 16, 'From the Ternary Model to the Binary Model: On the Reconstruction of Litigation Systems for Environmental Public Interest', Dr. Xiaobo Bo examines the relationship between the civil EPIL, administrative EPIL, and the newly introduced ecological and environmental damage compensation litigation (EEDCL). She points out that the overlap with the civil EPIL system exposed the shortcomings of the existing ternary litigation model. She argues that the basis of claims for ecological environmental damage compensation is not the national ownership of natural resources or the responsibility of environmental supervision, but the environmental public interest; and that there is no significant difference between these two kinds of litigation in terms of value objective, system function, basis of claim, object of relief, object of litigation, and the like. Accordingly, in her proposal to reform the current legal system relevant to EPIL in China, she suggests that the current ternary litigation model should be abandoned while coexistence of these forms of EPIL is unnecessary. To this end, a binary model is offered as a solution to resolve these problems.

In Chap. 17, 'Drawbacks in the System Design of the Environmental Public Interest Litigation and Ways for Improvement', Professor Xi Wang comprehensively discusses the shortcomings of the current EPIL system and proposes suggestions for improvement. The SPC, local courts and specialist environmental courts and tribunals are confronted with a wide range of challenges that will affect the coherence between different forms of EPIL. Wang argues that the responsibilities of the People's Procuratorate and social organizations are obviously different from those of administrative management. The duties of administrative agencies and the effective exercise of administrative power should be fully respected. The current provisions in relation to EPIL, e.g., Article 58 of the Civil Procedure Law and the related judicial Interpretation on public interest litigation, have led to procuratorial powers and judicial powers overruling administrative powers, falsifying administrative powers, acquiescing or laissez-faire administrative omissions, and neglecting administration. This chapter therefore calls for reform to the EPIL system within the broad political and economic context in China.

Overall, this book is written to provide a systematic review of Chinese experiences with EPIL in environmental matters, both with a view to gauging its success to date as well as discussing some more critical aspects. To this end, this book systematically examines the establishment and development of EPIL in China's legal, social, and political contexts. It particularly examines the significant role and functions of EPIL in China's environmental governance, and the far-reaching impacts on Chinese civil society and governments. It also offers readers an insider's perspective in terms of procedural and substantive issues with respect to EPIL, by reviewing the institutional designs, theoretical underpinnings and specific mechanisms, and the roles of various participants and stakeholders involved in this legal process. At the same time, it studies leading EPIL cases instituted to address environmental pollution, natural resource damage and ecological damage, and critiques the effectiveness of environmental adjudication that sustains EPIL as a new form of judicial instrument.

This highly topical work makes a novel and significant contribution to scholarship and practice in environmental law. We hope this book will prove an indispensable resource for law makers and regulators in the field as well as for judges, lawyers, academics, students, and anyone interested in tracing the legal history and legislative development of EPIL in China.

References

Gong G (2019) Reflections on the positioning of the nature of environmental civil public interest litigation. Chin J Law 41:127–147

Huang H (2018) The analysis of essential standards of judicial adjudication in environmental administrative public interest litigation instituted by procuratorial authorities. Law Sci Mag 39:107–113. https://doi.org/10.16092/j.cnki.1001-618x.2018.08.012

Li Y, Wu K (2016) Role of procuratorates in the environmental public interest litigation. J Renmin Univ China 30:2–13

Liu Y (2017) Judicial practice and theoretical exploration of procuratorial PIL. J Natl Prosecutors Coll 25:3–18+170

Liu J (2021) On the function orientation and procedure optimisation of environmental public interest litigation filed by the procuratorial authorities. J China Univ Geosci (Soc Sci Edn) 21: 28–40. https://doi.org/10.16493/j.cnki.42-1627/c.2021.04.004

Lv Z (2016) Judicial rationality on environmental protection should not be overridden by the pricey compensation. China Leg Sci 3:244–264

Lv Z (2017) Providing Legalized Plan for Ecological Damage Compensation System, Guangming Daily (22 December 2017)

Ma Y (2019) Vindicating environmental public interests in China: a balanced approach to institutional interaction in public interest litigation system. ELR 21(4):269–291. https://doi.org/10.1177/1461452919881342

Qin H (2019) An empirical study on the practice of administrative public interest litigation. Adm Law Rev 3:87–100

SPC (2020) White paper on environmental and resource adjudication in China 2019. https://www.court.gov.cn/zixun-xiangqing-228341.html. Accessed 6 July 2022

SPC (2021) White paper on environmental and resource adjudication in China 2020. https://www.court.gov.cn/zixun-xiangqing-307471.html. Accessed 6 July 2022

Wang M (2016a) On the development direction of Environmental Public Interests Litigation: analysis on administrative right and jurisdiction. China Leg Sci 1:49–68. https://doi.org/10.14111/j.cnki.zgfx.2016.01.003

Wang X (2016b) On the legislative order of environmental public interest litigation system. Tsinghua Univ Law J 10:101–114

Xie L, Xu L (2021) Environmental public interest litigation in China: a critical examination. Transnatl Environ Law 10:441–465

Chapter 2
Historical Development of Environmental Public Interest Litigation (EPIL) in China

Shijun Zhang

Abstract As early as the end of the last century, China gradually began to explore the EPIL System. The period from the beginning of the exploration to its formal establishment is more than 20 years. This can be generally summarized into three stages: early case exploration, local practice and legislative breakthrough, and national promotion. It can be expected that with the joint efforts of all sectors of the society and the successive introduction of relevant laws, regulations, and judicial interpretations and the continuous promotion of judicial practice, the EPIL System in China will become more developed. It will become an important litigation system to promote China's practice of the scientific outlook on development and the construction of anecological civilizationsociety.

Keywords Environmental Public Interest Litigation · EPIL · China · Civil Code

2.1 Introduction

*TheCivil Procedure Lawof the People's Republic of China*amended in 2012 explicitlyestablished a Public Interest Litigation System in its legislation for the first time, which provides in Article 55 that for acts that harm the public interest of society, such as the pollution of the environment, infringement of the lawful rights and interests of consumers, the organs and relevant organisations prescribed by law may file lawsuits with the people's courts. However, this Act is unclear as to which organs and relevant organisations as prescribed by law may bring public interest litigation as plaintiffs. Article 58 of the*Environmental Protection Law of the People's Republic of China*(EPL hereinafter), as amended in 2014, specifies thesocial organisationsthat are qualified to bringEPIL. It states that social organisations registered with the civil affairs department at or above the level of a district city and engaged in environmental public interest activities for 5 consecutive years or

S. Zhang (✉)
Law School of Shandong University, Shandong, China
e-mail: zhangshijun@sdu.edu.cn

X. Wang et al. (eds.), *Environmental Public Interest Litigation in China*,
https://doi.org/10.1007/978-3-031-26526-6_2

more without a record of violation of law may bring EPIL actions. TheEPIL Systemwas thus formally established and has been continuously improved since 2015 with the promulgation of supporting judicial interpretations and the piloting of EPIL.

As early as the end of the last century, China gradually began to explore theEPILSystem. The period from the beginning of the exploration to its formal establishment is more than 20 years. This can be generally summarized into three stages: early case exploration, local practice and legislative breakthrough, and national promotion.

2.2 Early Case Study Exploration Phase (Before 2007)

The introduction of theEPILSystemto China did not happen overnight but has gone through a certain stage of development. The relevant environmental judicial cases and local practices before 2007 have provided abundant practical experience for the establishment of the EPIL System.

2.2.1 Establishment of the Environmental Tribunal

Theenvironmental courtin China is a specialized court that mainly hears cases in relation to environmental and natural resource protection. It has played a positive role in improving the efficiency of thejudicial systemand promoting the rule of law inenvironmental protectionin the process of solving the environmental andecological damagethat has occurred in the course of China's economic and social development. As early as the end of the last century, the environmental court began operating. To effectively resolve environmental pollution disputes within itsjurisdiction, and in the face of the increasing number of environmental protection cases, in 1988, the IntermediatePeople's Courtof Wuhan City applied to the Supreme People's Court to set up an environmental protection court. This was thought to be the starting point of judicial specialization in environmental protection in China. Thereafter,environmental courtswere established in many other regions in China and started to play a very important role in the implementation of national environmental protection laws and policies.

In April 2002, the Environmental Protection Division of thePeople's Courtwas established in Dongling District, Shenyang City, Liaoning Province. Within 3 years of its establishment, it handled 253 cases involving various environmental administrative penalties. In April 2004, the Environmental Protection Circuit Court of the People's Court was set up in the Shahekou DistrictEnvironmental Protection Bureau, which is located in Dalian City, Liaoning Province. The Shahekou District People's Court specifically sent its staff to hear theenvironmental protectioncases at the Shahekou District Environmental Protection Bureau. In the same year, theEnvironmental Protection Courtwas established in Jinzhou City, which is located in Shijiazhuang City, Hebei Province. In 2006, the Environmental Protection Circuit Court of Renping County was established, which is located in Liaocheng City, Shandong Province. In June of the same year, the Environmental Protection Court of the People's Court of Tiexi District was set up in the Tiexi District Environmental Protection Bureau, which is located in Shenyang City, Liaoning Province. This court mainly accepts law enforcement cases of environmental protection and consultation on environmental laws and regulations, and the Administrative Tribunal of the District Court is responsible for the specific work, taking preliminary measures such as early intervention, prior mediation, and ordering rectification in the case handling process. Only the cases that refuse mediation enter the judicial process.

It is worth noting that these types ofenvironmental courts, established through cooperation between the People's Courts and local law enforcement agencies ofenvironmental protection, do not reflect the principle of judicial independence. Under this model,administrative powermight suspiciously interfere with the environmental judicial power. More importantly, there were no clear legal bases for the establishment of such courts, which eventually led to the abrogation of earlier environmental courts. In 2006, the HighPeople's Courtof Liaoning Province officially suspended the operation of all environmental courts in the Shenyang city. This event marks the end of China's early attempt at establishing environmental courts at the local levels. Nevertheless, as an attempt to innovate the judicial mechanism, the early model of environmental courts provided a wealth of practical experience for the later construction of environmental courts and the acceptance ofEPILcases.

2.2.2 *Public Interest Litigation Brought by the Public andLawyers*

Although there is no legal system for the public and lawyers to bringEPIL, in practice, there have been a number of cases brought by individuals, the public and lawyers to protect the environmental public interest, creating a "budding legal movement". For example, in the case of lawyer*Jin Kuixi v Hangzhou Municipal Planning Bureau*on February 25, 2003, theplaintiffargued that according to the*Hangzhou West Lake Scenic Spot Protection Management Regulations*, no unit is allowed to build or expand buildings unrelated to the scenic spots in the scenic area. However, the city's Planning Bureau issued a construction planning permit, allowing Zhejiang University for the Aged to build in the West Lake Scenic Spot. Consequently, Jin Kuixi filed a lawsuit for the environmental administrative public interest on the grounds that this Planning Bureau had issued a construction planning permit to allow the project to be carried out in the scenic spot, destroying the original appearance of West Lake and infringing on the public interest of citizens and tourists. ThePeople's Courtof West Lake District held that the construction permit issued by the Hangzhou Municipal Planning Bureau had no practical impact on the plaintiff and ruled against it on the grounds that the plaintiff was "not qualified".

Similarly, after thewater pollutionincident in theSonghua RiverBasin,[1]on 7 December 2005, three professors and three postgraduates from Peking University Law School filed a lawsuit in the Heilongjiang HighPeople's Court. This is the first environmental public interest lawsuit in China. The plaintiffs included some natural objects (sturgeon, Songhua River and Sun Island) as coplaintiffs. They asked the court to order the defendant to compensate 10 billion yuan for the establishment of the Songhua River Basin Pollution Control Fund to restore the ecological balance of this basin; to safeguard the rights of the sturgeon to survive; the rights of the Songhua River and Sun Island to maintain environmental cleanliness; and the rights of the natural person to travel and enjoy cases during the same period, including the major water pollution incident in the Tuo River, the case of*Shi Jianhui and Gu Dasong v Nanjing Municipal Planning Bureau*,[2]the "Environmental Impact Assessment (EIA) Storm" set off by the former State Environmental Protection Administration (SEPA), the Beijing Baiwang Family Court incident, and the Yuanmingyuan Lake bottom seepage prevention project case, etc. These plaintiffs who filed lawsuits to protect the environmental public interest were unable to have the matters heard by the court, as they were not qualified as plaintiffs.[3]There were, of course, cases in judicial practice where citizens have brought public interest litigation in other areas that have been accepted and won. In 1998, Mr Ge Rui, a native of Henan Province, went to the toilet in the waiting hall at Zhengzhou Railway Station. When he came out, the toilet manager asked him to pay 0.3 yuan for the toilet. However, he believed that this charge had been included in the first batch of payable items that were explicitly cancelled by theMinistry of Financeand theState Planning Commission; therefore, this charge was illegal. Mr Ge asserted that, after purchasing a ticket, passengers should enjoy the basic services such as sanitation, security, waiting rest, etc., which must be provided from entering to leaving the station. Thus, the railway station has infringed on the legal rights of consumers. Ge Rui then sued the Zhengzhou Railway Bureau for illegal toilet charges at the Zhengzhou Railway Station. After nearly 3 years of trial, the court eventually ruled in Ge Rui's favor, with the Zhengzhou Railway Bureau returning 0.3 yuan of toilet charges to Ge Rui and bearing the court fees—50 yuan each for the first and second trials. However, according to media reports, Zhengzhou Railway Station continued to charge the toilet fee even after losing the lawsuit. Although theplaintiffwon the case, his purpose of creating a ripple effect to maintain public welfare failed.

---

[1]On 13 November 2005, an explosion occurred in the first workshop of the Biphenyl Plant of Jilin Petrochemical Company. After this explosion, approximately 100 tons of benzene-like substances (benzene, nitrobenzene, etc.) flowed into theSonghua River, which seriously polluted the river water and affected the lives of millions of residents along the coast.

[2]Southeast University teachers sued the Planning Bureau for illegally approving the "observation deck", Nanjing IntermediatePeople's Courtof Jiangsu Province (2002).

[3]See Zhang (2007), pp. 10–15.

In general, the above cases can be divided into two categories: those in which the plaintiffs themselves do not have adirect interestin the case and those in which that interest pales in comparison to the cost of litigation. Although most of these cases were not heard, they all have value and social impact beyond the individual case. It is clear that public practice andpublic opinionare undoubtedly a huge driving force in the construction ofEPIL.

2.2.3 *Public Interest Litigation Brought by Environmental Protection Organisations and Environmental Protection Authorities*

Since the 1970s, when China began to attach great importance toenvironmental protection, China's civil environmental protection forces have gradually emerged and grown, gradually embarking on a path of exploration to solveenvironmental disputes.

On 12 November 2004, Greenpeace published the*Investigation Report on the Enclosure Deforestation in Yunan by Jinguang APP*.[4]The Zhejiang Hotel Association was informed and issued the*Notice on Boycotting APP Paper Products*, calling on the 417 star-rated hotels in the whole province to boycott the paper products of the Jinguang Group and its ancillary products. The Jinguang Group filed a lawsuit against the Zhejiang Hotel Association for infringement of its reputation, claiming 2.2 million yuan compensation. TheAll-China Environment Federationinvited environmental and legal experts to discuss this matter. After reviewing the relevant information, it was confirmed that the Jinguang Group had indeed caused damage to the ecology. In February 2005, the All-China Environment Federation formed an environmental legal aid team to support the Zhejiang Hotel Association in the defence of its rights. On the eve of the court hearing, the Jinguang Group withdrew its lawsuit.

Although the number ofEPILcases brought byenvironmental protectionorganisations at this stage was relatively small, these cases still played a positive role in safeguarding theenvironmental rightsand interests of society and the public. Environmental protection organisations accumulated rich practical experience for bringing EPILs at a later stage.

Meanwhile, environmental law enforcement agencies also actively explored the road toEPIL. In the case of the Tasman Sea oil spill in 2002,[5]the Tasman Sea vessel was involved in a shipping collision in the eastern waters of Dagukou, Tianjin City, causing a spill of crude oil from the vessel. This spilled crude oil formed a 2.5 nautical mile long and 1.4 nautical mile wide oil drift zone, causing serious damage

[4]Greenpeace (2007).

[5]*The Dagu Fishermen's Association of Tanggu District v Infinity Shipping Co., Ltd.*,Tianjin Maritime Court(2003).

to marine life and ecosystems. Subsequently, the Dagu Fishermen's Association of Tanggu District, where action was taken on behalf of 129 local fishermen, theTianjin Fishery Administration and Fishery Port Supervision and Management Officeand theTianjin Marine Bureaufiled separate lawsuits seeking compensation for damages from the shipping company involved in the accident, on the basis that the national fishery resources had been damaged and the marine environment had been polluted. TheTianjin Maritime Courtaccepted the case. The Dagu fishermen sought damages for the loss of marine fishing suspension, nets and beach shellfish farming caused by the pollution; the Tianjin Fishery Administration and Fishery Port Supervision and Management requested damages for the loss of fishery resources; the Tianjin Marine Bureau requested damages for the claim of ecological pollution damage and ecological restoration of the marine environment. There was no duplication of claims among the three. Therefore, the court not only upheld the fishermen's claim but also upheld the claims of the Fishery Administration and Fishery Port Supervision and Management Office and the Marine Bureau. This case was handled vaguely in a way that was reminiscent of EPIL, as the lawsuit brought by the Fisheries and Harbours Supervision and Management Office and the Marine Bureau was not to defend the interests of a particular individual or an entity, but for thepublic goodof the environment. Thus, the lawsuit actually fell within the scope of public interest litigation.

2.2.4 Public Interest Litigation Initiated byProcuratorial Authorities

At the end of the last century, a number of localprocuratorial authoritieshave conducted useful explorations of public interest litigation in different areas in China's judicial practice. Early cases in relation to the loss ofstate-ownedassetsexemplified the public interest litigation initiated by procuratorial authorities. In China, state-owned assets belong to all citizens, and state organs manage state-owned assets in accordance with the mandates of all citizens. However, for the loss of state-owned assets, citizens cannot directly file lawsuits, and state-owned asset management departments often refuse to file lawsuits on the grounds that they do not havelegal personality, which has given rise to a new institutional model of procuratorial organs filing public interest litigation. In 1997, the procuratorial authorities in Fangcheng, Henan Province, initiated the first public interest litigation against the loss of state-owned assets. This was followed by a case in Yueyang County, Hunan Province, where the victim, the Daily-use Miscellaneous Goods Company and its competent body, the County Supply and Marketing Society, explicitly expressed their unwillingness to bring a lawsuit, while the Yueyang County State Assets Office, one of the stated-owned assets management bureaus, stated that it did not have legal personality and it had the right to manage but not the right to sue. In the end, thePeople's Procuratorateof the Yueyang County had to bring a civil public action to recover state assets.

As of 2001, 141 similarcivil lawsuitshad been filed by procuratorial authorities across the country. In terms ofEPIL, in addition to the case of the*Leling CityPeople's Procuratoratev Fan Mou illegal small oil refinery project*in 2003, typical cases include the following. In November 2003, the LangzhongProcuratorateof Sichuan brought a civil action in court against the Qunfa Bone Meal Factory forinfringement of the environment. On May 12, 2004, in Ziyang, Sichuan Province, the Yanjiang District Procuratorate delivered procuratorial proposals to eight stone factories that seriously polluted the environment and warned these enterprises that if they did not actively control the pollution, they would be subject to civil prosecution according to law. In December 2004, the Civil and Administrative Prosecution Department of the Sichuan Provincial Procuratorate took the lead in proposing and establishing a "Public Interest Litigator" system that aimed to recover lost state assets and protect vulnerable groups. It also made relevant provisions to protect the national andcollective public interests, except for cases within the scope of the parties' right to dispose of and disputes between pure family members, and the procuratorial organs may support prosecution. These cases ofcivil EPILbrought by the procuratorial authorities ended in the procuratorial authorities' favor, which is a useful attempt to bring civil EPIL by the procuratorial authorities in China.[6]

In the early stages, both the establishment ofenvironmental courtsand the active exploration ofEPILby individual members of the public,environmental protectionorganisations and procuratorial authorities in practice laid a solid theoretical and practical foundation for the constructionof theEPIL Systemin China.

2.3 Local Practice and Legislative Breakthrough (2007–2014)

2.3.1 Local Practice

With the rapid development of the economy, the environmental situation in China has become increasingly serious, and the number of resultingenvironmental disputesand environmental mass incidents has soared. Theenvironmental courtswere created and developed during this period with the aim of "anticipating the better activation of judicial power to solve serious environmental problems in reality".[7]In November 2007, the Environmental Court of thePeople's CourtofQingzhenwas established in Qingzhen, Guizhou Province. On 3 July 2014, the Supreme People's Court officially announced the establishment of a special environmental resources court. According to incomplete statistics, as of July 2014, a total of 20 provinces,

[6]See Zhang and Xie (2007), pp. 107–109.

[7]See Zhang (2016), pp. 52–58.

including municipalities and autonomous regions, had set up 150 environmental resources tribunals, collegiate courts or circuit courts.[8]

Some of the more significant local practices include the Environmental Protection Trial Court of the IntermediatePeople's Courtof Guiyang and its subordinate Protection Court of the People's Court ofQingzhen, established in 2007; the Environmental Protection Trial Court of theIntermediate People's Courtof Wuxi, established in 2008; and the Environmental Protection Trial Court of the Intermediate People's Court of Kunming, established in 2008.

2.3.1.1 Practice of the Guiyang Environmental Protection Trial Court

On 20 November 2007, the IntermediatePeople's Courtof Guiyang established the Environmental Protection Trial Court and the Environmental Court in the People's Court ofQingzhen.

In December 2007, the IntermediatePeople's Courtof Guiyang issued the*Decision on Designation of Jurisdiction*,[9]specifying that all cases of the first instance involvingenvironmental protectionin Guiyang would be under the centralizedjurisdictionof theEnvironmental Protection Courtof the People's Court ofQingzhen, and that the Environmental Trial Division of the GuiyangIntermediate People's Courtwould be responsible for the second instance of appeals from the Environmental Protection Division of the People's Court of Qingzhen. According to the provisions of the Decision concerning the designation of jurisdiction, theEnvironmental Protection Court of Qingzhenaccepted the case of*"Two Lakes and One Reservoir" Administration v Tianfeng Chemical Company*.[10]It was for the first time that a government department sued asplaintiffin environmental pollution disputes and resolved the jurisdictional issue of cross-regional disputes by assigning jurisdiction.

In July 2009, theEnvironmental Protection CourtofQingzhenaccepted the case of the*All-China Environmental Protection Federation v Qingzhen Municipal Bureau of Land and Resources*,[11]which became the first environmental administrative public interest litigation case in China.

On 16 October 2009, the People's Congress of Guiyang passed the*Regulations on the Promotion of Ecological Civilization in Guiyang*,[12]which became the first local

[8] Ibid.

[9] Guiyang IntermediatePeople's Court(2007) Decision on Designation of Jurisdiction (Zhu Huanbao Zhizi No 1).

[10] *Guiyang "Two Lakes and One Reservoir" Administration v Guizhou Tianfeng Chemical Co., Ltd., theEnvironmental Protection CourtofQingzhen*(2008).

[11] *All-China Environmental Protection Federation vQingzhenMunicipal Bureau of Land and Resources, theEnvironmental Protection Courtof Qingzhen*(2009).

[12] Guiyang City's Regulations on Promoting the Construction of Ecological Civilization (2010).

regulation to promote the construction ofecological civilizationand the first national legislation onEPIL.[13]

2.3.1.2 Practice ofEPILin Kunming

The arsenic pollution of the Yangzong Sea in 2008,[14]the irregular construction of the Yunnan Ludi La and Longkou hydropower stations in 2009,[15]the ineffective reduction of sulfur dioxideemissionsby the Guodian Yangzonghai Power Generation Company in 2010,[16]and the chromium slag pollution in Qujing in 2011[17]have been the subject of extensive attention from all walks of life to the practice ofEPILin Yunnan. Yunnan was the first province in China to introduce normative documents on EPIL at the provincial and municipal levels.

On 5 November 2008, the KunmingIntermediate Court, theProcuratorate, the Municipal Public Security Bureau and the MunicipalEnvironmental Protection Bureaujointly issued the*Implementation Opinions on the Establishment of a Coordination Mechanism for Environmental Protection Law Enforcement.*[18]In December 2008, the Kunming and Yuxi Intermediate Courts set upenvironmental courts. On 13 May 2009, the Yunnan Provincial High Court launched the*Summary of the*

---

[13]These cases include, for example,*WangchengProcuratorateof Changsha, Hunan Province v Hunan Changsha Pingtang Cement Company Limited*(2008);*Haizhu Procuratorate of Guangzhou v Chen Zhongming, owner of Xin Zhongxing Water Washing Plant*(2008);*Guiyang Procuratorate v Xiong Jinzhi, Lei Zhang and Chen Tingyu*(2008);*Xinyu Yushui Procuratorate v Li Mou and Zeng Mou*(2008);*Guangzhou Panyu Procuratorate v Dongchong Dongtai Leather Factory*(2009);*Guiyang Procuratorate v Xiong Jinzhi, Lei Zhang and Chen Tingyu*(2009);*Wuxi Xishan Procuratorate v Li Mu and Liu Mou*(2009);*XingziPeople's Procuratoratev Xingzi Chenghui Stone Factory*(2009); *Guangzhou Panyu People's Procuratorate v Guangzhou Panyu Bolang Hardware Factory*; (2010); *Guangzhou Baiyun People's Procuratorate v Du Jiahua and Liang Zhongqiang*(2010);*Pinghu People's Procuratorate v Jiaxing Luyi Environmental Protection Service Co.*(2011). SeeTao (2013), pp. 93–110.

[14]Since the 1970s, affected by the development of biological resources in the lake area and the rapid development of local industry and agriculture, the water quality of the Yangzong Sea has deteriorated rapidly. In June 2008, the water quality of Yangzong Sea did not improve at all, and the arsenic content reached an astonishing 58 μg/L.

[15]The construction projects of Yunnan Ludila and Longkou hydropower stations failed to pass the EIA, but the company built dams without authorization in January 2009. On June 11, 2009, the Ministry of Environmental Protection ordered the two hydropower stations to stop construction.

[16]Due to the abnormal operation of the desulfurization facilities of Guodian Yangzonghai Power Generation Co., Ltd. and the ineffective reduction of sulfur dioxide, Chongqing Green Volunteers Federation filed a public interest lawsuit against Guodian Yangzonghai Power Generation Company in the Environmental Court of KunmingIntermediate Courtin 2010.

[17]In June 2011, more than 5000 tons of highly toxic chromium slag from Yunnan Luliang Chemical Industry Co., Ltd. was illegally dumped in the countryside of Qilin District, Qujing City, Yunnan Province, causing the death of 77 livestock and contamination of farmland in the nearby countryside, and seriously threatening the health and environmental safety of the surrounding people.

[18]Implementation Opinions on the Establishment of a Coordination Mechanism for Environmental Protection Law Enforcement (2008).

Symposium on the Construction of Environmental Protection Trials and the Trial of Environmental Protection Cases in the Province's Courts,[19]which not only mentioned the need to promote the construction of environmental courts throughout the province but also made clear provisions on the subjects ofEPIL, the scope of the case, the burden of proof, litigation procedures, litigation costs, etc. In October 2010, the Kunming Municipal Government promulgated the*Interim Measures for the Management of Special Funds for Environmental Public Interest LitigationRelief,*[20]which are mainly used for the investigation and collection of evidence, appraisal and assessment, litigation costs,environmental restorationand enforcement of EPIL.

In November 2010, the KunmingIntermediate Courtissued the*Opinions on Handling Civil EPILCases.*[21]In February 2011, the Yuxi Intermediate Court and the MunicipalProcuratoratejointly issued the*Opinions on Handling Environmental Resources Civil Public Interest Litigation Cases.*[22]As of September 2012, three intermediate courts and twelve municipal courts in Yunnan have set up specialenvironmental courts, ranking first in China in terms of the number of environmental courts.

2.3.1.3 EPILPractice in Wuxi

On 6 May 2008, the WuxiIntermediate Courtset up anenvironmental courtin response to the water ecology crisis caused by the cyanobacteria incident in Taihu Lake, with a view to promoting the pilot practice ofEPIL. In September of the same year, the Wuxi Intermediate Court and the MunicipalProcuratoratejointly formulated the*Trial Provisions on Handling Civil Environmental Public Interest Litigation Cases,*[23]which became the first local regulation on EPIL in China.

In December 2008, the WuxiIntermediate Court, the MunicipalProcuratorateand the Municipal Office of Legal Affairs jointly issued the*Opinions on the Provision of Evidence to the Procuratorate by Departments with Environmental Protection*

---

[19]Minutes of the Symposium on the Construction of the Environmental Protection Tribunal and the Trial of Environmental Protection Cases of the Yunnan Provincial HigherPeople's Court(2009).

[20]Interim Measures for the Administration of Special Funds for Environmental Public InterestLitigationRelief in Kunming (2010).

[21]Opinions on Several Issues concerning Handling of Environmental Civil Public Interest Litigation Cases (2010).

[22]Opinions of the IntermediatePeople's Courtof Yuxi City and thePeople's Procuratorateof Yuxi City on Several Issues concerning the Handling of Civil Public Interest Litigation Cases on Environmental Resources (for Trial Implementation) (2011).

[23]Provisions of Wuxi IntermediatePeople's Courtand WuxiPeople's Procuratorateon Handling Environmental Civil Public Interest Litigation Cases (for Trial Implementation) (2008).

Administrative Functions in Civil Environmental Public Interest Litigation.[24]However, due to inherent limitations in the institutional basis, the practice ofEPILin the Wuxi Environmental Protection Trial Court has always been limited, coupled with the fact that the function of the Environmental Protection Trial Court is primarily set to assist and support environmental administrative law enforcement. The main type of case it receives is administrative nonlitigation enforcement.

In June 2009, the Xishan DistrictProcuratoratefiled a civil public interest litigation against Li Huarong and Liu Shimi for the crime of "logging", initiating the judicial practice ofEPILin Wuxi as a "civil public interest litigant" rather than as aplaintiff. The two lawsuits, which could have been combined as criminal andcivil lawsuits, were deliberately divided into criminal and civil lawsuits to achieve a "zero breakthrough" incivil EPILin the Wuxi Environmental Protection Trial Court, which should be said to be more symbolic than practical. A month later, the WuxiEnvironmental Protection Courtaccepted the case of*All-China Environmental Protection Federation v Jiangyin Port Container Co., Ltd*,[25]which was the first case in China to recognize environmental associations as plaintiffs in EPIL.

2.3.1.4 Other Local Practices

As of April 2012, at least seven provinces and municipalities directly under the Central Government have issued at least 19 local documents to regulate and guide the local practice ofEPIL. In addition to the aforementioned, these include:*The Coordinated Opinions on Jointly Carrying Out EPIL Activities*[26]issued by Xinzhou District, Shangrao, Jiangxi Province (2010);*The Implementation Opinions on Establishing a Coordinated and Joint Working Mechanism for Environmental Protection Law Enforcement*[27]issued by the Chengyang District Court, the DistrictProcuratorateand the DistrictEnvironmental Protection Bureauof Qingdao (April 2010);*Several Opinions on Environmental Protection Public Interest Litigation*[28]issued by Jiaxing Procuratorate and Municipal Environmental Protection

[24]Opinions of WuxiIntermediate Court, the MunicipalProcuratorateand the Municipal Office of Legal Affairs on Evidence to the Procuratorate by Departments with Environmental Protection Administrative Functions in Environmental Civil Public Interest Litigation (2008).

[25]Shi Zhengmao,*All-China Environmental Protection Federation v Jiangyin Port Container Co., Ltd, IntermediatePeople's Courtof Wuxi City*, Jiangsu Province (2009).

[26]Collaborative Opinions of Xinzhou DistrictPeople's Procuratorateand Xinzhou DistrictEnvironmental Protection Bureauon Carrying out Environmental Public InterestLitigationActivities (2010).

[27]Opinions of Chengyang District Court, the DistrictProcuratorateand the DistrictEnvironmental Protection Bureauof Qingdao on the Establishment of a Coordinated and Linked Working Mechanism for Environmental Law Enforcement (2010).

[28]Several Opinions of Jiaxing CityProcuratorateand MunicipalEnvironmental Protection Bureauon Environmental Protection Public Interest Litigation (2010).

Bureau (May 2010);*Several Opinions on Establishing an Environmental Protection Public Interest Litigation System*[29]issued by Huzhou Central Court and Municipal Procuratorate (August 2010);*Opinions on the Active Use of Civil and Administrative Procuratorial Functions to Strengthen Environmental Protection*[30]issued by Zhejiang Provincial Procuratorate and Department of Environmental Protection (August 2010);*The Implementation Opinions on the Pilot Project of Civil Public Interest Litigation on Environmental Resources*[31]issued by the Hainan Provincial High Court (July 2011);*Interim Measures for the Management of Provincial EPIL Funds in Hainan*[32]issued by the Hainan Provincial High Court and the Provincial Department of Finance (September 2011);*Opinions on the Pilot Centralization of Environmental Protection Cases*[33]issued by the Chongqing Municipal High Court, the Municipal Procuratorate, the Municipal Public Security Bureau and the Municipal Environmental Protection Bureau (November 2011);*The Notice on Piloting Centralized Trial of Environmental Protection Cases in Some Courts in Jiangsu Province*[34]issued by the Jiangsu Provincial High Court (April 2012);*Regulations on the Promotion of the Construction of Ecological Civilization in Zhuhai Special Economic Zone*[35]issued by Zhuhai City, Guangdong Province (2014), among others.

2.3.2 Practice of ProcuratorialEPIL

According to incomplete statistics, from 2007 to 2012, there were 12 cases in which the procuratorate was theplaintiff, accounting for 40% of the total number ofEPILcases in China.[36]

These cases of civil environmental public interest litigation brought by procuratorial organs are a useful attempt to bring environmental civil public interest

[29]Several Opinions of Huzhou Central Court and MunicipalProcuratorateon the Establishment of an Environmental Protection Public Interest Litigation System (2010).

[30]Opinions of Zhejiang ProvincialProcuratorateand Department of Environmental Protection on Actively Using Civil and Administrative Procuratorial Functions to Strengthen Environmental Protection (2010).

[31]Implementation Opinions of the HigherPeople's Courtof Hainan Province on the Pilot Program of Civil Public Interest Litigation for Environmental Resources (2011).

[32]Measures of Hainan Provincial Department of Finance and Hainan Provincial HigherPeople's CourtInterim on the Administration of Provincial Environmental Public InterestLitigationFunds (2011).

[33]Opinions of Chongqing HigherPeople's Court, MunicipalPeople's Procuratorate, Municipal Public Security Bureau, and MunicipalEnvironmental Protection Bureauon the Pilot Centralized Handling of Environmental Protection Cases (2011).

[34]The Notice on Piloting Centralized Trial of Environmental Protection Cases in Some Courts in Jiangsu Province (2012).

[35]Regulations on the Promotion of the Construction of Ecological Civilization in Zhuhai Special Economic Zone §67 (2014).

[36]Ruan (2013), pp. 157–161.

litigation by procuratorial organs in China. It also shows that the courts around China have gradually recognized the procuratorate's right to bringEPILas an appropriate subject of theplaintiff.

2.3.3 Legislative Breakthroughs in the 2012 Amendments to theCivil Procedure Law

2.3.3.1 Legislative Breakthroughs

On 31 August 2012, the Standing Committee of theNational People's Congress (NPC) of China passed the amended*Civil Procedure Law*, which formally established the Civil Environmental Public Interest Litigation System. Article 55 of this law stipulates that "the authorities and relevant organisations provided for by law may bring a lawsuit to the people's court against acts that pollute the environment, infringe on the legitimate rights and interests of many consumers, and other acts that harm the public interest of society".[37]This is the first time that China's*Civil Procedure Law*has explicitly stated that in cases involving environmental pollution, infringement of the legitimate rights and interests of a cluster of consumers, and other cases that harm the public interest, the relevant authorities and organisations are authorized by law to initiate public interest litigation. Public interest litigation does not require that theplaintiffhave particular interests in the cause of action to bring the lawsuit, but rather a legal mandate to bring it.

The Civil Environmental Public Interest Litigation System added by the amended*Civil Procedure Law*gives the right to bring public interest litigation to third parties who have nodirect interestin the case, i.e., the two types of public authorities and relevant organisations as stipulated by the law, which provides the possibility for people who have no direct interest in the case to fileEPILfor the purpose of protecting the environment. It breaks through the traditionalcivil litigationrestrictions on the right to sue and is a substantial step forward in the development of EPIL. Although this Article only determines that specific organs and organizations are qualified to be plaintiffs in EPIL, it is significant that China has formally recognized the EPIL System at the national legislative level, which is a breakthrough in the construction of China's Public Interest Litigation System. Thus, this amendment has become a major highlight of the revision of the*Civil Procedure Law*and a new starting point for the construction of China's EPIL System.

[37]See Art. 55Civil Procedure Law.

2.3.3.2 Problems inEPIL

The 2012 amendments to the*Civil Procedure Law*further promoted the development ofEPIL, but in practice, since the law has come into force, it still suffers from many shortcomings and deficiencies. The provisions of Article 55 are overly simplified and general so that the provisions are reduced to the Principles of Public Interest Litigation and the practical operation is also poor, which eventually led to a downturn in EPIL actions being taken in 2013. The reasons are as follows:

First, the supporting substantive laws have not yet been amended, and the relevant judicial interpretation of the SupremePeople's Courthas not yet been issued. Thus, the institutional function ofEPILto broaden the avenues for citizens to defend theirenvironmental rightsand increase the efforts ofenvironmental protectionis greatly restricted and difficult to be effective.

Second, the new*Civil Procedure Law*only stipulates that organs and relevant organisation prescribed by law have the right to sue, but it does not specify which organs and organisations are referred to, and the definition of 'eligible organs' and 'relevant organisations prescribed by law' is not yet clear. What constitutes an organ under the law, and the scope of the organization in question, are two points that urgently require a relevant legal or judicial interpretation. Thus, it can either be interpreted restrictively to exclude thePeople's Procuratorateor expansively to include the People'sProcuratorate.

Between 2012 and 2014, four sets of amendments to the*EPL*were reviewed by the Standing Committee of theNational People's CongressofChina. China's new*EPL*, which came into force on 1 January 2015, has been widely praised as the strictestenvironmental protectionlaw ever and another milestone[38]in the history of China's environmental legislation for its significant innovations and breakthroughs in legislative concepts, legal principles and legal systems. This amendment to the*EPL*is of historic importance to the advancement of the rule of law in China.

The new EPL relaxed the eligibility of the scope of plaintiffs ofEPIL, explicitly expanding the scope of eligible plaintiffs tosocial organisations. Such social organisations should (1) have registered with the civil affairs department of the government at or above the district level, (2) have been continuously active for more than 5 years, and (3) have no disciplinary records. The relaxation of the restrictions on the participation of social organisations in EPIL resolved the shortcomings of the public interest litigation system under the*Civil Procedure Law*. This avoids the situation where the effectiveness of the EPIL system is greatly reduced due to a single litigant,[39]and is conducive to the better use of social coordination and public

[38]The Environmental Protection Law (for Trial Implementation) of 1979 was the first milestone in the history of China's environmental legislation, which stipulated the basic principles, tasks and policies of China'senvironmental protection.

[39]Article 55 of the NewCivil Procedure Lawstipulates that "organs and relevant organizations prescribed by law" also have the qualifications to initiate public interest lawsuits. The "relevant organizations stipulated by law" in this article generally only refer to someenvironmental protectionorganizations that are expressly stipulated by law, and in fact the scope of plaintiffs ofEPILis kind of narrow.

participation in the supervision of environmental damage and the protection of environmental public interests. Meanwhile, it also sets out the prerequisites for eligible plaintiffs, which avoided the problem of too many public interest litigation subjects and thus increased unnecessary judicial and social costs. Although individual citizens are unable to initiate public interest litigation, this is still a significant step, which shows that China has established the Public Interest Litigation System that strengthens the public's power to respond to environmental pollution with litigation.

2.4 Establishment and Full Promotion ofEPIL(2015–Present)

2.4.1 Pilot ProgramsEstablished from 2015 to 2018

In October 2014, the FourthPlenary Session of the 18th Central Committee of the Communist Party of China made a decision to explore the Procuratorial Public Interest Litigation System. In January 2015, the SupremePeople's Courtissued the*Interpretation on Several Issues concerning the Application of Law in Civil Environmental Public Interest Litigation*,[40]and in June of the same year, it issued the*Interpretation on the Application of Laws in Adjudicating EnvironmentalTortCases*,[41]which were used to guide the judicial adjudication of the second category of cases, focusing on solving the thorny issues of lack of basis for substantive adjudication and inadequate special procedural rules. On 1 July 2015, the 15th meeting of the Standing Committee of the 12thNational People's Congressauthorized the SupremePeople's Procuratorateto carry out a 2-year pilot public interest litigation in some municipal procuratorates in 13 provinces, autonomous regions and municipalities directly under the Central Government. The Standing Committee of the National People's Congress passed and published a decision authorizing the Supreme People'sProcuratorateto carry out pilot public interest litigation in the fields of ecological environment and resource protection, protection ofstate-ownedassets, transfer of state-owned land use rights, and food and drug safety. The pilot areas were identified as Beijing, Inner Mongolia, Jilin, Jiangsu, Anhui, Fujian, Shandong, Hubei, Guangdong, Guizhou, Yunnan, Shaanxi and Gansu. The pilot programs lasted for 2 years. This pilot reform was policy-driven, problem-oriented, and based on the long experience of the procuratorial authorities in judicial practice, and a comprehensive construction of the Procuratorial Public

---

[40]Interpretation of the SupremePeople's Courton Several Issues concerning the Application of Law in the Conduct of Environmental Civil Public Interest Litigations (2015).

[41]Interpretation of the SupremePeople's Courtof Several Issues on the Application of Law in the Trial of Disputes over Liability for Environmental Torts (2015).

Interest Litigation System was carried out.[42] On 2 July 2015, the Supreme People's Procuratorate announced the*SPPReform Plan on a Pilot Project for Instituting Public Interest Litigation by Procuratorial Organs*[43](hereinafter referred to as the*Pilot Program*). Since then, the procuratorate, as the "national force", has entered the stage of public interest litigation, which marks the most recent development of public interest litigation in China. Under the authority of the Standing Committee of the National People's Congress, the Supreme People's Procuratorate issued the*Implementation Measures on Pilot Projects of Public Interest Litigation*[44]and the Supreme People's Court issued the*Implementation Measures for Pilot Public Interest Litigation by the People's Courts*,[45]which have established the normative basis for environmental administrative public interest litigation by the procuratorial organs, and a systematic institutional practice has gradually begun.

During the pilot period, the national procuratorial authorities handled a total of 9053 public interest litigation cases, covering all authorized areas such as ecological environment and resource protection,state-ownedasset protection, state-owned land use rights transfer, and food and drug safety. In June 2017, the Standing Committee of theNational People's Congressadopted a decision to amend the*Civil Procedure Law*and the*Administrative Litigation Law*to formally establish a Public Interest Litigation Prosecution System, establishing a Chinese Scheme for the judicial protection of the public interest.

2.4.2 Full Promotion and Deepening ofEPILAfter 2018

Since 2018, ecological and environmental departments and procuratorial organs around China have actively collaborated, andEPILhas continued to develop. At present, China has basically established a pattern of EPIL that is jointly promoted by procuratorial organs andsocial organisationsand coordinated between the official and private sectors. From the perspective of practice, China's EPIL has deepened in the following four main directions since 2018.

First, the scope of litigation subjects continues to expand. First, procuratorial organs have become an important force in bringingEPIL. Since the pilot program of public interest litigation was launched in July 2015, procuratorial organs have performed their statutory duties and actively initiated environmental civil and administrative public interest litigation, and the number of EPIL cases initiated by

[42] See Liu (2017), pp. 3–18+170.

[43] Plan for the Pilot Project of Reform of Instituting Public Interest Litigations by the Procuratorial Organs (2015).

[44] Measures for the Implementation of the Pilot Program of Initiating Public Interest Litigation by People's Procuratorates (2015).

[45] Measures for the Implementation of the Pilot Program of Trial by People's Courts of Public Interest Litigation Cases Instituted by People's Procuratorates by People's Courts (2016).

procuratorial organs gradually exceeded that initiated bysocial organisations. As of September 2018, of the total 2041 EPIL cases that have been accepted by the courts, 1836 cases were filed by procuratorial organs, making up 90% of the total number of accepted cases. Second, the criteria for assessing the eligibility of social organisations have become clearer. In 2020, the SupremePeople's Court, in line with the implementation of the*Civil Code*,[46]revised and issued the*Interpretation on Several Issues Relating to the Application of Law in Hearing Civil Environmental Public Interest Litigation Cases*[47]and other relevant judicial interpretations in the field of environmental resources. This refined the criteria for assessing the social organisations entitled to bring environmental civil public interest litigation under the*Environmental Protection Law*. In addition, a number of local regulations have also explicitly provided criteria for the eligibility of subjects to bring EPIL. For example, on 1 October 2020, the first local legislation on EPIL in China, the*Shenzhen Special Economic Zone Regulations on Ecological and Environmental Public Interest Litigation*(hereinafter referred to as the*Regulations*),[48]came into force, providing an innovative and exploratory attempt at ecological and environmental public interest litigation. The*Regulations*clarify the division of labor between state organs, social organisations, and people's procuratorates. Administrative departments can become the subject of civil public interest litigation in relation to ecological andenvironmental protection, and clearly propose establishing an ecological and environmental public interest fund and an innovative Environmental Protection Ban System, which not only provides a stronger legal basis for procuratorial organs to carry out ecological and EPIL but also provides encouragement and support for social organisations to initiate EPIL.

Second, the composite nature of litigation continues to increase. With the establishment of the Procuratorial Public Interest Litigation System,EPILbegan to expand fromcivil litigationto administrative litigation and became associated with criminal litigation due to the criminal public prosecution function of procuratorial organs. On 2 March 2018, China's SupremePeople's Court(SPC) and SupremePeople's Procuratorate(SPP) jointly held a press conference in Beijing to jointly release the*Interpretation on SPC and SPP Interpretation on the Application of Laws in Procuratorial Public Interest Litigation*[49](hereinafter referred to as the*Interpretation*). This is the first judicial interpretation jointly issued by the highest judicial and procuratorial authorities of China since the Public Interest Litigation System was formally established on 27 June 2017. In response to the judicial practice of prosecuting public interest litigation, often leading from criminal cases,

---

[46]Civil Codeof the People's Republic of China (2020).

[47]Interpretation of the SupremePeople's Courton Several Issues concerning the Application of Law in the Conduct of Environmental Civil Public Interest Litigations (2020).

[48]Shenzhen Special Economic Zone Regulations on Ecological and Environmental Public InterestLitigation(2020).

[49]Interpretation of the SupremePeople's Courtand the SupremePeople's Procuratorateon Several Issues concerning the Application of Law for Cases regarding Procuratorial Public Interest Litigation (2018).

Article 20 of the*Interpretation*explicitly provides for the Procuratorial Criminal Incidental Civil Public Interest Litigation System, with the purposes of saving litigation resources, improving litigation efficiency, and properly determining the criminal and civilliabilityof criminal suspects. This*Interpretation*was also amended at the 58th meeting of the 13th Procuratorial Committee of the Supreme People'sProcuratorateon 28 December 2020. Criminal incidentalcivil EPILhas become an important form of EPIL, and specific litigation rules for it are now under development. In addition, some local people's courts have also accepted a number of civil public interest litigation cases incidental toenvironmental administration,and the composite nature of EPIL continues to grow.

Third, the ecologicalenvironmental protectionbecomes judicially specialized. China adopted a 'bottom-up' and 'point-to-point' approach when developing the specialization of environmental judiciaries. AsEPILcases are highly professional and complex, they must be heard by a specialized trial team in accordance with the law. Since the full implementation of the Procuratorial Public Interest Litigation System began in 2018, theSPPhas established the Eighth Procuratorial Office, taking charge of civil public interest litigation cases such as civil public interest litigation cases on damage to the ecological environment and resource protection; administrative public interest litigation cases in the field of ecological environment and resource protection; complaints involving public interest litigation under thejurisdictionof SPP, etc.

At the same time,SPPissued the*Notice on Promoting the Reform of the Internal Structure of People's Procuratorates below the Provinces*,[50]which requires local people's procuratorates to establish special institutions for public interest litigation. As of September 2019, public interest litigation procuratorates have been established separately in 25 provincial procuratorates. Special agencies and special case teams have also been set up in procuratorates at the city and county levels, with the organization of judicial specialization in ecologicalenvironmental protectiongradually improving.

For example, on 7 December 2018, the Eighth Procuratorial Department of Henan Province, which is specifically responsible for public interest litigation, jointly carried out a special action with theYellow River Commissionto clean up and protect the Yellow River. A new model for river governance, which is featured as 'river chief(河长) + procurator', has been explored by the special action. Since 2019, the Seventh Procuratorial Department of Jiangsu Province has been organizing special supervision activities such as "Guarding the Ocean", "Guarding the Yangtze River" and "Guarding Hongze Lake". The Department also issued the*Opinions on the Establishment of an Integrated Collaborative Mechanism for the Protection of Ecological Environment and Resources in the Yangtze River Economic Belt.*[51]The

[50]Notice of the SupremePeople's Procuratorateon Promoting the Reform of Internal Organizations of People's Procuratorates below the Province (2018).

[51]Opinions of the Jiangsu ProvincialPeople's Procuratorateon the Establishment of an Integrated Collaborative Mechanism for the Protection of Ecological Environment and Resources in the Yangtze River Economic Belt (2019).

objectives of this document are to strengthen the linkage between criminal and public welfare procuratorates between eight cities along the Yangtze River, and comprehensively promote the establishment of cross-provincial and cross-city collaboration in terms of ecological environment and natural resource protection.

Fourth, theecological damage compensation systemhas been continuously improved. Compensation for damage to the ecological environment is a direct manifestation and institutional expression of the principle that "the environment has a price, and the damage bears responsibility", and is essential to the construction ofecological civilization.

On 1 January 2018, the Central Committee of the Communist Party of China (CCP) and theState Counciljointly issued the*CCP Central Committee and State Council Reform Plan on the Compensation Regime for Eco-environmental Damage*[52](hereinafter referred to as the*Reform Program*). The*Reform Program*calls for the establishment of an ecologicalenvironment damagecompensation system that clearly addressesliabilityissues and technical standards. Moreover, the system must be easy to access in terms of protection and compensation, including effective steps forenvironmental restoration. However, the uncertain nature of eco-environmental damage lawsuits has led to difficulties in filing cases, and the unclear relationship between public interest litigation andcivil litigationmakes it difficult to determine thejurisdictionfor a particular suit.

As a response, on 8 February 2018, the SCC issued the*Main Points of the Work of the People's Courts,*[53]which clearly proposed exploring and improving the litigation system for compensation for ecological and environmental damages. On 5 June 2019, the Supreme Court issued the*Certain Provisions on the Trial of Ecological and Environmental Damage Compensation Cases (for Trial Implementation)*[54](hereinafter referred to as*Certain Provisions*), which formally established the rule that "ecological damagecompensation litigation takes precedence overEPIL" for both types of litigation. With the release and implementation of the*Certain Provisions*, three types ofcivil litigationrelating to the ecological environment including private litigation for environmental tort, civil public interest litigation for the environment, and litigation for compensation for ecological damage have basically been formed in China.

In addition, in May 2020, the*Civil Code*adopted the "green principle"[55]and added theliabilityclauses for ecological and environmental damage under the Chapter forTortLiability, with the eligible parties to bring the lawsuit being

---

[52] Plan for the Reform of the System of Damages for Harm to Ecology and Environment Issued by the General Office of the CPC Central Committee and the General Office of theState Council(2018).

[53] SPC(2018).

[54] Several Provisions of the SupremePeople's Courton the Trial of Cases on Compensation for Damage to the Ecological Environment (for Trial Implementation) (2019).

[55] The green principle spans the fields of Civil Law and Environmental Resources Law, which is an obligatory basic norm with universal binding force and value judgment. It is also the basic principle that carries the core value of eco-socialism and the concept of green development., for details, see Cai and Zhang (2018), pp. 1–8.

defined as "an organ prescribed by the State or an organization prescribed by law".[56]"Violation of state regulations" constitutes an element of liability and strengthens the government's responsibility for ecological andenvironmental protection.[57]As of November 2021, a total of 7600ecological damagecompensation cases have been handled nationwide, involving a compensation amount of more than 9 billion yuan (approximately US$1.3 billion), which promotes the effective repair of a number of damaged ecological environments including soil, groundwater, arable land, forest land, grassland, mining areas, grasslands, etc. and provides strong judicial protection and practical support for the comprehensive trial implementation of theEcological Damage Compensation System.

2.5 Summary

While the establishment of theEPILSystemis the ultimate means of judicial remedy against acts that harm the public interest, it is also the last resort for environmental governance in China. Its purpose is to avoid the embarrassing situation of public interests being damaged and the absence of litigation subjects. The continuous exploration and development of China's EPIL System has accumulated a wealth of experience in judicial practice and is playing an increasingly important role in safeguarding and promoting the construction ofecological civilization. However, this system is still facing the dual test of structural and coordinated refinement and new requirements of environmental governance. Issues such as the scope of the case, litigation procedures and the relationship between multiple lawsuits still need further study. It can be expected that with the joint efforts of all sectors of the society and the successive introduction of relevant laws, regulations, judicial interpretations and the continuous promotion of judicial practice, the EPIL System in China will become more developed. It will become an important litigation systemto promote China's practice of the scientific outlook on development and the construction of an ecological civilization society.

References

Cai SQ, Zhang Y (2018) Textual interpretation and system interpretation of green principle. J Gansu Univ Polit Sci Law5:1–8

Greenpeace (2007) The investigation report on the enclosure deforestation in Yunan by Jin-guang APP.https://www.greenpeace.org.cn/china/Global/china/_planet-2/report/2007/11/app-5.pdf. Accessed 18 July 2022

[56] Art. 1234Civil Code.

[57] Art. 1234Civil Code.

Liu Y (2017) Judicial practice and theoretical exploration of prosecutorial public interest litigation. J Natl Acad Public Prosecutors2:3–18+170
Ruan LJ (2013) Analysis of the judicial practice on the standing of plaintiffs in EPIL. Jiangxi Soc Sci12:157–162
SPC (2018) Main points of the work of the people's courts in 2018, Fafa [2018] No 3
Tao L (2013) The process of China's EPIL system and its outlook. Environ Rule Law China1:93–110
Zhang SJ (2007) Who can speak for the "non-speaking fish" - the construction of China's EPIL system with "plaintiff standing" as the core. J Kunming Univ Technol (Soc Sci Edn)3:10–15
Zhang SJ (2016) The dilemma of environmental courts and the way out — from the perspective of the scope of environmental courts. Law Forum2:52–58
Zhang SJ, Xie W (2007) A preliminary study on EPIL by the Procuratorate. Soc Sci5:107–109

Chapter 3
Defining Public Interest Under the Environmental Public Interest Litigation System

Shijun Zhang

Abstract Interest is of great social and economic diversity and complexity. On the one hand, interest demonstrates a variety of complexities in terms of the subject matter, contents and form of expression. It is therefore necessary for us to delineate the structural hierarchy of interests for a better understanding of the interests to be protected, how they are to be protected and to what extent they are to be protected. The holistic nature of the environmental public interest lies in the fact that it meets the need for a healthy, clean, green and harmonious environment in which to live. EPIL involves multiple, as well as competing, interests. The adjudication process is essentially the process by which judges balance, remedy, or redistribute interests. As some scholars have argued, at a macro level, there should be a hierarchy of rights and interests in legal value judgments. EPIL protects the public interest by weighing the interests to a certain extent and protecting the rights and interests of greater value at the expense of the rights and interests of lesser value.

Keywords Classification of legal interests · Environmental public interest litigation · Private interest · Public interest

3.1 Public Interest and the Environmental Public Interest

3.1.1 The Notion of Interest

3.1.1.1 Definition of Interest

The word "interest" originally comes from the Latin word "*interesse*". It originally meant "in the middle of something", "to participate in something", or "to exist in something". In the ancient texts of China, "interest" was originally composed of two Chinese characters ("LI"(利) and "YI" (益)) with separate meanings. According to the interpretation of *Shuowen Jiezi (Chinese:说文解字)*, an ancient Chinese

S. Zhang (✉)
Law School of Shandong University, Shandong, China
e-mail: zhangshijun@sdu.edu.cn

X. Wang et al. (eds.), *Environmental Public Interest Litigation in China*,
https://doi.org/10.1007/978-3-031-26526-6_3

dictionary from the Han dynasty, "LI" is related to harmony. This means that things or people develop in harmony with each other to have good benefits. For example, *The I Ching (《易经》), or Book of Changes,* says that one must seek unity with morality to get the benefit.[1] "YI" (益) means being affluent. In addition, affluence means happiness. Later, people put the Chinese charters ("Li" (利) and "YI" (益)) together and made a new word "interest" due to the similarity between "LI" and "YI" in the sense of good things in our daily life, such as food, clothing, housing, and transport.[2] By this token, the basic semantic meaning of "interest" is deemed to be benefit or utility.[3] The encyclopedia of standard Mandarin Chinese, *Cihai (《辞海》),* essentially defined the word Interest or 'Li Yi' in the same way, stating that "interest is the same as benefit, such as collective interest and individual interest".[4] *The Oxford Dictionary of Law* defines 'interest' as "a claim, demand, wish or need of an individual or a group of individuals seeking to be satisfied and protected".[5]

Definitions of interest vary in different disciplines. For example, from a philosophical point of view, interest is expressed as a particular mental or material object that has meaning for the subject. This object is directly perceived, reasonably assumed or recognized by the subject itself or other evaluators as valuable (useful, necessary, or worth pursuing) for the existence of the subject in question. In the field of law, it has been argued that interests represent the various needs of people for certain objects, which are governed by objective laws and arise to satisfy their own survival and development.[6] The environmental social relationship regulated by environmental law is also essentially a relationship of interest. According to Roscoe Pound, "law cannot create interest"[7] but can identify, safeguard, and balance it. As "the different needs of people expressed through social relations",[8] interest is an important social phenomenon in social life. The interest of human beings is the original driving force for their survival and development, and "everything that human beings struggle for is related to their interest".[9] Human activity is so closely related to interest that Holbach even saw it as the sole motivating force of human action.[10] Engels pointed out that "the economic relations of every society manifest themselves above all as interest".[11]

---

[1] *The Book of Changes* is an ancient Chinese philosophy book which constitutes a common source for both Confucianist and Taoist philosophy. It has exerted a living influence in China for more than 3000 years. It represents one of the first efforts of the human mind to place itself within the universe.

[2] See Zhang (2002), p. 1.

[3] See Cihai (1988), p. 188.

[4] See Xia (1989), p. 1955.

[5] See Walker (1988), p. 454.

[6] See Li (2001), pp. 57–59.

[7] See Pound (1984), p. 36.

[8] See Li (2001), p. 57.

[9] See Xu (1999), pp. 366–367.

[10] See Zhang (1996), p. 129.

[11] Marx and Engels (1973), p. 537.

3.1.1.2 Classification of Legal Interests

Interest is of great social and economic diversity and complexity. On the one hand, interest demonstrates a variety of complexities in terms of the subject matter, contents and form of expression. It is therefore necessary for us to delineate the structural hierarchy of interests for a better understanding of the interests are to be protected, how they are to be protected and to what extent they are to be protected.

There are multiple ways in which 'interests' have been classified by scholars in a range of disciplines, sometimes leading to complexity and confusion when this issue is being discussed. Some examples of the different ways in which 'interests' can be classified are as follows:

(a) The subjects of interest, where the classification is individual interest, group interest and social interest.
(b) The object of the interest, they can be divided into economic interest and non-economic interest.
(c) The timeframe to realize the interest: long-term, short-term, and immediate interests.
(d) The spatial scope of the activities of the area of the interest: overall interest, local interest and individual interest.
(e) The way of obtaining interests and the impact of interests on the subject, there can be direct interest and indirect interest.
(f) The relationship between law and interests, interests can be divided into personal interest and property interest. Such interests are not limited to economic interests. Interests of non-economic value, such as aesthetics, recreation and comfort of life, are of paramount importance to people as well.

Traditionally, legal interests are divided into individual interests, group interests and social interests. In China, the term "social interest" is often referred to as the "social public interest" and the state sectors are entrusted to protect it. This proposition has been prefunded impacted by the theories of modern legal philosophers.

The British philosopher, utilitarian jurist Jeremy Bentham (1748–1832) advocated that individual interest should be paramount.[12] He believed that the public interest of society is a fictitious entity that represents nothing more than the 'aggregate of individual interests'.[13] Thus, increasing private interest means increasing the interest of society as a whole. The German jurist Rudolf von Jhering (1818–1892), on the other hand, emphasized social interest, or the combination of social interest and personal interest, and advocated striving to balance individual principles and utilitarian principles.[14] Jhering inherited Bentham's utilitarian tradition. He regards rights as the purpose and fundamental sign of law, and rights are interests protected by law. The difference between Jhering and Bentham is that Jhering does not

[12] See Bentham (2011).
[13] Bentham et al. (1970), p. 12.
[14] See Jhering (1997).

emphasize personal interests, but emphasizes social interests or the combination of social interests and personal interests, and strives to balance personal principles and utilitarian principles. Jhering's theory had a considerable influence on the legislation after the reunification of Germany at the end of the nineteenth century, and promoted the transformation of western law from individual-based to social-based. His "social interest" theory directly constitutes the ideological source of interest law.[15]

Moreover, as a communist country, the modern philosophy of law in China has been greatly impacted by Karl Marx and Friedrich Engels' classics. When the nature of legal interests is concerned, Karl Marx and Friedrich Engels stated that, "Over against the spheres of civil law and personal welfare, the family and civil society, the state is on the one hand an external necessity and their superior authority, and both their laws and interests are subordinate to and dependent upon the nature of this authority".[16] Individuals can only have their own interests if they maintain "public order, public safety, and the public interest".[17]

3.1.2 Public Interest

The term 'public interest', also known as the public good, is a multifaceted, multilayered, variable and flexible concept, generally considered to be the well-being and interests of the public in question, or the interests belonging to the community. In theory, it usually refers to an uncertain interest of the majority. In the literal sense, it is a concept opposite to the individual, private interest. In contrast, the term private interest refers to the interests of individual social agents, while the public interest focuses on the common overall interests of all social agents in society.

According to the *Oxford Advanced Learner's English-Chinese Dictionary*, public means "of or relating to the public" or "for the public, (especially if provided by the central or local government)".[18] The essence of public interest is the common interest of a certain group. It is neither the interests of individual members of society, nor a simple aggregate of the interests of individual members of society. Instead, it is a synthesis of the interests of all members of society.

The public interest should have two major forms: the first is the public interest of society, i.e., the interest enjoyed by all or some members of society, and the second is the interest of the state.[19] The purpose of public interest litigation is to protect the state and the public interest of society, to stop administrative subjects from abusing

[15] Fu (2001), p. 32.
[16] See Marx and Engels (1975), p. 5.
[17] See Marx and Engels (1972), p. 609.
[18] See Horby (1997), p. 1196.
[19] See Zhang and Tian (2009), pp. 1438–1448.

their power and endangering the state and society, to promote the formation of good social order and to pursue social justice and fairness.[20]

From a developmental perspective, the public interest refers to the common need of a certain range of citizens to seek environmental benefits and create social order, in accordance with the contemporary cultural and social contexts and the original constitutional intent.[21] The public interest represents the interests of the majority of individuals and is clearly individual in nature, but it is not simply the aggregate of individual interests.[22]

Along with the classification of interests that have been discussed in the previous section, public interest may fall within some general categories from the perspective of the subject and content of interest. First, public interest has been defined as the aggregate of individual interests, and this was the major argument of Jeremy Bentham.[23] Second, public interest is an alternative to the social interest. Western scholars of social contract theory believe that the public interest and the state interest are consistent and that the state interest here refers specifically to the interest in the sense of political domination, i.e., the interest of the government.[24] Third, public interest refers to the state interest. General will, according to Rousseau, evolved into the public interest after the formation of the state. There is also the view that the public interest is the social interest, the interest of the majority of society, or the interest of society as a whole, and these views can be summarized as the Theory of the Social Interest.

In terms of content, interest can be divided into one-level theory, two-level theory and three-level theory. In "One-Level Theory", scholars restrict the subject of the public interest to the public only. The Japanese scholar Takeshi Kojima argues that the public interest refers only to the interest of the unspecified majority, or the public interest of society.[25] In "Two-Level Theory", scholars believe that the public interest is the public interest of society and the state interest and that the public interest is the interest enjoyed by all or some members of society.[26] This doctrine considers the national interest and the public interest of society (the interest of all or some of its members) as elements of the public interest. In "Three-Level Theory", scholar Han Bo believes that the public interest consists of three levels: first, the national interest, which is the core of the public interest, such as state-owned assets; second, the interest of the unspecified majority, which is the permanent form of the public interest, such as the interest of the unspecified majority of consumers, victims of environmental pollution, and those who suffer from monopolistic operations; third, the interest that requires special protection, which is the special form of the public

[20] See Li (2004), pp. 12–20.

[21] See Guo (2015), pp. 9, 16, 174.

[22] See Wu (2017), p. 5.

[23] See Zhang (2011), p. 5.

[24] See Yu (2010), pp. 87–98.

[25] See Kojima (2001), pp. 40–41.

[26] See Yan (2019a), p. 19.

interest that must be protected for balanced and sustainable social development, such as the interest of the elderly individuals, children, women, and the disabled.[27]

In common law systems, the public interest refers primarily to "principles and standards regarded by the legislature or the courts as fundamentally relevant to the state and society as a whole, which require that the general public interest and the good of the community be taken into account so as to give the courts reason to refuse to recognize the legal effect of certain transactions or other acts of the parties".[28] According to the Anglo-American concept of constitutional democracy, the public interest is based on individual interest, not directly from the "people", but in the plural, or private individuals. In common law countries, there is also the concept of "Public Interest Law", which refers to a general term for laws of public interest such as Civil Rights Laws, Poverty Relief Laws, Environmental Protection Laws and Health Care Laws. According to this interpretation, public policy principles allow for restrictions to be placed on the freedom of contract or private dealings of the parties and set the necessary limits on the performance of any act that may cause harm to the general public interest.[29] In civil law countries, the concept associated with "Public Policy" is that of public order, also known as public decency. In the civil law tradition, the public interest is defined in substantive terms, as something that benefits as many people as possible and is as beneficial as possible to the lives of the beneficiaries.

In summary, the public interest is the common good of a not-well-defined majority, whereas a "not-well-defined" is different from the common good of an "identifiable" majority. The subjects of public interest include the whole population, the whole or the majority of the people in a given territory, special groups of people and unspecified majorities.

3.1.3 Environmental Public Interest

The public interest in the environment, or the environmental public good in environmental law, is the interest of the public in the environment.[30] The environmental public interest is a public interest based on the ecological services of the natural environment, i.e., the ability of the natural environment to meet certain needs of people.[31] By its nature, the environmental public interest is not a personal or property interest but rather an interest arising from people's certain needs from the environment, a human interest alongside personal and property interests.[32] In terms

[27] See Han (2013), pp. 31–37.
[28] See Xue (2003), p. 1117.
[29] See Han (2005), pp. 5–9.
[30] See Fu (2017), p. 14.
[31] See Wu (2017), p. 14.
[32] See Wang (2011), pp. 50–57.

of expression, the public interest in the environment is shared by all members of society and is not characterized by separateness, exclusivity or competition.[33] It is a kind of public interest. Some scholars believe that the "public interest" in environmental public interest litigation should only refer to the pure environmental public interest,[34] the subject of which cannot be determined and which arises from the unique function of the environment. This is not the sum necessarily of the interests of the majority, but rather the interests of a specific subject in a particular region or country who has suffered environmental damage. The environmental public good is both unified and layered. The unity lies in the fact that humans are a part of the whole ecosystem, and that living and nonliving things exist in the system, interdependent and limited. A disruption to any one of these could endanger the whole system and create a crisis in human life. The public interest in the environment, as a unified and integral ecological interest, is indivisible and is not enjoyed exclusively because of certain qualities. The hierarchical nature is characterized by geographical differences, and geographical differences require attention to the characteristics and requirements at different levels when interpreting the environmental public interest.

3.1.4 Links and Differences Between Environmental Public Interest and Public Interest

3.1.4.1 Links Between Environmental Public Interest and Public Interest

The concept of public goods is directly related to the public interest, and Samuelson's explication of the concept of public goods provides a good illustration of the scope of the public interest. He does not treat public goods as an outcome of government action but as a class of goods with specific properties. Examples he uses to illustrate public goods are national defence, security, lighthouses, rule of law and order, transport facilities, health care, the environment, and scientific research and education. Samuelson analyses the concept of public goods by using private goods as its counterpart.[35] Pound argues that the public interest includes the social interest, which is subdivided into six categories: "...Fourth, the interest in the protection of the resources of society. This consists of the protection of natural resources, i.e., the protection of forests, energy sources, etc. ...".[36]

The environment has the attributes of public goods and covers certain public interests, so the environmental interest is still a kind of public interest.[37] The public

---

[33] See Wu (2017), p. 14.

[34] See Yan (2019b), p. 56.

[35] See Samuelson (1964), pp. 1192–1203.

[36] See Pound (1959); Shen (1996), pp. 291–295; Zhang (1997), pp. 300–303.

[37] See Wu (2017), p. 13.

interest is holistic,[38] universal in its subject matter, and nonexclusive and noncompetitive in its content. There are similarities between the environmental public interest and the public interest in terms of their characteristics.

The holistic nature of the environmental public interest lies in the fact that it meets the need for a healthy, clean, green and harmonious environment in which to live. Here, the environment is a whole that contains the natural ecosystem, not a partial composition or a specific part of it. The environment consists of the natural environment, such as air, water and soil; and the human environment, such as cultural artefacts and relics. These elements form a large living whole, and people and other elements are in this community, each of which is an important link and cannot be separated from the others. Integrity is also reflected in the needs of all humans and the unanimous pursuit of environmental interests by all humans reflects the integrity of environmental interests. It is not the pursuit of a beautiful environment by people in a particular region that is sufficient; their local pursuit is ultimately not the environmental health of the whole. Local satisfaction of environmental interests cannot realize the overall interests. We should advocate more care for the environment when enjoying the public benefits of the environment, so that everyone can enjoy the benefits of the environment and future generations can enjoy these benefits as well.

The universality of the public interest in the environment is reflected in the subject and content of the interest. People of all regions and times have the desire to pursue a healthy and clean environment, which is also the desire for a better life. The subjects of interest are diverse, whether they are natural persons, groups of social organizations or state organs, all of whom can pursue the public interest in the environment. This is what distinguishes it as a 'public interest' from a 'collective interest'. The plurality of non-specific subjects of interest allows everyone to enjoy the benefits and to pursue a healthy and beautiful environment. Regardless of status, age or geographic location, all humans in the environmental community have the opportunity to do so. The scope of environmental public interest includes direct and indirect interests, including economic interest, health interest, etc., including the interests of the present and future generations and the environmental interests of creatures other than humans. Its content is universal, which is different from the private interest protected in traditional civil litigation. This is because private interests are more focused on economic interests without considering long-term indirect interests.

The non-exclusive nature of the public interest in the environment means that it is difficult to exclude anyone from the enjoyment of the public interest in the environment. Each subject benefits from the public interest without being subject to the will of others. Generally speaking, no natural person enjoys the public benefits of the environment to the same extent as others. The non-competitive nature of the public interest in the environment means that the enjoyment of the public interest in the environment by one person does not diminish the degree of satisfaction of the public

[38] See Jiang (2014).

interest in the environment by other persons. Every subject enjoys the public interest in the environment to the same extent.[39]

3.1.4.2 Distinction Between Environmental Public Interest and Public Interest

The environmental public interest belongs to a category of public interests, but due to the special nature of the environment, the content of the environmental public interest is correspondingly different from other public interests.

Firstly, the environmental public interest is a fundamental human interest and has priority among all public interests. The public interest meets people's needs for resources and conditions for survival and development. The environmental public interest provides for the satisfaction of a healthy and clean environment and does not include other resources and conditions. It is fundamental to the maintenance of human life and the continuity of generations and is more important than other public interests, which must give way to human survival and health when they conflict with it.[40]

Secondly, the content of environmental public interest has gone beyond the property of material ownership and is reflected in the quality of the environment suitable for human survival and development. The essence of environmental public interest is the interest in the healthy survival of human beings, and once destroyed, the environmental public interest is irreversible and difficult to repair. Therefore, it requires special protection.

Thirdly, the environmental public interest has a longer time dimension than other public interests. It is needed not only by the present generation, but also by future generations, and it needs to be maintained in perpetuity. Both the natural and human environments change with time, and in pursuing and defending the environmental public interest, there is a greater need to be mindful of the trends of the times and the changes in the general environment. With the progress of the times, interests that were not previously part of the environmental public interest will be incorporated with new content.

Finally, the environmental public interest is trans-regional in its spatial dimension, and its subjects of interest and scope of influence are far greater than those of other public interests. In the earth's environment, all environmental materials and energy flow over a considerable area, and the scope of influence of many events transcends national boundaries. For example, environmental pollution is highly mobile, with water and air pollution exhibiting a strong transregional dimension and a wider range of environmental public interests that they can infringe upon. Today, global interests are tightly linked to form a global community of interests.

---

[39] See Wang (2011), pp. 50–57.

[40] Fu (2017), p. 15.

3.2 Weighing the Interests of the Environmental Public Interest Litigation System

3.2.1 Private Interests Versus Public Interests

From an economic point of view, the characteristics of public goods are distinct from those of private goods in three ways: (1) the non-exclusivity of consumption; (2) the non-exclusivity of benefits; and (3) the high or even unquantifiable private transaction costs of producing public goods.[41] By comparing the characteristics of public goods with those of private goods, it is clear that the natural environment is a typical public good and environmental benefits are typically public benefits. Public goods such as the environment are holistic in nature and cannot be divided into parts and supplied separately to different environmental consumers.

However, not all public goods are in fact purely public goods, the characteristics of which have been listed above. Public goods also cover a wide range of objectives and are hierarchical in nature. Generally, all goods for public consumption have the characteristics of public goods. Goods for entirely public consumption without private consumption are purely public goods. Otherwise, they are not purely public goods. There are great differences between the various types of public goods, and even the same public goods may have different uses in different situations. The environment, as a kind of public good, has a variety of characteristics and can therefore be further classified into two categories, depending on its nature. For example, land, forests, minerals, water and other natural resources are owned by the state or the collective (they can also be contracted by units or individuals who have the right to use and benefit from them) and have a certain degree of exclusivity. However, the atmosphere and ecological benefits cannot be exclusive to environmental users.

The significance of distinguishing between the two types of environmental public goods lies in the fact that different measures of protection can be taken for environmental public goods of different natures. For the former, market-based instruments can be used to protect environmental public goods through the clarification of property rights. For the latter, market-based instruments can hardly work. If we look at the issues of environmental public goods from the perspective of environmental litigation, the former can be brought on the basis of proprietary rights or through the EPIL system. However, when environmental public goods that are not exclusive or exclusive but have been infringed upon, according to the existing legal provisions, environmental litigation cannot be brought on the basis of proprietary rights. The only way to protect the environmental public interest is to introduce the EPIL by legal means.

EPIL is "a litigation system whereby any person who believes that the environment has been directly or indirectly damaged or is in danger of being damaged,

[41] Mao (1996), pp. 97–98.

brings a lawsuit in his or her own name on behalf of the state or the public against an environmental actor, seeking an order to stop the environmental damage that is occurring or is likely to occur, or to compensate for the environmental damage that has been caused."[42] EPIL aims to protect the public interest, not just private interests. It expands its function from environmental dispute resolution to the protection of the public interest. It has the effect of stopping environmental infringements and safeguarding the environmental public interest; at the same time, it supervises the administrative actions of the government in terms of the environment and natural resource protection. In this sense, EPIL is essentially a reflexive benefit, i.e., it will bring some benefits to some individuals while safeguarding and promoting the public interest.[43]

Scholars have different views on whether the two interests should be traded off, which can be divided into the 'private interest priority' theory, the 'public interest priority' theory and the 'intertwined public and private interest' theory. The private interest priority theory holds that private interest involves the right to life of the parties. According to the principle that private interest protection is superior to public interest protection, when private interests and public interests are harmed, private interests should be prioritized. If public interest litigation is given absolute priority, it will inevitably delay the relief of private interests, given the number of people involved, the scope of impact, and the time and effort involved in the trial.[44]

In contrast, the public interest priority theory, which is the prevailing view, believes that public interest litigation protects the environmental public interest. The traditional forms of civil remedies by Chinese courts that aim to protect the environmental public interest, such as to stop the infringement, remove obstructions, and eliminate danger, can realize the environmental private interest at the same time; thus there is no need to bring separate environmental private interest litigation based on the same set of facts.

Generally, environmental cases damaging private interests have the following features: a large number of victims; complex case scenarios; high litigation costs; long duration of the litigation process, etc.[45] The intertwined public and private interest theory is said to help create a synergy between the public and private interests, thereby balancing the preservation of the public and private interests. In most cases, the realization of public interests also means the realization of individual interests.[46]

[42] See Zhang (2011), p. 31.

[43] See Lv (2000), pp. 112–113.

[44] See He (2019).

[45] Ibid.

[46] See Ding and Guo (2020), pp. 205–210.

3.2.2 *Protection of the Public Interest*

EPIL in China is designed to protect the public interest, not just the private interest. More importantly, it only requires the claimed interest to be related to the public interest, no matter whether the litigant has a "direct interest" in the environmental harm.[47] In EPIL, anyone can be a potential beneficiary of environmental protection as well as a potential victim of the environmental damage. Therefore, the Chinese legislatures gradually removed certain restrictions on standing to sue in the EPIL, as a way to enhance public interest protection. Eligible litigators are not limited to those with a direct interest but include those with a potential interest to sue for environmental damage.[48]

Damage in the context of EPIL can either be real or potential. One important feature of the environmental public interest is that, once it has been infringed upon, it will be difficult to remedy or to restore it to the original state.[49] Due to the irreversibility of most environmental problems, EPIL must be able to prevent environmental risks or damage. For the same reason, prohibitive litigation, a preventive rights protection mechanism, can be brought against potential environmental administrative actions and environmental civil actions that may give rise to environmental damage.[50]

EPIL involves multiple, as well as competing, interests. The adjudication process is essentially the process by which judges balance, remedy, or redistribute interests. As some scholars have argued, at a macro level, there should be a hierarchy of rights and interests in legal value judgments.[51] EPIL protects the public interest by weighing the interests to a certain extent and protecting the rights and interests of greater value at the expense of the rights and interests of lesser value.

The distinctive measures of the EPIL system highlight its unique function in public interest protection. The EPIL helps to achieve the checks and supervision of the judicial power over the administrative power. On the one hand, EPIL mainly targets the environmental actions of the government; on the other hand, the EPIL system can also play the role of supervisor of the general public. EPIL can not only remedy environmental damage in the way that the traditional litigation systems do, but more importantly, it can effectively prevent environmental damage. The establishment of the EPIL system provides institutionalized access for the public to participate in environmental protection in China.[52]

---

[47] See Wu (2017), p. 19.

[48] See Zhang (2011), p. 31.

[49] See Wu (2017), pp. 19–20.

[50] See Zhang (2011), p. 32.

[51] See Chen (2000), p. 240.

[52] See Xu et al. (2009), pp. 52–54.

References

Bentham J (2011) The principles of morals and legislation. Commercial Press, Beijing

Bentham J, Burns JH, Hart HLA (1970) An introduction to the principles of morals and legislation. Athlone Press

Chen QS (2000) Sustainable development and legal change. Law Press, Beijing

Ding GM, Guo SJ (2020) The problem of defining public interest private interests in environmental tort litigation and the choice of strategy. Soc Sci Front 11:205–210

Fu Z (2001) Legal interpretation of interests. The Jurist 2:11–32

Fu JQ (2017) On environmental public interest damage relief — from "tragedy of the commons" to "relief of the commons". China Social Science Press, Beijing

Cihai (1988) Hunan and Henan Dictionaries Revision Group and the Editorial Department of the Commercial Press. The source of words. Commercial Press, Beijing

Guo JH (2015) A trial discussion on the legal basis for the establishment of plaintiffs in China's environmental public interest civil litigation and its influencing factors. J Chongqing Second Normal Coll 1: 9, 16, 174

Han DY (2005) Normative analysis of "Public Interest" in constitutional texts. Leg Forum 1:5–9

Han B (2013) The power combination of the public interest litigation system. Contemp Jurisprud 1: 31–37

He J (2019) Harmonization of environmental public interest litigation procedures. Chongqing University

Horby (1997) Oxford advanced English-Chinese dictionary. Commercial Press, Beijing

Jhering RV (1997) The struggle for law, 2nd edn. The Lawbook Exchange, Ltd., New York

Jiang SQ (2014) On environmental public interest and its related institutional construction. Chongqing University

Kojima T (2001) Jurisprudence and demonstration of litigation system reform. Law Publisher, Beijing

Li SY (2001) The theory of value orientation of administrative law. J Soc Sci Shanxi Higher Sch 11: 57–59

Li KY (2004) The construction of administrative public interest litigation system – the plaintiff of administrative public interest litigation. J East China Univ Polit Sci Law 5:12–20

Lv ZM (2000) A new vision of environmental law. China University of Political Science and Law Press, Beijing, pp 112–113

Mao SL (1996) An economic analysis of the functions of the Chinese government. China Radio and Television Press, Beijing, pp 97–98

Marx K, Engels F (1972) The complete works of Marx and Engels, vol 2. People's Publishing House, Beijing

Marx K, Engels F (1973) Selected works of Marx and Engels, vol 2. Lawrence & Wishart, p 537

Marx K, Engels F (1975) Marx-Engels collected works, vol 3. Lawrence & Wishart, p 5

Pound R (1959) Jurisprudence, vol 3. Western Publishing Company, Racine

Pound R (1984) Social control through law: the task of law. Commercial Press, Beijing

Samuelson PA (1964) Economics. Commercial Press, Beijing, pp 1192–1203

Shen ZL (1996) Modern western jurisprudence. Peking University Press, Beijing, pp 291–295

Walker DM (1988) The Oxford dictionary of law. Guangming Daily Press, Beijing

Wang XG (2011) On the interests and rights basis of environmental public interest litigation. J Zhejiang Univ (Humanit Soc Sci Edn) 3:50–57

Wu YJ (2017) A study on the diversity of environmental public interest litigation subjects in China. China Procuratorate Press, Beijing

Xia Z (1989) Dictionary of the sea (reduced version). Shanghai Dictionary Press, Beijing

Xu XM (1999) Jurisprudence tutorial. China University of Political Science and Law Press, Beijing, pp 366–367

Xu XM, Hu ZH, Mei H et al (eds) (2009) A study of environmental public interest litigation - centered on institutional construction. China Legal Publishing House, Beijing, pp 52–54

Xue B (2003) Yuanzhao dictionary of common law. Law Press, Beijing
Yan YQ (2019a) A study on the theory and system of ecological environmental public interest litigation delivery with Chinese characteristics. China University of Political Science and Law Press, Beijing
Yan YQ (2019b) Research on the concept and practice of public interest litigation. Law Publishing House, Qingdao
Yu SX (2010) What is public interest – analysis of the concept of public interest in western legal philosophy. Jianghuai Tribune 2:87–98
Zhang WX (1996) A study of western philosophical thought in the twentieth century. Law Press, Beijing
Zhang NG (1997) An outline of the history of western philosophy (revised edition). China University of Political Science and Law Press, Beijing, pp 300–303
Zhang JH (2002) Interest and politics. Peking University Press, Beijing
Zhang SJ (2011) Research on the qualification of plaintiffs in environmental public interest litigation. Shandong Wenyi Publishing House, Jinan
Zhang SJ, Tian J (2009) Basic concept and scope definition of environmental public interest litigation and setting of plaintiff types. In: Ecological civilization and environmental resource law – 2009 National symposium on environmental resource law (annual conference) paper, pp 1438–1448

Chapter 4
From Compensation to Prevention: Expanding the Function of EPIL in China

Kaijie Wu

Abstract China's current environmental public interest litigation (hereafter "EPIL") laws and cases focus on remedying environmental damage instead of risk prevention. Social organizations have filed a few preventive civil litigations but they have encountered difficulties. The existing preventive civil EPIL cases are in the nature of tort litigation, and face limitations in terms of scope, adjudication basis and litigation effect in preventing environmental damage. Preventive EPIL in China should be enforcement litigation, rather than tort litigation, to achieve a preventive function by supervising and assisting administrative organs to implement preventive regulations. In order to strengthen its preventive function, the EPIL system should expand the scope of cases, avail of the independent value of pre-litigation procedures, evaluate the performance of duties with behavioral standards, and strengthen the support of environmental professional capacity. These improvements could be realized by revising judicial interpretations in the short term and formulating special legislation in the long term.

Keywords Environmental public interest litigation · Civil public interest litigation · Administrative public interest litigation · Tort litigation · Enforcement litigation

4.1 Introduction

The current practice of environmental public interest litigation (hereafter "EPIL") in China is characterized by ex post facto relief. In most EPIL cases, environmentally harmful behaviors have already caused damage to the environment. For example, an energy company built a wind farm in the Bagong Mountain Scenic Area in Anhui Province. Although it would damage the environment of the scenic area and violate relevant laws, it was approved by the local government due to the demand for

K. Wu (✉)
Peking University Law School, Beijing, China
e-mail: wukaijie@pku.edu.cn

X. Wang et al. (eds.), *Environmental Public Interest Litigation in China*,
https://doi.org/10.1007/978-3-031-26526-6_4

investment.[1] After inspection by the central environmental protection inspection team, the company finally dismantled the completed wind farm, causing economic losses amounting to more than 300 million RMB, not including the cost of environmental damage. Since then, Friends of Nature (FON), one of the oldest environmental NGOs in China, has filed a civil EPIL against the energy company for environmental damage.[2] Post-event remediation is necessary, but often costly or even irreversible. If the wind farm project could be stopped by EPIL before construction, it would not only prevent extensive ecological and environmental damage, but also avoid huge economic losses.

Preventive EPIL is in line with the basic principles of environmental law and environmental justice in China. The preventive principle of environmental law in China requires EPIL to pay attention to prior prevention. Based on the profound understanding of environmental problems, Article 5 of the *Environmental Protection Law* (as amended in 2014) stipulates that "environmental protection shall adhere to the principle of prevention",[3] requiring that "in the whole process of environmental management, prevention beforehand shall be combined with remediation afterwards, and priority shall be given to prevention before it is too late".[4] As an important part of the *Environmental Protection Law*, the EPIL system should follow the principle of prevention. Preventive EPIL is also a reflection of the concept of preventive justice. As early as 2014, the Supreme People's Court (SPC) issued *Opinions of the Supreme People's Court on Fully Strengthening Environmental Resources Trial Work to Provide Powerful Judicial Safeguards for Promoting Eco-Civilization Construction*, which set "promoting the improvement of the strictest source protection system" and "insisting on prevention" as the guiding principles of environmental judicial work".[5] The Opinion further required courts at all levels to actively take judicial measures to prevent environmental damage, to reduce environmental risks and damage through preventive measures beforehand.

On October 31, 2019, the Decision of the Fourth Plenary Session of the 19th Communist Party of China (CPC) Central Committee clearly stated that "the scope of EPIL cases should be expanded".[6] On November 16, 2020, General Secretary Xi Jinping stressed that "we should improve the preventive legal system" and "we should actively promote legislation in important areas such as ecological

---

[1] See Yue and Ma (2016).

[2] See Friends of Nature (2019).

[3] Art. 5 of the Environmental Protection Law (1989, revised 2014).

[4] Xin (2014), p. 17.

[5] Opinions of the Supreme People's Court on Fully Strengthening Environmental Resources Trial Work to Provide Powerful Judicial Safeguards for Promoting Eco-Civilization Construction (No. 11 [2014] of the Supreme People's Court).

[6] Decision of the Central Committee of the Communist Party of China on Several Major Issues on Adhering to and Improving the Socialist System with Chinese Characteristics and Advancing the Modernization of the State Governance System and Governance Capability (Adopted at the Fourth Plenary Session of the Nineteenth Central Committee of the Communist Party of China on October 31, 2019).

civilisation".[7] In March 2021, the National People's Congress (NPC) approved the 14th Five-Year Plan for *National Economic and Social Development and the Outline of Vision 2035*, setting "improving the EPIL system" as an important part of "improving the modern environmental governance system".[8] The 14th Five-Year Plan also emphasized the importance of prevention, including "strict prevention and control of environmental risks," "prioritizing the protection of the people's health and adhering to the policy of prevention," and "to establish a system of public safety and security prevention and control". It can be seen that preventive EPIL is an important direction for the future development of the EPIL system.

With this background, this chapter first summarizes the current status of preventive EPIL laws and practice, then argues for law enforcement as an intrinsic part of preventive EPIL. Finally, it provides suggestions to improve the legislation around preventive EPIL.

4.2 Current Status of Preventive EPIL Laws and Practice

In China, EPILs are classified into civil EPILs and administrative EPILs governed by different laws. At present, EPIL laws at the national level have not explicitly allowed for preventive EPIL, in either civil EPILs or administrative EPILs. Judicial interpretations by the SPC, however, have confirmed the preventive function of civil EPIL by expansively interpreting relevant laws. However, there is no such expansive interpretation of administrative EPIL. On the basis of the national EPIL laws, some provinces have begun to explore preventive administrative EPIL filed by means of local legislation.

4.2.1 Preventive EPIL in Laws

4.2.1.1 Preventive EPIL in National Laws and Judicial Interpretations

In terms of civil EPIL, the Civil Procedure Law and Environmental Protection Law do not explicitly deal with preventive cases. The Civil Procedure Law stipulates that "the authorities and relevant organizations prescribed by law may file lawsuits in the people's courts against acts that harm the public interests of society, such as pollution of the environment and infringement of the legitimate rights and interests of consumers."[9] The Environmental Protection Law only allows qualified social

---

[7] Xi (2020).

[8] Outline of the Fourteenth Five-Year Plan for National Economic and Social Development of the People's Republic of China and Vision 2035 (approved at the Fourth Session of the Thirteenth National People's Congress on March 11, 2021).

[9] Art. 55 of the Civil Procedure Law (1991, amended 2007, 2012, 2017 and 2021).

organizations to file lawsuits in the people's courts for "acts that pollute the environment, damage the ecology, or harm the public interest of society".[10] The law is not clear as to whether "acts that harm the public interest" include "acts that have not yet caused damage but have a risk of damage".

Although the current laws on EPIL have not explicitly provided for preventive EPIL, the SPC has adopted a judicial interpretation that allows certain entities to prevent environmental damage by filing civil EPIL. The Interpretation of the Supreme People's Court on Several Issues Concerning the Application of Law in Environmental Civil Public Interest Litigations (hereafter "Civil EPIL Interpretation") stipulates that the authorities and relevant organizations as provided for by law may file lawsuits against "acts of environmental pollution and ecological damage that pose a significant risk of harming the public interest".[11] In line with this, the Judicial Interpretation further stipulates that the materials to be submitted for filing a civil EPIL include "the defendant's behavior has harmed the public interest or has a significant risk of harming the public interest".[12] Thus, the Civil EPIL Interpretation has expanded the interpretation of "damage to public interest" in Article 55 of the Civil Procedure Law and Article 58 of the Environmental Protection Law to include "significant risk of damage to public interest", which shows that the SPC is open to filing preventive EPIL in cases where damage has not occurred yet.[13]

In terms of administrative EPIL, both the Administrative Litigation Law and judicial interpretations fail to explicitly cover preventive cases. Article 25 of the Administrative Litigation Law provides that "where the people's procuratorate finds, in the performance of duties, that any administrative authority assuming supervision and administration duties in such fields as the protection of the ecological environment and resources …… exercise duties in violation of any law or conducts nonfeasance, which infringes upon national interest or public interest, it shall issue procuratorial opinions to the administrative authority, and urge it to perform duties in accordance with the law. If the administrative authority fails to perform duties in accordance with the law, the people's procuratorate shall file a lawsuit with the people's court in accordance with the law."[14] The *Interpretation of the Supreme People's Court and the Supreme People's Procuratorate on Several Issues concerning the Application of Law for Cases regarding Procuratorial Public Interest Litigation* (hereinafter "Procuratorial PIL Interpretation") further stipulates that the materials to be submitted by the procuratorates for filing administrative EPIL include, "materials proving that the defendant has exercised its power or failed to act in violation of the law, resulting in the infringement of state interests or social public

[10] Art. 58 of the Environmental Protection Law (1989, revised 2014).

[11] Art. 1 of the Interpretation of the Supreme People's Court on Several Issues concerning the Application of Law in Environmental Civil Public Interest Litigations (2015, revised 2020) (hereafter "Civil EPIL Interpretation").

[12] Art. 8 of the Civil EPIL Interpretation.

[13] See Cheng and Wang (2016), p. 39.

[14] Art. 25 of the Administrative Litigation Law (1989, amended 2014 and 2017).

interests".[15] Accordingly, the scope of administrative EPIL only includes acts "causing the infringement of national interests or social public interests", and requires the provision of "materials proving the infringement of national interests or social public interests" when prosecuting, without explicitly covering the situation of "causing the national interest or social public interest to be infringed".

4.2.1.2 Preventive EPIL in Local Legislation

Compared to the national legislature, local legislatures in China are more actively exploring preventive EPILs filed by procuratorates. As of 2021, 25 standing committees of the provincial People's Congress have issued special decisions to strengthen the PILs filed by procuratorial organs.[16] Some local legislation has allowed procuratorial authorities to issue pre-litigation opinions[17] or initiate public interest litigation in cases where the national interest and social public interests are at risk of being infringed upon. These local authorities include Guangdong, Shaanxi, Zhejiang, Shanghai, Yunnan and other provinces and municipalities, but with different expressions such as "the risk of suffering serious damage" (please see Table 4.1).

4.2.2 Preventive EPIL in Practice

4.2.2.1 Civil EPIL Cases

With the authorization by the Civil EPIL Interpretation, some qualified social organizations have attempted to initiate preventive EPIL against enterprises carrying out environmentally harmful projects since 2015. At least three of these cases have been accepted by courts (please see Table 4.2).

Among these preventive EPIL cases, the green peafowl case is the first case of preventive EPIL for wildlife protection in China, and has received the most attention. On July 12, 2017, FON filed a lawsuit over the impact of the Jiasha River Stage I Hydropower Station (hereafter "Jiasha Hydropower Station") on endangered species such as the Green Peacock in the main stream of the Red River (Yuanjiang) in

---

[15] Art. 22 of the Interpretation of the Supreme People's Court and the Supreme People's Procuratorate on Several Issues concerning the Application of Law for Cases regarding Procuratorial Public Interest Litigation (2018, revised 2020) (hereinafter "Procuratorial PIL Interpretation").

[16] See Zhang (2021), p. 18.

[17] The Administrative Litigation Law stipulates that procuratorates shall "issue procuratorial opinions to the administrative authority and urge it to perform duties in accordance with the law" first before filing litigations. Art. 25 of the Administrative Litigation Law (1989, amended 2014 and 2017).

Table 4.1 Provisions Regarding Preventive Procuratorial PILs by Local Legislatures

NO.	Place	Provisions Regarding Preventive Procuratorial PILs
1	Guangdong Province (2020)	If procuratorial organs find that there is *a risk that the national interest and social public interests may suffer serious damage*, or if they find other general problems, they may issue procuratorial opinions to the relevant state organs, so as to enhance *the preventive function* and governance effectiveness of procuratorial EPIL.[a]
2	Shaanxi Province (2020)	If the relevant administrative organs exercise their powers illegally or fail to act in a way that *may harm the national interest and social public interest*, a case shall be opened for investigation, ... if, after supervision by the procuratorial organs, the administrative organs still fail to perform their duties in accordance with the law, and the state interests or social public interests are still in a state of infringement *or in great danger*, the procuratorial organs shall file administrative EPIL in accordance with the law.[b]
3	Zhejiang Province (2020)	If it is found that there is *a potential risk of serious infringement of the national interest and social public interests*, procuratorial opinions may be sent to the administrative organ to urge it to take measures to eliminate the potential problems.[c]
4	Shanghai Municipality (2020)	Courts shall, in accordance with the law, hear litigation requests made by the procuratorial authorities around confirming violations of law against the public interest, *preventing damage to public interests*, and promoting the relief of damaged public interests, as well as to promote the application of punitive damages in PIL in relevant areas in accordance with the law.[d]
5	Yunnan Province (2019)	In the process of PILs, procuratorial organs may propose to the departments with regulatory responsibilities to improve their work and governance in cases where problems exist in the system and management of relevant units, and where *damage to the national interest or social public interests may result.*[e]

[a]Decision of the Standing Committee of the Guangdong Provincial People's Congress on Strengthening Prosecutorial Public Interest Litigation (Adopted at the Twenty-second Meeting of the Standing Committee of the Thirteenth People's Congress of Guangdong Province on July 29, 2020)
[b]Decision of the Standing Committee of the Shaanxi Provincial People's Congress on Strengthening Prosecutorial Public Interest Litigation (adopted at the 16th meeting of the Standing Committee of the 13th People's Congress of Shaanxi Province on March 25, 2020)
[c]Decision of the Standing Committee of the Zhejiang Provincial People's Congress on Strengthening Prosecutorial Public Interest Litigation (Adopted at the Twenty-first Meeting of the Standing Committee of the Thirteenth People's Congress of Zhejiang Province on May 15, 2020)
[d]Decision of the Standing Committee of the Shanghai Municipal People's Congress on Strengthening Prosecutorial Public Interest Litigation (Adopted at the 22nd Meeting of the Standing Committee of the 15th Shanghai Municipal People's Congress on June 18, 2020)
[e]Decision of the Standing Committee of the People's Congress of Yunnan Province on Strengthening the Public Interest Litigation Work of the Procuratorial Organs (adopted at the 13th meeting of the Standing Committee of the 13th People's Congress of Yunnan Province on September 28, 2019)

Table 4.2 Preventive EPIL Cases Filed by Social Organizations

Case Name	Date	Basic Facts	Claims
Green Development Council v. Yalong River Basin Hydropower Development Co., Ltd ("Acer Pentaphyllum Case")[a]	2015.9	The Acer pentaphyllum is a critically endangered wild plant, with fewer than 500 plants remaining in the world. The development of the Yagen II hydropower plant would threaten the survival of this endangered species.	The defendant should immediately suspends the construction of the Asagen II Hydropower Station and its auxiliary facilities until the defendant takes effective measures to eliminate the threat to the survival of the Acer pentaphyllum.
Friends of Nature v. China Hydropower Consultants Group Xinping Development Co. and others ("Green Peacock Case")[b]	2017.7	The Green Peacock is a national endangered wildlife, and the construction of the Jiasha River Hydropower Plant would inundate the habitat of the Green Peacock, seriously threatening the survival of the species.	The defendant should immediately stop the construction of the hydroelectric power station and not cut down the vegetation in the area flooded by the hydroelectric power station, such as Chen's Sutra.
Friends of Nature v. Yunnan Huarun Power (Xishuangbanna) Co. and others[c]	2018.3	Most of the vegetation in the construction and inundation areas of the Huilongshan Hydropower Station is tropical rainforest with high ecological value. The Luosuo River, where the hydropower station is located, is the only migratory channel for fish in the Lancang River Basin.	The defendant should stop cutting and destroying vegetation such as national protected plants and tropical rainforests and eliminate the danger to the environment caused by the construction of Huilongshan Hydropower Plant, etc.

[a]China Biodiversity Conservation and Green Development Foundation (2019)
[b]See ClientEarth (2021), p. 1
[c]See Du (2018)

Yunnan Province.[18] The plaintiff argued that the construction and operation of the Jiasha Hydropower Station would cause serious damage to the largest green peafowl habitat in China and would most likely result in the regional extinction of the green peafowl population. The plaintiff also stated that the project would threaten the survival of many protected species, such as Chen's sundew (a national-level protected plant) and the black-necked long-tailed pheasant (a national-level protected animal), as well as cause significant damage to the only remaining and still relatively well-preserved dry hot river valley monsoon rainforest in the Red River Basin. In view of such significant risks to the public interest, the plaintiff requested that the two defendants work together to eliminate the danger of the construction of the hydropower station on the Jiasha River environment, to

[18] See Diao (2017).

immediately stop the construction of the hydropower station, and not to intercept the flow of water for storage and not to cut down the vegetation in the area flooded by the hydropower station. On the morning of August 14, 2017, the FON received a notice of filing from the court.

After the trial, the court of first instance recognized that the inundation area of the Jiasha hydropower station is the habitat of the green peafowl, and once inundated, it was likely to cause serious damage to the survival of the green peafowl.[19] At the same time, the EIA Report of the hydropower station failed to evaluate Chen's Sutra, and the defendant did not take any protective measures for Chen's Sutra. If the construction of the hydropower station continued, it would expose the survival of rare flora and fauna in the region to significant risks. On March 20, 2020, the court of first instance ruled that the defendant immediately stoped the hydropower station construction project under the existing EIA. The defendant was instructed to conduct a post-evaluation of environmental impact in accordance with the requirements of the Ministry of Ecology and Environment (MEE), to take improvement measures and report to the MEE for the record. The relevant administrative departments were to make a decision according to the law relevant to the specific situation. Both parties appealed after the verdict was announced on December 31, 2020, and the court of second instance affirmed the verdict.[20]

The common features of the three preventive EPIL cases that have been accepted for trial by the courts thus far, as represented by the green peafowl case, are remarkable. To protect wildlife and their habitats, the plaintiffs filed litigations against the hydropower plant developers, trying to stop the damage before it happened. In such cases, social organizations are reluctant to wait until the damage has been done before asking the developer to compensate, repair or apologize, because once the habitat of endangered species is flooded, ecological restoration is almost impossible. Faced with the risk of irreversible environmental damage, social organizations believe that the only way to achieve the goal of environmental protection is to initiate EPIL.

These three cases do not represent all of the attempts by social organizations to initiate preventive EPIL, as some of the cases are politically sensitive and therefore difficult to be filed in court. For example, on October 27, 2015, the FON filed a lawsuit with the Kunming Intermediate Court, demanding that PetroChina Yunnan Petrochemical Company stop the construction of the Yunnan oil refinery project, which they submitted would cause serious water, air and solid waste pollution.[21] On December 25, 2015, the plaintiff, at the court's request, resubmitted evidence indicating that the Yunnan refinery project imposed a significant risk of harm to the public interest. However, it was not until January 3, 2017 that the court issued a ruling of inadmissibility on the grounds that "the EIA report of the project has been

[19] See the Civil Judgment of Kunming Intermediate People's Court ([2017] Yun 01 Min Chu No. 2299).

[20] See the Civil Judgment of Yunnan High People's Court ([2020] Yun Min Zhong No. 824).

[21] Friends of Nature (2017).

approved by the MEE, so the materials do not indicate that the defendant's actions have harmed the public interest or have a significant risk of harming the public interest". In another case filed by the China Biodiversity Conservation and Green Development Foundation, a hydropower company planned to carry out large-scale transplantation of an ancient tamarisk colony for construction of the Yellow River Yangqu Hydropower Plant. The plaintiff requested that the hydropower company should stop transplanting and suspend the hydropower project. After receiving the petition, the court repeatedly asked the plaintiff to correct the relevant information, but it had not been filed.[22] These cases show that social organisations face considerable resistance to bringing preventive civil EPIL in practice.

4.2.2.2 Administrative EPIL Cases

At present, only procuratorial organs are explicitly authorized by law to file administrative EPILs for remedying actual damage. The number of preventive EPILs initiated by procuratorial organs against administrative agencies is relatively small. Most administrative EPILs were initiated after environmental pollution and ecological damage had occurred. Very few cases were initiated in the planning, EIA or permitting stages beforehand. In a typical case of biodiversity protection released by the SPP in 2021, the Wenchang City Procuratorate issued opinions to the Wenchang City Bureau of Natural Resources and Planning, stating that the ecological damage caused by illegal reclamation and construction in the Qilin Cai Nature Reserve, and requesting that measures be taken to eliminate impact on the environment in accordance with the law. The administrative authorities concerned actively implemented rectification after receiving the procuratorial opinions.[23] Ecological restoration was carried out in the protected area, and illegal structures occupying the sea area and surrounding buildings in the protected area were removed. The Wenchang City procuratorate deserves recognition for effectively urging the administrative authorities to carry out ecological restoration work. However, it is worth reflecting on why the nature reserve has long failed to formulate a zoning plan and why the authorities have failed to approve project engineering permits in accordance with the law. The nature reserve would have been protected from environmental damage if supervision through EPILs had been carried out at the prior stage of planning and permitting.

Local procuratorial authorities should have been aware of the need for prevention; however, they face obstacles in filing preventive EPILs. First, as set forth above, there is no clear legal basis for filing administrative EPILs without actual damage, and the scope of preventive enforcement duties subject to litigation is not

[22] China Foundation for Biodiversity Conservation and Green Development (2017).

[23] Supreme People's Procuratorate (2021) Fourteen Typical Cases of Public Interest Litigation on Biodiversity Protection (Oct. 9), https://www.spp.gov.cn/spp/xwfbh/wsfbt/202110/t20211009_531433.shtml#2.

clear. As preventive cases often involve major decisions of local governments, procuratorates face greater resistance and confrontation.[24] In the absence of a clear legal basis, local procuratorial organs are conservative in this regard. Second, it is difficult for procuratorial authorities to obtain clues, investigate and collect evidence during the preventive stage. Procuratorates are not capable of assessing environmental risks before actual damage emerges due to a lack of professional staff and equipment. Third, preventive public interest litigation easily crosses the boundary between judicial power and administrative power. Preventive laws often give administrative organs wide discretionary powers, allowing them to make decisions based on their professionalism and flexibility in responding to complicated and dynamic issues. As a legal supervisory body, procuratorial authorities need to abide by the modesty of justice and therefore need to be more cautious.

4.2.3 Preventive PIL in Other Legal Areas

Although the laws at the national level have not explicitly allowed for preventive EPIL, they allow for preventive EPIL in other specific areas of governance. In addition to environmental protection, Chinese procuratorates have also been authorized to file PILs in other fields including food and drug safety, state-owned property protection, assignment of the right to use state-owned land, production safety and personal information protection. For example, the *Production Safety Law* provides that "if a work safety violation causes a potential risk of major accidents or causes a major accident, inflicting harm on the national interest or the public interest, the people's procuratorate may initiate public interest litigation under the relevant provisions of the *Civil Procedure Law* and the *Administrative Litigation Law*".[25] These words of "causing risks of major accidents" have a preventive connotation, which clearly authorizes the procuratorate to initiate preventive EPIL in the field of workplace safety.

For example, in a typical case released by the SPP, a local procuratorate found that a building construction planning permit violated relevant railway safety regulations and posed safety hazards, so it issued a pre-litigation opinion to urge the responsible agency to revoke the illegal permit before the construction began.[26] After receiving the pre-litigation opinion, the agency made a written reply to the procuratorate, saying that it had set up a task force, held several special study sessions, developed corrective measures, and revoked the illegal construction planning permit.

[24] Li and Wu (2016), p. 10.

[25] Art. 74 of the Safety Production Law (2002, revised 2009, 2014 and 2021).

[26] The Supreme Prosecutor and the State Railway Group jointly issued 10 typical cases of public interest litigation in the field of railroad safety production (December 24, 2020).

The legislation and judicial practice of preventive PIL in the field of workplace safety helped prove the necessity and feasibility of preventive EPIL.

4.3 Arguments for Enforcement Litigation Nature of Preventive EPIL

The legal nature of EPIL must be identified first to serve its preventive function.[27] While EPILs are now divided into civil and administrative EPILs by law, such a division does not reflect their functional nature. In accordance with their different functions in protecting public interests, EPILs are theoretically comprised of two distinctive types: the first type uses a tort liability mechanism to remedy the damaged environmental public interest, which can be referred to as "tort litigation"; the second type of EPIL supervises and assists administrative agencies in enforcing environmental law, which can be called "enforcement litigation".

At present, the preventive EPILs filed by social organizations are by nature tort litigations, which have insurmountable limitations in preventing environmental damage. To max out the preventive function, preventive EPILs should have an enforcement dimension.

4.3.1 Different Nature of EPIL and its Preventive Function

The differences between these two types of EPIL mainly involve the function, scope, defendant, and form of liabilities. First, tort litigation is mainly aimed at restoring and recovering environmental public interests, while enforcement lawsuits are tasked with maintaining the objective environmental legal order. Second, tort litigation is filed against environmental damage caused by the use of the environment, without the violation of legal provisions as a prerequisite, while the plaintiff of the enforcement litigation can only file a lawsuit against a violation of environmental laws and regulations. Third, tort litigation only prosecutes enterprises polluting or damaging the environment, while law enforcement litigation prosecutes both enterprises and administrative organs. Finally, tort litigation is based on the liability to repair the damage, while enforcement litigation is based on the liability to detain or correct the wrongdoing.

Current civil EPIL only has the nature of tort litigation, but not enforcement litigation.[28] The tort litigation nature of civil EPIL is a unique feature of China's civil

[27] Environmental public interest litigation is neither a civil nor an administrative litigation. See Lv (2008), p. 133.

[28] China's current environmental civil public interest litigation has the nature of tort litigation and should be transformed to the positioning of public law litigation. See Gong (2019), p. 127.

EPIL legal system. Some countries, such as the US, only have civil EPILs, which are the nature of enforcement litigation.[29] Some other countries do not have civil EPIL against enterprises, but only administrative EPIL against administrative agencies, such as Germany. Such tort litigation nature of EPIL was designed to remedy environmental damage in China. As the relief mechanism led by administrative organs had long been ineffective, there was an urgent need for EPIL to undertake the task. To perform the relief function, the EPIL system borrows its main rules from tort liability laws, and makes reforms and innovations with the goal of protecting the environmental public interest. Although the main function of the EPIL system is to remedy ecological and environmental damage, it has also been expanded for prevention. The Civil EPIL Interpretation allows statutory authorities and relevant organizations to file lawsuits for "pollution and ecological damage that poses a significant risk of harming the social public interest of the society", and correspondingly stipulates the preventive liability of eliminating danger, stopping infringement and removing the nuisance.[30]

Unlike civil EPIL, China's administrative EPIL has the nature of enforcement litigation, but has not been explicitly assigned a preventive task.[31] According to the Administrative Litigation Law, the conditions for the procuratorial authorities to initiate administrative EPIL include not only the "illegal exercise of authority or inaction of the administrative organ with enforcement responsibilities", but also the "infringement of national interests or social interests".[32] In practice, the procuratorates hold the view that they have to fulfil both requirements to satisfy the standing requirement to file an administrative EPIL.[33] Since the expression "infringement of the national interest or social public interest" does not explicitly include "significant risk", it is doubtful whether the procuratorial authorities can file an EPIL before environmental damage occurs. In addition, an internal rule issued by the SPP to local procuratorates requires that "in handling administrative EPIL cases, the circumstances under which the environment has been damaged should be identified first" and "the fact that the national interest or social public interest has been infringed, i.e., the process, fact and extent of environmental pollution".[34]

[29]The U.S. citizen suits filed directly against illegal enterprises is clearly positioned as a kind of law enforcement litigation, the scope of the case is the violation of specific environmental laws and regulations, liability is limited to the "injunction" and "civil fine" for the correction of errors. It has the same effect as administrative enforcement, and avoids duplication with administrative enforcement through procedures such as the pre-litigation notice and diligent enforcement. See Gong (2017), p. 173.

[30]Art. 18 of the Civil EPIL Interpretation.

[31]Administrative public interest litigation is not a subjective litigation, but an objective litigation. See Liu (2018b), p. 39.

[32]Art. 25 of the Administrative Litigation Law (1989, amended 2014 and 2017).

[33]Liu (2018a), p. 21.

[34]The Guidelines for Handling Administrative EPIL Cases by Procuratorial Organs (for Trial Implementation) (2018).

In conclusion, as previously stated, the existing civil EPIL is tort litigation by nature, but can be filed in circumstances where environmental damage has not yet occurred; the existing administrative EPIL is enforcement litigation by nature, but its preventive function has not been clearly confirmed by law. It is worth reflecting on whether tort litigation meets the requirement of preventing damage to the ecological environment. Furthermore, what kind of nature should EPIL have to effectively perform its preventive function? The following sections will analyze the current practice of EPIL in China.

4.3.2 Limitations of the Preventive Function of Tort Litigation

The current Civil EPIL Interpretation takes into account the need to prevent environmental damage and modifies the traditional tort litigation system to include the preventive function. The current preventive EPIL cases filed by social organizations, however, have shown that the preventive function of tort litigations is inherently limited in terms of scope, adjudication basis and litigation outcome.

First, there is a risk where the uncertainty of consequences may hardly fall into the scope of tort litigation. The Civil EPIL Interpretation only provides for liability for the purpose of "eliminating a danger".[35] Since that liability must be statutory, i.e., clearly provided by law, the "significant risk" should therefore be interpreted as "significant danger", which requires an obvious possibility of foreknowledge.[36] Accordingly, the scope of tort litigation only includes the significant risk of specific environmental damage that can be fully ascertained based on the evidence available in the litigation and the current level of science and technology. This will result in a large number of cases with great uncertainty but catastrophic consequences being excluded from the scope of EPILs. For example, in the case of *FON v State Grid Gansu Electric Power Company* (hereinafter referred to as the "Gansu Wind and Light Abandonment Case"), the Lanzhou Intermediate Court dismissed the case because of the complexity and uncertainty of the causal relationship between renewable energy generation and climate change.[37] Although the Gansu High Court finally directed the Gansu Mining District People's Court to hear the case, other courts may still uphold the Lanzhou Intermediate Court's position in the face of similar risks of uncertainty.[38]

[35] Art. 19 of the Civil EPIL Interpretation.

[36] Zhang (2017), p. 165.

[37] See civil ruling of the Lanzhou Intermediate People's Court ([2017] Gan 01 Min Chu 434); Diao Fanchao. Lanzhou Central Court rejects lawsuit by Friends of Nature against State Grid Gansu for abandoning wind and light: conditions do not fulfiled [N]. The Paper, 2018-8-20.

[38] See Civil Ruling of the Gansu Provincial High People's Court ([2018] Ganmin Final 679).

Second, it is difficult to reach a consensus on environmental risks without contradicting the "statutory" basis. "Significant risk of harm to the public interest" is a public policy issue that itself is subject to scientific judgment and the balancing of interests. It is necessary to consider not only the negative impact of development on the environment, but also the risk trade-off between this negative impact and the positive economic and social benefits. For example, in the Green Peafowl case, the plaintiff and the defendant argued that the construction of the dam, the clearing of the reservoir area and the inundation of the area posed significant risks to the endangered plants and animals and their living environment. This involved a high degree of scientific professionalism and balancing of interests. In the absence of a "statutory" basis, the court, as a neutral adjudicator, lacked the relevant professional knowledge and value judgment standards, making it difficult to come to a convincing decision. For the same reason, the process of tort litigation is slow. The heavy burden of proof and lengthy litigation processes makes it difficult for plaintiffs to respond quickly to widespread environmental risks because of the high cost of litigation. For instance, the Acer Pentaphyllum case was filed in 2015 and only went to trial on October 30, 2019, taking nearly 4 years. The green peafowl case was not adjudicated until April 2020, although it went to trial as early as August 2018.

Third, tort litigation lacks a mechanism to interface with administrative enforcement. The professionalism and flexibility of administrative organs can help them to effectively respond to environmental risks. In tort litigation, administrative organs can only support prosecution or provide reference advice on request but have no obligation to take the initiative to perform their regulatory functions. As a result, the court has to make various professional judgments in order to protect the public interest, while the administrative organ, which is responsible for protecting the public interest and has professional knowledge, is not involved. For example, in the Green Peafowl case, FON and five other social organizations wrote to the former Ministry of Environmental Protection (hereafter "MEP"), suggesting that the environmental impact assessment (hereafter "EIA") approval for the Jiasha River hydropower plant be revoked in accordance with Article 69 of the *Administrative License Law*.[39] However, the request was not taken seriously due to the lack of standing. In its final decision, the court required the defendant to stop the construction of the hydropower plant "based on the existing environmental impact assessment", but the implementation of the EIA system is the responsibility of the administrative agencies, so the court was essentially overstepping its role vis a vis an administrative agency.[40]

---

[39] Art. 69 of the Administrative License Law (2003, amended 2019).

[40] See the civil judgment of Kunming Intermediate People's Court ([2017] Yun 01 Min Chu No. 2299).

4.3.3 Suitability of Law Enforcement Litigation in Prevention

Compared with tort litigation, enforcement litigation is more suitable for performing a preventive function.[41] Enforcement litigation allows the plaintiff to file a lawsuit against illegal behavior or administrative agencies' failure to perform their duties. The United States, Europe and other countries with well-developed EPIL systems have given certain plaintiffs the right to file preventive enforcement lawsuits, supervising and assisting administrative organs in risk regulation and enforcement. In the United States, most federal environmental laws provide for citizen suit provisions, authorizing plaintiffs to file lawsuits against actions with the risk of environmental damage according to corresponding legal provisions.[42] For example, in *Tennessee Valley Authority v Hill*, a U.S. federal court held that the construction of a multi-million dollar dam violated the Endangered Species Act by endangering an endangered fish and therefore required the construction to stop.[43] In Europe, the Aarhus Convention authorizes the public to sue for decisions, acts or omissions that violate national laws relating to the environment.[44] The advantages of enforcement litigation in playing a preventive function are also reflected in the scope, the basis for adjudication, and the litigation outcome.

First, enforcement litigation can prevent environmental risks in an earlier and broader way. After four decades of development, China has established an environmental legal system comprising more than 30 laws, 60 administrative regulations, 600 administrative rules and 1200 national environmental standards,[45] which weaves a tight institutional net for preventing environmental damage. EPIL plaintiffs can supervise various types of behaviors with environmental risks based on basic regulatory instruments such as environmental planning, environmental impact assessment, environmental standards, and environmental permits. For instances, they can require environmental impact assessment at the time of planning and before the construction project begins, obtain emission permits before an enterprise discharges pollutants, and install specific emission equipment to prevent excessive emissions. Plaintiffs can also file enforcement lawsuits based on other legal norms with preventive functions. For example, in the aforementioned case of wind and light abandonment in Gansu, the plaintiff could have claimed that the power grid company violated its full acquisition obligation under the *Renewable Energy Law*, which requires it to "fully acquire the grid-connected power of renewable energy projects within the coverage of its power grid".[46] In fact, the plaintiff, the FON, has already raised the fact that the defendant had violated the Renewable Energy Law and other

---

[41] Wu (2017), p. 153.

[42] Adelman and Glicksman (2020), p. 388.

[43] See *TVA v Hill*, 437 U.S. 153 (1978).

[44] Art. 9(3) of the Convention on Access to Information, Public Participation in Decision-making and Access to Justice in Environmental Matters (Aarhus Convention).

[45] Lv (2020), p. 6.

[46] Art. 14 of the Renewable Energy Law (2009).

environmental laws in the litigation process.[47] But due to the nature of tort litigation, the plaintiff could not directly demand the defendant correct the violation. If the existing environmental regulations could be fully implemented, the goal of preventing environmental risks would be achieved to a large extent.

Second, enforcement litigation allows the court to avoid dealing with public policy issues, but makes use of its expertise in the application of law. Taking the EIA system as an example, China has promulgated laws and administrative regulations such as the Environmental Impact Assessment Law, the Regulations on the Administration of Environmental Protection of Construction Projects, and the Regulations on Environmental Impact Assessment of Planning, as well as numerous and continuously updated regulations and other normative documents such as the Measures on Public Participation in Environmental Impact Assessment and the List of Environmental Impact Assessment Classification Management of Construction Projects, which establish a detailed system of substantive and procedural rules. In view of the current preventive EPIL cases, it would help to reduce the difficulty of litigation if the plaintiff could directly invoke preventive laws. For example, in the Green Peafowl case, the plaintiff FON pointed out that the EIA Report of the Jiasha River Hydropower project was incomplete and factually incorrect. Not only did it fail to mention the habitat of endangered animals that would be submerged, but it also ignored the large number of sapsuckers and did not mention the tropical rainforest. Consequently, the plaintiff argued that "the environmental impact assessment of the hydropower station construction project has serious procedural and substantive problems".[48] If the plaintiff could bring enforcement litigation, invoking the relevant procedural and substantive laws to review the preparation and approval of EIA, it would help reduce the plaintiff's burden to prove the existence of "significant risk", and thus prevent environmental risks more quickly and effectively.

Finally, enforcement litigation can fully stimulate the professional advantages of administrative organs, improving their preventive mechanism. In the face of enforcement litigation, administrative agencies can no longer stay out of it, but need to take the initiative to supervise the environmental behavior in accordance with the preventive regulations; otherwise, they will face the loss of the lawsuit. For example, in the Green Peafowl case, although the court recognized the fact that the EIA Report of the hydropower project was inaccurate and ill-considered, it failed to review the legality of the agency's approval due to the nature of the tort lawsuit. The court only requested the defendant to "complete the post-evaluation of the environmental impact according to the requirements of the Ministry of Ecology and Environment and take improvement measures".[49] In contrast, an enforcement lawsuit can urge the

---

[47] See Gansu Mining District People's Court Public Interest Litigation Notice ([2019] Gan 95 Min Chu No. 7).

[48] Ge (2018), p. 61.

[49] See the civil judgment of Kunming Intermediate People's Court ([2017] Yun 01 Min Chu No. 2299).

administrative authorities to review the EIA of the hydropower station according to EIA laws and supervise the law implementation. In addition, an enforcement litigation may prompt administrative organs to establish a long-term mechanism for environmental risk regulation, while tort litigation is only a case-by-case relief.[50] For instance, both of the Green Peafowl case and the Acer Pentaphyllum case were caused by a hydropower project threatening endangered species and their living environment. If they are stopped separately through tort lawsuits, that would not help prevent other similar risks. However, if the administrative organs are prompted to improve the enforcement mechanism of the EIA system through enforcement lawsuits, the impact will go beyond the individual lawsuits themselves and prevent more potential risk behavior without needing repeated prosecutions.

4.4 Paths Toward the Improvement of Preventive EPIL Legislation

In order to establish a preventive EPIL system with an enforcement factor, the Chinese legislature should improve current EIPL laws in terms of the scope, pre-litigation procedures, duty evaluation and professional capacity.

4.4.1 Proper Expansion of the Scope

In order to build a preventive EPIL system, it is necessary to expand its scope through legislation, not limited to cases that have caused damage to an environmental public interest. The scope of a case could be appropriately expanded in the following three aspects.

First, the scope of administrative and civil EPIL should include "significant risk of damage to an environmental public interest". Due to the enforcement nature of preventive EPIL, the judgment of whether it constitutes a "significant risk" is based on the objective environmental laws and regulations rather than subjective rights.[51] The plaintiffs, whether they file an administrative EPIL to promote law enforcement or a civil EPIL to subrogate law enforcement, should be able to use the preventive obligations under environmental law as the basis for litigation. The courts should not judge the environmental risks separately from their statutory basis.

Second, the scope of the preventive regulations that can be used as the basis for litigation is limited. Limiting the scope of the preventive regulations can help focus the limited supervision power on the key areas that need to strengthen prevention, and avoid excessive interference with the priority judgment of the administrative

[50] Gong (2019), p. 141.

[51] See Wu (2021), p. 39.

body. In the U.S., only certain provisions of federal environmental laws can be used as the basis for citizen suits. For instance, the *Clean Air Act* and the *Clean Water Act* only allow citizen suits for "violations of emission standards and limitations, and specific permits and their terms, as provided in this Act".[52] In determining the specific scope, it is necessary to consider not only the preventive function of the specific environmental regulations, but also the degree of comprehensiveness of the legal provisions, the maturity of the administrative enforcement mechanism, and the feasibility of supervision through EPIL mechanisms.

Third, priority should be given to the non-discretionary acts of administrative organs. Considering the balance between administrative power and judicial power, priority can be given to the non-discretionary acts that clearly violate the substantive or procedural provisions of the law. Such cases can fully utilize the legal interpretation and application expertise of the judicial branch, requiring less environmental expertise and not excessively interfering with the discretionary power of administrative organs. Illegal approval of construction projects in nature reserves is a typical non-discretionary act. For instance, the Regulation on Scenic and Historic Areas stipulates that "the following activities shall be prohibited within a scenic and historic area; (1) cutting into a mountain, opening a quarry or mine, opening up wasteland, building a tomb or tombstone, or conducting any other activity that would destroy the landscape, vegetation or landforms; (2) building a facility to store explosive, combustible, radioactive, poisonous or corrosive substances……".[53] In practice, many local governments illegally approve the prohibited activities in order to attract investment and promote economic development, and should be subject to supervision by preventive EPILs.

4.4.2 Independent Value of Pre-Litigation Procedures

Pre-litigation procedures are not just a subsidiary of the litigation process but have independent value in protecting the public interest.[54] The *Administrative Litigation Law* stipulates that procuratorates "shall make recommendations to the administrative organs to urge them to perform their duties in accordance with the law".[55] Pre-litigation procedures are not compulsory and are designed to request the administrative organ to reconsider its decision of action or inaction. Unlike the rigid litigation procedure, pre-litigation procedures are more flexible in respecting administrative discretion and saving judicial resources. If a pre-litigation procedure can make administrative organs perform their enforcement duties according to law, plaintiffs will not need to go through the time-consuming and labor-intensive

[52] Clean Air Act, 42 U. S. C. 7604 (a)(1), (3); Clean Water Act, 33 U. S. C. 1365 (a)(1).

[53] Art. 26 of the Regulation on Scenic and Historic Areas (2006, revised 2016).

[54] See Qian (2011), p. 160.

[55] Art. 25 of the Administrative Litigation Law (1989, amended 2014 and 2017).

litigation process, which is conducive to filling the regulatory gap in time and avoiding the consequences of environmental damage. To this end, the current pre-litigation procedures should be improved to better perform a preventive function.

First, introducing public hearings, consultations and other mechanisms should be introduced to enhance the interactivity of the pre-litigation process. With the backing of litigation, administrative organs are inclined to act in compliance with the requirements proposed in the pre-litigation opinions. More than 90 percent of all administrative EPIL cases have been ended in the pre-litigation stage.[56] In order to identify the true facts, clarify regulatory responsibilities, and promote consensus among all parties, EPIL legislation should change the one-way nature of pre-litigation procedures, and enhance interaction by introducing mechanisms such as public hearings and consultations. These mechanisms would assist administrative organs in understanding their preventive responsibilities and taking measures to fulfil them.[57] In practice, procuratorial authorities have actively explored public hearings and other mechanisms. The SPP has issued 12 typical cases of EPIL hearings on July 22, 2021.[58]

Second, a pre-litigation injunction system should be established. To effectively prevent damaging results, plaintiffs should also have the right to apply for an injunction that requires administrative organs to take temporary measures or suspend the making or execution of administrative acts with environmental risks.[59] For example, procuratorial authorities may request that the administrative authorities to suspend the approval of an EIA permit for a construction project. Due to the uncertainty of the risk and the irreversibility of damage, no one can be sure what the consequences of a moment's delay will be, so it is necessary to take immediate measures when the potential risks are discovered. European countries, in accordance with the requirement of "effective" prevention of environmental damage under Article 9(4) of the *Aarhus Convention*, attach importance to the plaintiff's right to seek a temporary injunction from domestic courts, and thus effectively avoid the outcome of "winning in court, but losing in practice".[60] Some local legislations have also explored the establishment of pre-litigation preservation systems, such as injunctions, to prevent environmental damage in a timely and effective manner, which are worthy of reference by national legislation (please see Table 4.3).

---

[56] Supreme People's Prosecutor (2021).

[57] See Wang and Qiang (2021), p. 13.

[58] The Supreme People's Procuratorate Releases 12 Typical Cases of Public Interest Litigation Prosecution Hearings (July 22, 2021).

[59] See Wang (2018), p. 56.

[60] Darpö (2018).

Table 4.3 Provisions Regarding Pre-Litigation Injunction by Local Legislatures

NO.	Place	Provisions Regarding Pre-Litigation Injunction
1	Yunnan Province (2019)	*Timely preservation measures* in cases that may lead to damage to national interests or social public interests.[a]
2	Anhui Province (2020)	In the case of PIL, if it is found that acts damaging national interests or social public interests still exist, *preservation measures such as injunction* shall be taken in a timely manner in accordance with the law.[b]
3	Hubei Province (2019)	If it is found during the trial of a PIL case that acts damaging national interests and social public interests continue to exist, *measures such as injunction to preserve the behavior* shall be taken in a timely manner in accordance with the law.[c]
4	Jiangsu Province (2020)	Before filing civil EPIL, the procuratorial authorities may urge the infringer to take *measures such as stopping the infringement, eliminating the danger and restoring the original state of affairs.*[d]

[a]Decision of the Standing Committee of the People's Congress of Yunnan Province on Strengthening the Public Interest Litigation Work of the Procuratorial Organs (adopted at the 13th meeting of the Standing Committee of the 13th People's Congress of Yunnan Province on September 28, 2019)
[b]Decision of the Standing Committee of the People's Congress of Anhui Province on Strengthening Prosecutorial Public Interest Litigation (Adopted at the Twenty-second Meeting of the Standing Committee of the Thirteenth People's Congress of Anhui Province on November 13, 2020)
[c]Decision of the Standing Committee of the People's Congress of Hubei Province on Strengthening Prosecutorial Public Interest Litigation (Adopted at the Tenth Meeting of the Standing Committee of the Thirteenth People's Congress of Hubei Province on July 26, 2019)
[d]Decision of the Standing Committee of the People's Congress of Jiangsu Province on Strengthening Prosecutorial Public Interest Litigation (Adopted at the Nineteenth Meeting of the Standing Committee of the Thirteenth People's Congress of Jiangsu Province on November 27, 2020)

4.4.3 Duty Evaluation with Behavioral Standards

Requirements of standing are important to avoid preventive EPIL from deviating from its supervisory position. The standing requirements are the requirements that the procuratorial organs should meet when filing an EPIL, in order to avoid improperly interfering with the operation of administrative power. The *Administrative Litigation Law* provides that after the pre-litigation procedure, "if the administrative organ does not perform its duties in accordance with the law, the people's procuratorate shall bring a lawsuit to the people's court in accordance with the law".[61] Accordingly, the procuratorial authorities need to examine whether the administrative organs "perform their duties in accordance with the law" before filing EPIL, and the courts need to decide whether to accept the case based on this standard. Disagreement exists on the interpretation of "performing duties in accordance with the law" in terms of adopting the behavioral standard, outcome standard, or both.[62]

[61] Art. 25 of the Administrative Litigation Law (1989, amended 2014 and 2017).
[62] See Wang (2020), p.129.

In preventive EPIL, the judgment of whether the administrative organ has "performed its duties in accordance with the law" should be based on the behavioral standard of "exhaustion of administrative means" rather than the outcome standard.[63] As long as administrative organs have exhausted all enforcement measures, such as administrative orders, administrative penalties and administrative coercion, even if they fail to effectively stop environmental violations or eliminate environmental risks, administrative organs should be considered as performing their duties according to law. On the one hand, if the administrative organ has "exhausted administrative means" but the environmental risks remain, it will not help file the lawsuit. On the other hand, the court lacks the ability to make substantive judgments on whether environmental risks have been effectively controlled, making the outcome standard difficult to operate. Judgment of the "exhaustion of enforcement measures" requires consideration of objective factors and other constraints on administrative enforcement, as well as legal procedures such as administrative review and litigation.

If administrative organs fail to "exhaust enforcement measures", EPIL plaintiffs should decide whether to file a lawsuit to urge enforcement or a lawsuit to subrogate enforcement depending on the specific circumstances. If there is a lack of will and motivation to enforce the law, due to local protectionism or regulatory capture, plaintiffs should file a lawsuit to urge enforcement. If there is a lack of resources and ability to enforce the law, due to limited funding for administrative agencies or excessive enforcement tasks, plaintiffs should consider filing a lawsuit to subrogate enforcement.

4.4.4 Support of Professional and Information Capacity

Due to the enforcement nature of preventive EPIL, procuratorial organs have no need to make policy judgments on whether a "significant risk of harming the environmental public interest" exists, even though, they still need to make professional determinations on relevant facts of the case. Considering the limited environmental expertise of the procuratorates under the current system, legislation is needed to provide this necessary support.

First, we establish a communication and cooperation mechanism with environmental social organizations. From the existing preventive EPIL cases, we can see that environmental social organizations are very concerned about the prevention of environmental risks and have the professional ability to find the necessary data. For example, in the Green Peafowl case, the environmental organization Wild China was the first to come across the issue during its fieldwork activities and then transferred

[63] Zhang Xueqiao, deputy procurator-general of the Supreme People's Procuratorate, expressed this view in an interview. See Chang (2019).

the case to the FON to file a lawsuit.[64] As only procuratorial authorities are qualified as plaintiffs to bring administrative EPIL, environmental social organizations are willing to work with the procuratorial authorities to urge administrative agencies to fulfil their enforcement responsibilities.

Second, the obligations of government and enterprises should be refined in terms of environmental information disclosure. At present, many local procuratorial organs take the initiative to implement environmental monitoring and other activities in order to gain evidence in environmental pollution cases.[65] These initiatives not only increase costs, but also make it difficult to ensure the evidential validity of monitoring results due to the lack of environmental expertise and qualifications. In fact, existing environmental laws and regulations already require enterprises and the government to disclose environmental information such as the type, quantity, concentration and method of pollution discharge. If procuratorial organs can supervise relevant subjects to fulfil the obligation of information disclosure in accordance with the law, and compare the disclosed environmental information with the legal obligations of enterprises and administrative organs, they can obtain factual evidence and prevent the occurrence of environmental damages in time.

4.4.5 Short-Term and Long-Term Legislative Paths

In the short term, the SPP and the SPC should revise the Procuratorial PIL Interpretation to specify that the scope of public interest litigation under the current legislation includes cases with a "significant risk of harming social public interests", and to implement the aforementioned improvement suggestions. While revising its judicial interpretations, the SPP may first revise its operational rules to include preventive EPIL. Such operational rules of the SPP do not have legal effects but can clarify the understanding of preventive EPIL among procuratorial organs nationwide and provide guidelines for handling cases.

In the long term, the legislative path to establish a preventive EPIL system can be divided into three stages. First, the Standing Committee of the NPC could establish a preventive EPIL system in special environmental laws. At present, the *Environmental Impact Assessment Law, Marine Environmental Protection Law, Wildlife Protection Law, National Park Law, Nature Reserve Law, Wild Plant Protection Regulations*, and other environmental laws and regulations are in the process of being formulated or revised. It is possible to take advantage of the current legislative opportunity to include provisions on preventive EPIL and explore preventive EPIL in certain areas of environmental protection first. Second, the legislature could revise

[64] See Rongzhen (2018).

[65] For example, procuratorial authorities in Hubei and Zhejiang procured drones to take aerial photographs of ecological damage sites in order to collect evidence. See Wenbin (2018) and Mei (2018).

the general legal basis of the EPIL system. At present, Article 25 of the *Administrative Procedure Law*, Article 55 of the *Civil Procedure Law* and Article 58 of the *Environmental Protection Law* are the main legal bases of the EPIL system. The preventive EPIL system could be established by amending the corresponding articles of these three laws, which are applicable to all environmental cases. Third, the legislature could consider formulating a special law on EPIL. By enacting a specialized EPIL law, the rules of civil EPIL and administrative EPIL, which are scattered in different legislations, can be integrated. The scope of EPIL could then include acts with environmental risk, and litigation rules can be established to adapt to the characteristics of preventive EPIL.

4.5 Conclusion

China's EPIL system focuses on remedying environmental damage instead of risk prevention, and social organizations have encountered difficulties in filing preventive civil EPIL cases. The existing preventive civil EPIL cases are of a nature of tort litigation, facing limitations in terms of scope, adjudication basis and litigation effect in preventing environmental damage. Preventive EPIL should not adopt the nature of tort litigation, but the nature of enforcement litigation, achieving preventive function by supervising and assisting administrative organs to implement preventive regulations. To strengthen preventive function, the EPIL system should expand the scope of cases, give full play to the independent value of pre-litigation procedures, evaluate the performance of duties with behavioral standards, and strengthen the support of an environmental professional capacity. These improvements could be realized by revising the judicial interpretation in the short term and formulating special legislation in the long term.

References

Adelman DE, Glicksman RL (2020) Reevaluating environmental citizen suits in theory and practice. Univ Colorado Law Rev 91(2):385–452

Chang T (2019) Public interest litigation, not recourse litigation: A conversation with Zhang Xueqiao, deputy procurator-general of the Supreme Procuratorate, Southern Weekend (Mar. 21)

Cheng D, Wang C (2016) On the interface between environmental damage compensation system and environmental public interest litigation. Environ Protect 2:39–42

China Biodiversity Conservation and Green Development Foundation (2019) Court Tomorrow: the First National Case of the Protection of Endangered Plants Acer Pentaphyllum Preventive Public Interest Litigation Case Public Hearing, Oct.29, https://mp.weixin.qq.com/s/WnhHt5JUHb3bLIKngrpDAA. Accessed 16 Mar 2022

China Foundation for Biodiversity Conservation and Green Development (2017) Qinghai Ancient Tamarisk Case Filed or not still Unknown, Yellow River Yangqu Hydropower Plant Project has been Confirmed Illegal Construction (Sep. 11). http://www.cbcgdf.org/newsshow/4856/2948.html. Accessed 16 Mar 2022

ClientEarth (2021) 10 Landmark Cases for Biodiversity. https://mp.weixin.qq.com/s/m6oZJDW8eE79AsSrApUt6g. Accessed 16 Mar 2022

Darpö J (2018) Principle 10 and access to justice. In Elgar Encyclopedia of environmental law. Edward Elgar Publishing

Diao F (2017) Yunnan Green Peacock Habitat Threatened by Hydropower, Preventive Environmental Public Interest Litigation Filed. The Paper (Aug. 15). http://www.thepaper.cn/newsDetail_forward_1762645. Accessed 16 Mar 2022

Du Y (2018) Seriously Wounding China's Last Rainforest". Caixin Weekly. https://weekly.caixin.com/2018-03-17/101222542.html. Accessed 16 Mar 2022

Friends of Nature (2017) Environmental Public Interest Lawsuit against PetroChina's Yunnan Refinery Project not Accepted. Friends of Nature to Appeal (Jan. 4), https://mp.weixin.qq.com/s/1IXCHErhIb6BvRVdD96e5A

Friends of Nature (2019) Case filed! Huainan Central Court accepts ecological damage case filed by Friends of Nature on Bagong Mountain (Oct. 21). https://mp.weixin.qq.com/s/ZCHsqnOhfMnEPovPlAQsCg. Accessed 16 Mar 2022

Ge F (2018) The history and typical cases of environmental public interest litigation in China: the practice of "Friends of Nature" environmental public interest litigation as an example. Soc Gov 2:51–63 p.61.

Gong G (2017) The standing of the United States environmental citizen suits and its inspiration. Legal Commer Res 5:171–182

Gong G (2019) Reflections on the positioning and nature of environmental civil public interest litigation. Legal Stud 3:127–147

Li Y, Wu K (2016) Role of procuratorates in the environmental public interest litigation. J Renmin Univ China 2:2–13

Liu C (2018a) Pre-litigation procedure of environmental administrative public interest litigation. J Law 1(114–123):21

Liu Y (2018b) Constructing an objective litigation mechanism for administrative public interest litigation. Legal Stud 3:39–50

Lv Z (2008) Analysis of environmental public interest litigation. Legal Commer Res 6:131–137

Lv Z (2020) On the communication and coordination mechanism of environmental law - a perspective of modern environmental governance system. Law Forum 1(5–12):6

Qian S (2011) The value goal and institutional design of setting up administrative public interest litigation. Chinese Soc Sci 1:151–163

Rongzhen M (2018) A "Green Peacock Defense Battle" Started by Multiple Environmental NGOs. Southern Weekend (Sep. 30).

Supreme People's Prosecutor (2021) Key Case Handling Data from January to September for National Prosecutorial Organs (Oct. 18). https://mp.weixin.qq.com/s/jMgPHoyXeUt6BRGYkk088A. Accessed 16 Mar 2022

Wang C (2018) On "Preventive" administrative public interest litigation by procuratorial organs. Zhejiang Soc Sci 11:51–58

Wang C, Qiang J (2021) Judicial design of pre-litigation procedures for administrative public interest litigation. People's Procurat 7(13–16):13

Wang Q (2020) Benchmarks for review of administrative inaction in environmental administrative public interest litigation. Tsinghua Jurisprud 2:129–142

Wu K (2017) On preventive environmental public interest litigation. Theory Reform 3:146–161

Wu K (2021) The nature and positioning of preventive procuratorial environmental public interest litigation. J China Univ Geosci (Soc Sci Edn) 1(30–44):39

Xi J (2020) Xi Jinping Delivers Important Speech at Central Conference on Comprehensive Rule of Law (Nov. 17). http://www.gov.cn/xinwen/2020-11/17/content_5562085.htm. Accessed 16 Mar 2022

Xin C (2014) Interpretation of the environmental protection law of the People's Republic of China. Law Press, Beijing

Xiong W (2018) Get off the ground! Fancheng Procuratorate Releases Drones to Help Environmental Public Interest Litigation. Jingchu Fair Justice Network (Mar. 26) http://www.hbjc.gov.cn/xxkj/201803/t20180326_1266672.shtml. Accessed 16 Mar 2022

Xu M (2018) Our Provincial Procuratorate Uses Drones to Obtain Evidence for Public Interest Litigation. Zhejiang News Network (Jan. 17). https://zj.zjol.com.cn/news.html?id=851855. Accessed 16 Mar 2022

Yue H, Ma Z (2016) Huainan Bagong Mountain Scenic Area Illegal Construction Wind Power Project: Repeatedly Ordered to Dismantle Still Put into Use. The Paper (June 26). https://m.thepaper.cn/newsDetail_forward_1489205. Accessed 16 Mar 2022

Zhang C (2021) Interview of Zhang Xueqiao on how to improve the legal system of procuratorial public interest litigation. Democracy Legal Syst 31:17–22

Zhang X (2017) Thinking about the procedural rules of preventive environmental civil public interest litigation. Legal Sci (J Northwest Univ Polit Sci Law) 4:164–172

Chapter 5
The Standing to Sue of NGOs in Environmental Public Interest Litigation in China: A Doctrinal Analysis of Laws and Cases

Qi Gao, Quan Ke, and Xiaobo Zhao

Abstract Environmental public interest litigation filed by NGOs is an important legal breakthrough to improve the private enforcement of environmental law. Different from traditional standing rules, NGOs are allowed to sue even if their rights or legal interests are not affected by the acts or omissions in question. Since individual rights are considered as superfluous in actions vindicating environmental public interests, certain threshold requirements are applied to NGOs in Article 58 of Environmental Protection Law to ensure the professionalism of NGOs and prevent frivolous lawsuits. The year 2022 marks the 8th year of its implementation in practice and the judiciary has applied and interpreted this provision in numerous cases. Notably, the status quo of practice does not match expectations. Instead, NGOs are increasingly marginalized in this court-centred environmental movement. Although the Chinese courts have generally adopted a relaxed interpretation of Article 58 in individual cases, the standing to sue of the administrative agencies and procuratorates in environmental public interest litigation has a significant impact on the participation of NGOs. Moreover, it should be admitted that difficulties still exist in providing a sound doctrinal explanation for the application of relevant provisions. This is the dilemma faced by the rapid development of environmental judicial innovations driven by legal policies.

Keywords Environmental public interest litigation (EPIL) · NGOs · Standing to sue · Ecological and environmental damage compensation litigation (EEDCL)

Q. Gao (✉)
Koguan Law School, Shanghai Jiao Tong University, Shanghai, China
e-mail: gaoqifxy@sjtu.edu.cn

Q. Ke
School of Law, South China University of Technology, Guangdong, China

X. Zhao
School of Law and Justice, University of Southern Queensland, Springfield, QLD, Australia
e-mail: bob.zhao@usq.edu.au

X. Wang et al. (eds.), *Environmental Public Interest Litigation in China*,
https://doi.org/10.1007/978-3-031-26526-6_5

5.1 Introduction

China has formally recognized environmental public interest litigation (EPIL) since the revision of the Civil Procedure Law in 2012. As a major legal breakthrough, it allows certain kinds of organizations and authorities to bring EPIL even if their rights or legal interests are not affected by the acts or omissions in question. Despite doctrinal controversies,[1] this is considered as a response to the criticism that the civil society and the judiciary have not been involved in environmental governance as they should be in China.[2] The NGOs are therefore expected to bring more EPIL cases to court to provide opportunities for the judiciary to exert more influence on environmental governance.

Nevertheless, the reliance on NGOs to initiate a court-centred environmental movement has proven to be unsatisfactory in practice. Between 2015 and 2019, the EPIL cases brought by NGOs fluctuated around 60 cases per year.[3] Some claim that the legal standing requirements on NGOs are overly strict, especially considering China's fledgling civil society.[4] However, according to statistics, there are approximately 700 NGOs alleged to have legal standing in the EPIL in China. Ironically, only 25 of them have actually filed such lawsuits between 2015 and 2017.[5] To reveal the reasons behind this phenomenon, it is necessary to explore how the judiciary interprets relevant legislation in the quasi-legislative interpretation documents and individual cases. Doctrinal analysis is the main methodology used in this research. In particular, representative cases are selected to analyse Article 58 of the Environmental Protection Law (EPL) from the aspects of activity performance, activity area and activity period. Attention is then paid to the impact of the standing of the administrative agencies and procuratorates in EPIL on the participation of NGOs.

Through the above analysis, this chapter attempts to answer the following key legal questions: (1) What is the rationale for expanding the standing to sue beyond individual rights? (2) Are the threshold requirements on the standing of NGOs in EPIL overly strict in the context of China? (3) Should administrative agencies and procuratorates be allowed to bring EPIL as a complement?

---

[1] See Sect. 5.2.

[2] Only less than 1% of environmental disputes were settled through judicial process in 2013. Environmental victims usually prefer filing petitions or complaints to administrative authorities as the primary tool for environmental dispute resolution. Chaofei Yang, *Several Issues on Environmental Protection and Environmental Legal System* (23 November 2012) <http://www.npc.gov.cn/npc/c541/201211/63ba26f719454bf1a673ac1762850c85.shtml> (in Chinese).

[3] Li (2017), p. 1; The State Council Information Office (2019b); SPC (2020).

[4] Huang (2020).

[5] Ministry of Civil Affairs, Hongyu Liu (2019).

5.2 The Evolution and Theoretical Rationale of EPIL

5.2.1 *The Evolution of the Standing Rules in EPIL*

The relaxation of standing in environmental cases is a result of both judicial interpretation and legislative enactment. Although judges cannot make laws in China, they can enforce them via judicial interpretation on a case-by-case basis (with no binding force) or on a general basis (issued by the Supreme People's Court (SPC) with the effect of quasi-legislation).[6] The latter is represented by the *Judicial Interpretation on EPIL* (2015) and *Judicial Interpretation on Civil Procedure Law* (2015). Even before the formal recognition of public interest litigation in the *Civil Procedure Law* in 2012, individual rights and powers granted to public authorities were stretched by the judicial branch in several cases to relax standing for those who claim to represent the public interest. Since the implementation of the *Civil Procedure Law* in 2013, both liberal and conservative interpretations of standing have been observed in practice.

Recent years have also witnessed the revision of some important statutory laws that constitute the current legal framework for EPIL in China. In particular, Article 55 of the *Civil Procedure Law* (as revised in 2012) grants standing to 'relevant organizations and authorities prescribed by law' to enhance environmental protection and consumer protection. The EPL was further amended in 2014 to set threshold requirements on the standing of NGOs. Moreover, Article 89 of the *Marine Environmental Protection Law* has been considered as the statutory basis for public interest litigation initiated by administrative authorities since 1999, although the nature of this provision is in fact disputable.[7] As a response to relevant disputes, the *Judicial Interpretation on EPIL to Protect Marine Natural Resources and Environment* was adopted in 2022.[8] It clarifies that procuratorates can also bring EPIL to protect marine natural resources in case the administrative authorities in charge of marine environment protection fail to do so after receiving notification from the procuratorates. In contrast, judges still preclude NGOs from bringing EPIL to protect marine natural resources. This results in the marginalization of NGOs.[9]

The phenomenon is also reflected in the authorisation of procuratorates to bring EPIL. The Supreme People's Procuratorate (SPP) has actively advocated for the expansion of standing to procuratorates since 2000.[10] After losing anti-corruption authority to the National Supervision Committee system in 2018,[11] EPIL is in fact considered as an important approach for procuratorates to expand their power and influence. As a result, in July 2015, the Standing Committee of the National People's

[6]Wang (2006), p. 1.

[7]See Sect. 5.4.

[8]*Judicial Interpretation on EPIL to Protect Marine Natural Resources and Environment* (2022).

[9]Zhang (2019), p. 5.

[10]Xin (2014), p. 201.

[11]Smith (2018).

Congress authorized a pilot practice in 13 provinces that allowed procuratorates to bring public interest litigation to promote environmental protection. While NGOs are only qualified to initiate EPIL against polluters (mostly private persons), procuratorates in pilot areas can also challenge the legitimacy of administrative decisions in courts. This eventually led to the revision of both the Civil Procedure Law and Administrative Litigation Law on 27 June 2017. Procuratorates across China are now authorized by statutory laws to bring EPIL.

Based on the above observation, the current standing requirements for EPIL can be concluded as follows in Table 5.1.

5.2.2 *Theoretical Rationale for the Relaxation of the Standing Rules*

The above evolution of the standing rules can be analysed from several perspectives: First, environmental law enforcement in China heavily relies on administrative authorities, while the potential of civil society and the judiciary in environmental governance in China remains largely unexplored. The relaxation of the standing rules is expected to facilitate a court-centred environmental movement and enhance the public's right of access to environmental justice. However, as will be discussed in detail later, this movement is increasingly evolving into another kind of public enforcement that is more powerful but less restricted than traditional administrative approaches.

Moreover, the EPIL considers individual rights to be superfluous in actions vindicating environmental public interests. Traditionally, plaintiffs without individual rights do not have access to court.[12] With regard to public or collective interests, it is believed that the remedy lies in the political process, not in the courts.[13] A sharp distinction exists between law and politics, rights and policies used to gain most doctrinal support from mainstream legal scholars.[14] This 'individual rights' model, however, has been criticized for being human centric rather than ecocentric; hence, it is unable to provide enough protection for environmental values that exceed the boundaries of individual rights, particularly with regard to nature conservation.[15] Increasing scientific uncertainty has also made it difficult for individuals to bring an act or omission to court based on the violation of their rights or legal interests.[16] It is therefore argued that 'in the environmental context, a move towards less formalistic and individualistic justiciability doctrines might thus have appeared not as a

[12]Greve (1989), p. 201; Rabkin (1998), pp. 179–183; Henry (2001), pp. 233–235.

[13]Greve, above note 16, p. 213.

[14]Ibid 232.

[15]Ibid.

[16]Ibid.

Table 5.1 Standing Requirements for EPIL in China (by author)

Public Interest Litigation as a Civil Action	Article 55(1) of the Civil Procedure Law (2017) grants standing to '**relevant organizations** and **authorities prescribed by law**' to bring lawsuits against environmental pollution activities Article 55(2) of the Civil Procedure Law (2017) grants standing to **procuratorates** as a complementary measure to promote public interest litigation	**NGOs:** NGOs that fulfil prerequisites stipulated by Article 58 of the EPL are allowed to bring public interest litigation against polluters and destroyers (further interpreted by Articles 2–5 of the Judicial Interpretation on EPIL).
		Administrative Authorities: Article 89 of the Marine EPL authorizes administrative authorities in charge of marine environment protection to seek compensation from polluters and destroyers *on behalf of* the State.
		Procuratorates: Article 55(2) of Civil Procedure Law allows procuratorates to initiate public interest litigation against polluters and destroyers[a] if no organizations and authorities stipulated in Article 55(1) existed or they refused to file such cases (Article 3 of the *Judicial Interpretation on EPIL to Protect Marine Natural Resources and Environment* clarifies that this provision is applicable to EPIL to protect marine environment).
Public Interest Litigation as an Administrative Action	The amendment of the Administrative Litigation Law in 2014 did not include any provision on public interest litigation against administrative authorities. This was changed in 2017. Article 25(4) of the Law currently opens the door for **procuratorates** to bring public interest litigation against administrative authorities.	**Procuratorates**: Article 25(4) of the *Administrative Litigation Law* allows procuratorates to initiate public interest litigation against administrative authorities in charge of environmental protection and natural resources preservation.[b] Pre-trial prosecutorial suggestions should be given to administrative authorities to urge them to comply with legal requirements. Only if they refused to correct the alleged violation, can procuratorates initiate public interest litigation.

[a]Procuratorates are also allowed to bring public interest litigation against private persons to protect consumer interests on the safety of food and pharmaceuticals

[b]Procuratorates are also allowed to bring public interest litigation against administrative authorities in charge of the protection of state-owned properties, the transfer of right to use state-owned land and the safety of food and pharmaceuticals

revolution but as a natural adaptation of the legal system to more complex technologies and social realities'.[17]

Under such circumstances, the 'objective legality' model has gradually gained momentum. Compared to the protection of individual rights, a system of comprehensive judicial oversight has been proposed to improve adherence to environmental legal norms (objective concept of judicial review).[18] The shift towards objective legality provides justification for the relaxation of standing.[19]

This can be identified in the United States during the 'environmental decade' of the 1970s. The term 'private attorney general' refers to one who brought an action to vindicate the public interest.[20] It is argued that 'the role of private litigation is not simply to secure compensation for victims, but is at least equally to generate deterrence, principally by multiplying the total resources committed to the detection and prosecution of the prohibited behaviour'.[21] Government agencies were accused of being 'unable or unwilling to enforce regulatory laws as they should be enforced' due to bureaucratic and political constraints.[22] The gap between promise and performance added momentum to the support of a citizen suit as a vital complement to the public enforcement of environmental law.[23]

Unlike China, the Supreme Court in the United States expanded standing not by abandoning the foundation of individual rights.[24] Rather, it is the categories of injury that were expanded via judicial interpretations.[25] Moreover, organizations were allowed to sue on behalf of their members if certain criteria were met.[26] In contrast, the standing of 'relevant organizations and authorities prescribed by law' in EPIL is directly established by statutory laws. NGOs and procuratorates do not need to prove that their interests or their members' interests have sustained or will sustain direct injury or harm in order to gain their standing in EPIL. This largely resembles the

---

[17] Ibid 223.

[18] The law and right are considered as two different phenomena, but in German law, there is only one corresponding word '*Recht*', which is used in both subjective sense and objective sense. The subjective sense refers to individual rights, while the objective sense indicates rules and norms. In the early days, there were rights only. 'Law as the order of the State was superinduced with the purpose of sanctioning and protecting rights which, independently from this order, had come into existence'. Ibid 236, n.15; Kelsen (1945), p. 78.

[19] Ibid 220; Degenhart (1981), pp. 154–155.

[20] Coffee (1986), p. 669.

[21] Coffee (1983), pp. 215–218.

[22] Boyer and Meidinger (1985), pp. 834, 836–837.

[23] Ibid 838.

[24] *Sierra Club v Morton (1972)*, 405 U. S. 727, 738.

[25] In *Sierra Club v Morton (1972)*, the categories of injury that may be alleged in support of standing were relaxed from 'legal interest' to 'injury in fact'. The court found that injuries to recreational and aesthetic interests would be sufficient grounds for establishing standing. Ibid 733, 738, 753.

[26] See *Hunt v Washington State Apple Advertising Commission (1977)*, 432 U.S. 333.

"association suits" (Verbandsklage) in Germany.[27] Article 89 of the Marine EPL, however, does not quite fit in the objective model. Administrative authorities in charge of marine environment protection are allowed to sue 'on behalf of' the State. It is therefore argued that the standing of administrative authorities is based on the state ownership of marine natural resources. Nevertheless, Article 89 is usually considered as a special kind of EPIL. But an inconsistency exists between Article 89 and Article 55, which raises the question of whether NGOs and procuratorates can bring EPIL to protect the marine environment and what is the relationship between EPIL and the more recently emerged ecological and environmental damage compensation litigation (EEDCL). Further discussion will be made in Sect. 5.3.

Since individual rights are generally considered as superfluous in EPIL, it is argued that the standing of the plaintiffs should be subject to certain restrictions to ensure professionalism, prevent vexatious litigation and prevent the intrusion of individual rights in the name of public interests, which is also the case for the association suit in Germany.[28] This sets the keynote for the following discussion on specific threshold requirements on the standing of NGOs. It is particularly interesting to observe how the court balances public interests and individual rights on a case-by-case basis.

5.3 The Threshold Requirements on the Standing of NGOs

According to Article 58 of the EPL, NGOs that fulfil the following requirements are considered as eligible plaintiffs: (1) registered with the state at the municipal level or above in accordance with law; (2) specialized in environmental protection public interest activities and have no record of administrative or punitive penalties for their activities in the past five consecutive years. This can be analyzed from several perspectives.

5.3.1 Registration and Legal Personality

Environmental NGOs in China should be formally registered at the municipal level or above in order to meet the first threshold requirement set by Article 58. Therefore, NGOs that registered at the county level do not have standing to bring EPIL. But according to statistics in 2009, 63.05% of NGOs in China are registered at the county

---

[27] Most states (provinces) in Germany have adopted association suit in the field of nature conservation since the early 1980s. Rehbinder (2014); Bundesnaturschutzgesetz (BNatSchG) [Federal Nature Conservation Act], Section 63, 1 March 2010, Federal Law Gazette, 2009, Part I, No. 51, 2542, an unofficial English version is available at: https://www.bmu.de/fileadmin/Daten_BMU/Download_PDF/Naturschutz/bnatschg_en_bf.pdf.

[28] Rehbinder, above note 32.

level.[29] Notably, the initial drafts of the amendment were even more conservative, restricting the standing to sue to only one or a few nation-wide NGOs.[30] It was criticized for a lack of trust on NGOs and eventually led to a relaxation of the standing rules in the final version.

Local NGOs are usually more familiar with local circumstances. However, given the fledging civil society in China, they are more likely to face challenges in the professionalism and resources needed to bring EPIL. Alternatively, renowned NGOs, such as Friends of Nature, have cooperated with local NGOs in aspects of information exchange and evidence collection in practice.[31] In this way, local NGOs can still participate in the EPIL process.

In addition, the registration authority is restricted to civil affairs departments. Therefore, foreign NGOs registered at the police department and for-profit organizations registered at the industry and commerce department are not eligible to bring EPIL. However, given the strict and complicated registry system,[32] grassroots organizations are likely to choose to be registered at the industry and commerce department to gain legal personality.[33] Under such circumstances, they do not have standing to bring EPIL.

Of course, for-profit organizations can engage in activities that promote public interests, and even in a more effective way. Moreover, traditional lawsuits brought to protect individual rights can also contribute to public welfare.[34] However, the standing of EPIL is expanded beyond individual rights, which justifies certain restrictions on the plaintiffs. The requirement of the non-profit purpose of the organization can help save judicial resources and prevent vexatious litigation. Even for NGOs, it is difficult to prevent the abuse of EPIL for blackmail and unfair competition.[35]

Notably, the standing of NGOs in EPIL is not restricted to those with legal personalities. Corporations and foundations for non-profit purposes and the so-called 'civil non-enterprise institutions' (social service organizations that are

[29] Due to statistical reasons, there is no way to calculate how many environmental NGOs are registered below the county level. Huang (2011), p. 5.

[30] Jost Wübbeke (25 April 2014); National People's Congress (17 July 2013); Dongyu Peng (12 December 2013).

[31] Feng (2017).

[32] Not only should an NGO be registered at the civil affairs department, it also has to find another authority to supervise its operation in order to be qualified for registration. In reality, the latter is often very difficult to find, since government authorities also lack incentives to take on such responsibility and do not trust grass-root organizations. Even if an NGO manages to find a supervisory authority, this relationship can be easily ended unilaterally by the latter. *Regulations on Social Groups Registration Management*, State Council, 6 February 2016, art 3.

[33] 4.4% of environmental grassroot organizations are registered at the industry and commerce department in 2008. All-China Environmental Protection Federation, Report on the Status of Development of Environmental Social Organizations in China (2008).

[34] Lin (2006), pp. 48–149.

[35] The Paper (3 April 2021).

not necessarily registered as legal persons)[36] are all allowed to bring EPIL as long as they are legally registered at or above the municipal level and meet other criteria set by Article 58. This is different from the case in Germany, which requires the plaintiff to be corporations for non-profit purposes.[37] Full voting rights and open membership are highlighted by German law to reflect the democratic nature of corporations.[38] It is a response to the criticism that NGOs are 'self-appointed guardians' of the public interest, while the government represents the public interest and gains its legitimacy through a representative democratic process.[39] It is the lack of the democratic element that foundations are excluded from eligible NGOs in Germany to bring association suit.[40] However, given the status quo of civil society development in China, it is not necessary to raise the standard in this aspect. The traditional theory of representative democracy itself has also been shaken by the popularity of deliberative democracy. In fact, it is the deficiencies of representative democracy and a sense of distrust in the government to protect the public interest that add momentum for deliberative democracy and civil society participation.

5.3.2 *Objective and Performance*

Article 58 of the EPL further requires the NGOs to specialize in public interest activities that promote environmental protection. Article 4 of the Judicial Interpretation on EPIL (2015) further elaborates this requirement from the aspects of objective, scope of work and public interest activities. It is interpreted that the public interest issue raised in the EPIL should be relevant to the NGO's objective and scope of work.

5.3.2.1 The NGO's Objective and Scope of Work: To Protect Environmental Public Interest

'Environmental public interest' is not a legal terminology. Legal interpretation methodologies should be applied to clarify its meaning in EPIL. In practice, Guiding Case No. 75 is selected by the SPC as an example.[41] In *China Biodiversity Conservation and Green Development Foundation v Ningxia Ruitai Technology Co., Ltd.*,

---

[36]Interim Regulations on Registration Administration of Private Non-Enterprise Units, State Council, Order No. 251, 25 October 1998, art 2 (in Chinese).

[37]Rehbinder, above note 32.

[38]Ibid.

[39]Ibid.

[40]Ibid.

[41]Guiding cases should be referred to by the court when similar cases are heard in the future. The SPC Rules on Guiding Cases, No.51 [2010].

the plaintiff's standing was denied in the first and second instances for failing to explicitly mention 'environmental protection' as its objective, although biodiversity protection and green development were listed as the major objectives in its charter.[42] This rigid interpretation was fiercely criticized and the previous civil rulings were finally overruled by the SPC in January 2016.[43] In the final ruling, the SPC referred to the definition of biodiversity in Article 2 of the Convention on Biological Diversity and Article 30 of the EPL and the definition of environment in Article 2 of the EPL.[44] It is ruled out that the protection of biodiversity is an important aspect of environmental protection and that the interpretation should not focus on a literal expression.[45]

5.3.2.2 Actually Engages in Activities to Protect Environmental Public Interests

Not only should NGOs state environmental protection as their objective and scope of work in their charters, they should also actually engage in relevant activities. After all, Article 58 emphasizes the evaluation of a performance record in order to ensure the professionalism of NGOs. Initially, rather conservative rulings can be identified in judicial practice. For example, in *Friends of Nature v Taizhou Woaite Chemical Co., Ltd.* et al. (2015), the Intermediate Court of Taizhou refused to recognize the standing of Friends of Nature in the first instance. It was considered as a mere 'research institute' on the environment and was not specifically engaged in activities to promote environmental public interest.[46] This interpretation was later overturned by the High Court of Jiangsu in the appeal.[47]

Chinese courts now tend to make a broad interpretation on what counts as activities to protect environmental public interests. In Guiding Case No. 75, the SPC ruled that activities to protect environmental public interests 'not only include

[42] China Biodiversity Conservation and Green Development Foundation, *Introduction of CBCGDE* (2018); The Paper (29 January 2016); *China Biodiversity Conservation and Green Development Foundation v Ningxia Ruitai Technology Co., Ltd.*, Intermediate Court of Zhongwei, Civil Ruling, Public Interest Litigation, First Instance, No.6, 19 August 2015 (in Chinese); *China Biodiversity Conservation and Green Development Foundation v Ningxia Ruitai Technology Co., Ltd.*, High Court of Ningxia, Civil Ruling, Public Interest Litigation, Appeal, No. 6, 6 November 2015 (in Chinese).

[43] The Paper, above note 48.

[44] *China Biodiversity Conservation and Green Development Foundation v Ningxia Ruitai Technology Co., Ltd.*, Supreme People's Court, Civil Ruling, Retrial, No. 47, 28 January 2016 (in Chinese).

[45] Ibid.

[46] *Friends of Nature v Taizhou Woaite Chemical Co., Ltd.* et al., Intermediate Court of Taizhou, Environmental Public Interest Litigation, First Instance, No.00001, 15 January 2015 (in Chinese). According to Article 9 of the Charter of Friends of Nature, its scope of activities includes research and promotion of solid waste disposal and environmental protection.

[47] Ibid.

activities that directly restore the environment, such as afforestation, but also involve activities that improve environmental governance and raise environmental awareness, such as environmental education, training, academic communication, legal aid and EPIL'.[48]

5.3.2.3 The Relevance Between the Public Interest Issue Raised in the EPIL and the NGO's Objective and Scope of Work

The standard of relevance required by the SPC also aims to ensure the professionalism of NGOs.[49] Diverse rulings, however, can be identified in judicial practice. For example, in *Friends of Nature v Jiangxi Pohu Low Carbon Environmental Protection Co., Ltd.* (2019), the Intermediate Court of Nanchang ruled that the scope of work of the plaintiff is relevant to solid waste, therefore it is not qualified to bring an EPIL against water pollution.[50] In the appeal, however, Friends of Nature referred to the Guiding Case No. 75 and argued that as long as the issue brought to the court is relevant to the environmental elements or ecosystems that the NGOs intend to protect, the standard of relevance should be considered as fulfilled.[51] This relaxed standard was supported by the court, and they found in favour of Friends of Nature.[52] Of course, solid waste could have a negative impact on the aquatic environment. In this way, it is not difficult for NGOs to meet the standard of relevance.

In the same year, however, the standing of an NGO was denied for failing to meet the standard of relevance.[53] The defendant, Hebei Changheng Energy Technology Co., Ltd., was accused of illegal discharging of air and water pollutants, which led to severe pollution. The plaintiff, the Henan Centre to Promote Corporate Social Responsibility, however, is devoted to corporate social responsibility. Corporate social responsibility is generally understood to mean that corporations have a degree of responsibility not only for the economic consequences of their activities, but also for the social implications in the aspects of labour, consumer, community and environment.[54] The charter of the plaintiff also listed the promotion of ecological

---

[48] Ibid.

[49] Ibid.

[50] *Friends of Nature v Jiangxi Pohu Low Carbon Environmental Protection Co., Ltd.*, Intermediate Court of Nanchang, First Instance, Civil Division, No. 510, 25 July 2019 (in Chinese).

[51] *China Biodiversity Conservation and Green Development Foundation v Ningxia Ruitai Technology Co., Ltd.*, Supreme People's Court, Civil Ruling, Retrial, No. 47, 28 January 2016 (in Chinese); *Friends of Nature v Jiangxi Pohu Low Carbon Environmental Protection Co., Ltd.*, High Court of Jiangsu, First Instance, Civil Division, No. 526, 28 August 2019 (in Chinese).

[52] *Friends of Nature v Jiangxi Pohu Low Carbon Environmental Protection Co., Ltd.*, High Court of Jiangsu, First Instance, Civil Division, No. 526, 28 August 2019 (in Chinese).

[53] *Henan Center to Promote Corporate Social Responsibility v Hebei Changheng Energy Technology Co., Ltd.*, Intermediate Court of Shijiazhuang, First Instance, No. 818, 15 July 2019.

[54] Seven key themes are listed in the Guidance on Social Responsibility, which include environmental protection. General Administration of Quality Supervision, Inspection and Quarantine and

civilization and sustainable development as one of its objectives.[55] Following the logic of Guiding Case No. 75, the plaintiff should be able to pass the relevance test. In fact, the same plaintiff was granted standing in other EPIL cases, such as the *Henan Center to Promote Corporate Social Responsibility v Tongren Tongxin Mercury Co., Ltd.* (2017).[56]

5.4 Duration of Activity

Article 58 requires NGOs to have engaged in environmental protection activities in the past five consecutive years. Simple and clear as it seems, disputes have occurred in judicial practice. The typical case here is *Friends of Nature and Fujian Green Home Environmental Friendly Center v Zhijin Xie* et al. *(2015).* As one of the very first Chinese environmental NGOs, it was formally known as 'Green Culture College' and was registered as a branch of the International Academy of Chinese Culture (an already existing NGO) in 1993.[57] This was an expedient to circumvent strict NGO registration requirements.[58] It has since then engaged in environmental protection activities in China under the name of 'Friends of Nature'. But it did not gain legal personality until its independent registration on 18 June 2010.[59] The defendants therefore argued that it failed to meet the five-year existence standard at the time of filing this lawsuit (1 January 2015).[60]

The court, however, recognized its standing. It ruled that although Friends of Nature was not independently registered until 2010, it has engaged in environmental protection activities for more than 5 years. This ruling was later sustained on appeal.[61] Namely, the court is in favour of the argument that the five-year standard can be traced back to even before the NGOs are independently registered. This issue can be further elaborated in several aspects:

Standardization Administration, *Guidance on Social Responsibility*, GB/T 36000-2015 (in Chinese).

[55] Charters of Henan Center to Promote Corporate Social Responsibility (2022).

[56] *Henan Center to Promote Corporate Social Responsibility v Tongren Tongxin Mercury Co., Ltd.* (2017). High Court of Henan, Civil Division, Appeal, No. 232, 15 September 2017 (in Chinese).

[57] Southern Weekly (17 November 2007).

[58] Ibid.

[59] *Friends of Nature and Fujian Green Home Environment Friendly Center v Zhijin Xie* et al., Intermediate Court of Nanping, Civil Division, Appeal, No.2060, 29 October 2015 (in Chinese).

[60] Ibid.

[61] Ibid.

5.4.1 NGOs That Have Registered for Over Five Years

Article 58 requires NGOs to have engaged in environmental protection activities in the past five consecutive years. This is more a standard based on actual experience. Simply, per se., being registered for over 5 years does not necessarily meet the standard. Attention should also be paid to its actual performance. The word 'consecutive' here can be understood as year by year. The annual report of NGOs can be considered as evidence in this aspect. With regard to what constitutes as environmental protection activities, it is suggested that a broad interpretation should be adopted.

At present, no plaintiff has been challenged on this issue. However, there could be potential risks for NGOs that have a rather broad scope of work and just begin to show interest in participating in EPIL practice. Those NGOs should devote more resources to enhancing their record on environmental protection activities per se.

5.4.2 NGOs That Have Not Registered for Over Five Years

Although the court supports the interpretation that the five-year standard can be traced back to even before the NGOs are independently registered, it failed to elaborate the reasons behind this ruling. In addition, it seems necessary to explore whether the logic of this ruling can be applied to other NGOs that have not registered for over 5 years.

From the perspective of semantic interpretation, Article 58 is ambiguous on the duration standard. On the one hand, it can be understood as a requirement on the duration of existence of a particular NGO. The registration date is the starting point of calculation. On the other hand, the five-year standard is included as part of the performance evaluation requirement, which seems to place emphasis on actual experience. If it truly intends to consider it as a requirement on the duration of existence, it should be stated together with the requirement on registration, which is the case in German law on association suit for environmental purposes.[62] In the *Friends of Nature and Fujian Green Home Environmental Friendly Center v Zhijin Xie* et al. *(2015)*, the court obviously took the second opinion.

From the perspective of teleological interpretation, as previously mentioned, the standing of the plaintiffs is restricted by Article 58 to ensure professionalism and prevent vexatious litigation. Meanwhile, excessive restrictions should also be avoided. Otherwise, the initial intention of promoting civil society participation through EPIL cannot be achieved. Friends of Nature is the first grassroots

[62] Rehbinder, above note 32; Bundesnaturschutzgesetz (BNatSchG) [Federal Nature Conservation Act], Section 63, 1 March 2010, Federal Law Gazette, 2009, Part I, No. 51, 2542, an unofficial English version is available at: https://www.bmu.de/fileadmin/Daten_BMU/Download_PDF/Naturschutz/bnatschg_en_bf.pdf.

environmental NGO registered at the Ministry of Civil Affairs and one of the first few NGOs that have actively initiated EPIL cases in China. It has been commonly known as the 'Friends of Nature' and has participated in environmental protection activities since 1994. But it was initially registered as an affiliated group of the International Academy of Chinese Culture and only gained independent legal personality since June 2010. Consistency of its activities can be identified before and after its independent registration in 2010, and it has been widely acknowledged by the Chinese society. If NGOs like this are precluded from initiating EPIL, it is against the legislative intention of Article 58. Given the status quo of civil society development in China, a relatively relaxed interpretation on the five-year standard should be preferred. In fact, the Chinese standard is already stricter than that in Germany, where environmental NGOs are only required to have existed for 3 years.[63] Overall speaking, considering the five-year standard as a standard on actual experience will give more flexibility and leverage to the court to evaluate the performance of NGOs on a case-by-case basis.

Moreover, this issue also relates to the legal effect of registration and the division of a legal person. In the case of *Friends of Nature and Fujian Green Home Environmental Friendly Center v Zhijin Xie* et al. *(2015)*, Friends of Nature was divided from the International Academy of Chinese Culture in 2010. According to Article 19 of the Regulations on Social Groups Registration Management, Friends of Nature as an affiliated group should operate within the scope of authorisation by the International Academy of Chinese Culture.[64] Correspondingly, the legal effects of its behaviour belong to the latter. However, in case the division of a legal person occurs, Article 67 of the Civil Code requires that the rights and obligations of the legal person shall be enjoyed and assumed jointly, unless otherwise agreed upon by its creditors and debtors.[65] Although it does not address the issue of standing, this provision does indicate the relevance of the rights of the legal person in the case of division. Of course, it is unlikely for NGOs to make arrangements on the standing to sue in advance, it is therefore necessary for the court to decide. The consistency of its activities, objectives and scope of work before and after the division should be the key factors to consider.

Another possible situation is that an NGO has been newly established for less than 5 years. In China, most types of non-profit legal persons are considered established from the moment of registration.[66] Under such circumstances, a newly established NGO does not even exist before the registration. In addition, NGOs are not allowed to operate in the name of social organizations without legal registration.

---

[63] Ibid.

[64] *Regulations on Social Groups Registration Management*, State Council, 6 February 2016, art 3 (in Chinese).

[65] *Civil Code*, National People's Congress, 28 May 2020, art 67 (in Chinese).

[66] Zhu (2016), p. 438; *Regulations on Social Groups Registration Management*, State Council, 6 February 2016, art 3 (in Chinese); *Regulations on the Management of Foundations*, State Council, 11 February 2004, art 11 (in Chinese); *Interim Regulations on Registration Administration of Private Non-Enterprise Units*, State Council, 25 October 1998, art 3 (in Chinese).

If the activities before its registration are illegal, there is no way for the court to recognize them when considering the standing to sue.[67] In contrast, Article 60 of the Swedish Civil Code considers an NGO established once the charter is issued.[68] Registration here is just a declaration. Under such circumstances, the legal effect of its activities before the registration certainly belongs to the NGO itself.

In addition, in case NGOs are merged for less than 5 years, Article 67 of the Civil Code states that the rights and obligations of such legal persons shall be enjoyed and assumed by the surviving legal person. To apply by analogy, it is reasonable to recognize the standing of a merged NGO if at least one of the predecessors has participated in environmental protection activities for more than 5 years.

Furthermore, in case NGOs changed their objectives for less than 5 years, it is suggested that it should not result in the recount of the duration standard. The objectives could be enlarged, narrowed or shifted. For example, one may expand their objective from the protection of one particular wild animal to a category of wild animals, narrow their objective from the protection of water resources to the protection of water resources in a specific river basin, or change their objective from the prevention of solid waste pollution to water pollution. The change of objectives and the scope of work should go through a registration process in China, but it has no impact on the legal personality.[69] In addition, the court already established a rather relaxed rule on the relevance between the dispute raised in the EPIL and the NGO's objective and scope of work. The change of objectives will be less likely to affect the standing of NGOs.

If the objective was changed from for-profit to non-profit, however, the duration of activities would need to be recalculated. As previously mentioned, many grass-roots NGOs in China choose to be registered as a for-profit organization to circumvent stringent requirements on the registration of social organizations. However, the change of objective from for-profit to non-profit is in fact a procedure to establish a new NGO. The EPL clearly restricts the standing to NGOs, instead of all legal persons who have been engaged in environmental public interest activities. Therefore, it is prudent to preclude the standing of NGOs under such circumstances.

---

[67] *Provisional Measures on Banning Illegal NGOs*, Ministry of Civil Affairs, 6 April 2000, art 2 (in Chinese).

[68] Zhu, above note 72.

[69] *Regulations on Social Groups Registration Management*, State Council, 6 February 2016, art 16 (in Chinese); *Regulations on the Management of Foundations*, State Council, 11 February 2004, art 10 (in Chinese); *Interim Regulations on Registration Administration of Private Non-Enterprise Units*, State Council, 25 October 1998, art 12 (in Chinese).

5.5 The Issue of Compliance

Article 58 also requires the NGOs to have no record of administrative or punitive penalties for their activities in the past five consecutive years. Article 5 of the Judicial Interpretation on EPIL further clarifies that only administrative or punitive penalties applied for their operational activities will affect the standing to sue. Again, it is obvious that this provision aims to evaluate the professionalism of NGOs and ensure that they comply with government supervision rules. In contrast, if an NGO were fined for violation of fire safety regulations in the workplace, it should not be considered as a record that affects the standing to sue in EPIL.

In addition, penalties applied to NGO members, staff and principals do not affect the standing to sue. For example, two staff members of the China Biodiversity Conservation and Green Development Foundation were arrested as suspects of racketeering in April 2021.[70] If they are convicted, this will not have an impact on the standing of the NGO. However, it will expose the NGO to the risk of potential administrative penalties.

In practice, the same NGO also raised a dispute as to whether the failure to pass an annual inspection should count as a record of breaking the law. In *China Biodiversity Conservation and Green Development Foundation v Zhejiang Taobao Network Co., Ltd. (2019)*, the plaintiff failed an annual inspection in 2015 and was instructed to rectify its behavior in 2017. The main problem was the defendants investing money against its non-profit purposes and being in violation of the rules on holding council meetings.[71] The defendant therefore argued that it is not an eligible plaintiff. The court, however, ruled that the failure of annual inspection does not constitute as an illegal record.[72] After all, an annual inspection is not a kind of administrative or punitive penalty. But according to Article 42 of the *Regulation on Foundation Administration*, foundations that failed an annual inspection should be given a warning or ordered to stop their default. In the case of a severe violation, the registration could be withdrawn. All these measures, if applied, can be considered as an illegal records for the China Biodiversity Conservation and Green Development Foundation. Interestingly, the National Social Organizations Management Bureau has not applied any sanction on it, which risks the Bureau being charged for an administrative omission. The defendant could also consider bringing administrative litigation against it to challenge its nonfeasance.

---

[70] The Paper, above note 41.

[71] *China Biodiversity Conservation and Green Development Foundation v Zhejiang Taobao Network Co., Ltd.*, High Court of Zhejiang, Civil Division, Appeal, No. 863, 14 October 2019 (in Chinese).

[72] Ibid.

Table 5.2 Number of Court-resolved EPIL Cases 2015–2020[a]

Year	NGOs	Procuratorates
2015	53	12
2016	68	77
2017	58	1304
2018	65	1737
2019	58	1895
2020	103	3454

[a]Supreme People's Procuratorate (2016); Li (2017), p. 1; The State Council Information Office (2 March 2019a); SPC (2020)

5.5.1 *The Influence of Other Eligible Plaintiffs on the Standing of NGOs*

Overall, the Chinese courts have generally taken a relaxed position on the threshold requirements of Article 58. Nevertheless, environmental NGOs in China are not as active as expected in initiating EPIL. Some may argue that Article 58 is too stringent on NGOs compared to its counterpart in Germany, especially with regard to the five-year standard. In addition, there remains a long way to go to build capacity for NGOs. China, however, is eager to explore the potential of the judiciary in environmental law enforcement. This leads to the authorization of procuratorates to bring EPIL and the introduction of EEDCL to allow local governments to bring lawsuits against polluters. Procuratorates and local governments are considered more powerful and resourceful to make polluters pay for environmental damages through the judicial process. Moreover, as a partial response to the criticism that the EPIL only targeted polluters, the procuratorates are also allowed to bring an EPIL to challenge the legitimacy of administrative decisions. Meanwhile, disputes have occurred in judicial practice regarding whether NGOs can bring EPIL against marine pollution. Compared to the threshold standing requirements on NGOs, other eligible plaintiffs of EPIL could have a more significant impact on the standing of NGOs in practice. EPIL in China has eventually evolved from private enforcement into another kind of public enforcement that is more powerful and less restricted than traditional administrative authorities.

5.5.1.1 Procuratorates

As previously mentioned, Article 55(2) of the *Civil Procedure Law* (2017) grants standing to procuratorates as a complementary measure to promote EPIL. Namely, they are allowed to bring EPIL against polluters if no organizations and authorities stipulated in Article 55(1) existed or they refused to file such cases. Nevertheless, NGOs are increasingly marginalized in the EPIL. As the diagram below (Table 5.2) shows, the number of EPIL cases brought by procuratorates rapidly increased since 2017,[73] while the number of EPIL cases brought by NGOs stagnated at a much lower level. In 2018, the Judicial Interpretation on Public Interest Litigation Initiated by Procuratorates formally confirmed that procuratorates can bring an incidental civil EPIL as part of a criminal procedure, which eventually became the mainstream practice.[74] In addition, a large number of EPIL cases brought by procuratorates are against administrative authorities,[75] which NGOs are not yet allowed to do so.

In order to reflect the complementary role of procuratorates, Article 13 of the Judicial Interpretation on Public Interest Litigation Initiated by Procuratorates requires a 30-day pre-trial notification process. But this notification process has also raised disputes in practice.

5.5.1.2 The Scope of Application of the Pre-Trial Notification Procedure

As previously mentioned, procuratorates can bring an incidental civil EPIL in a criminal procedure. What is controversial, is whether the procuratorates should also follow the 30-day notification rule if the EPIL process is brought incidental to a criminal procedure. Judicial practice is divergent on this issue. Opponents argue that the legal basis for bringing an incidental civil litigation in the criminal procedure is not Article 55(2) of the *Civil Procedural Law*, but the Article 101 of the *Criminal Procedure Law*. It states that in a case where state-owned and collectively-owned properties suffered loss, procuratorates can bring a civil litigation in the criminal procedure.[76] Therefore, it is different from the standing to sue granted by Article 55(2) and this provision has been referred to many times in judicial practice.[77] Moreover, incidental civil litigation in the criminal procedure is designed to improve litigation efficiency and save judicial resources. If the 30-day notification rule still

---

[73] The sharp increase of EPIL cases brought by procuratorates in 2017 was mostly motivated by political pressure from top-down to deliver a good performance result at the end of the two-year pilot period. It was used to add momentum for the revision of relevant legislation to formally recognize the standing of procuratorates in public interest litigation, which eventually came into being in July 2017. Zhang (2017).

[74] SPC (2019).

[75] The number of administrative EPIL cases brought by procuratorates is almost 8 times more than that of the civil EPIL cases. Supreme People's Procuratorate (2022).

[76] Article 101 is referred to in *People's Procuratorate of Lianyun District v Baoshan Yin* et al. (2016). See *People's Procuratorate of Lianyun District v Baoshan Yin* et al., Intermediate Court of Lianyungang, Criminal Division, Appeal, No. 99, 2016 (in Chinese); Liu (2019a), pp. 84–86.

[77] Liu, above note 85. Article 101 and Article 55(2) are different on the scope of issues allowed to sue. Article 55(2) targets at the activities that are harmful to the public interests on the protection of environment and natural resources. Article 101, however, applies to all criminal activities that injured properties that are state-owned and collectively owned. From the perspective of environmental protection, some environmental interests cannot be protected by state-ownership, such as air quality. Only Article 55(2) can be considered as the legal foundation for the standing of procuratorates.

has to be applied, the judicial process will be delayed and difficulties will arise for the coordination of civil and criminal procedures.[78]

Supporters, however, insist that incidental civil litigation brought by procuratorates in the criminal procedure is still a kind of EPIL. Article 55(2) should also be applied under such circumstances. Otherwise, NGOs will be completely excluded from the procedure, especially considering that the incidental civil EPIL is in fact the mainstream practice of procuratorates. For example, in *People's Procuratorate of Wuqia v Qingquan Zhao* (2019), the court of the appeal withdrew the ruling at first instance for failing to comply with the pre-trial notification requirement.[79] In 2019, the SPC and the SPP issued a joint written reply on this dispute and confirmed that pre-notification should be made in incidental civil EPIL brought by procuratorates in the criminal procedure. Article 101 of the *Criminal Procedure Law*, however, was not mentioned as the legal basis.

Although the decision is largely commendable, the written reply still cannot provide a sound doctrinal explanation of the relationship between Article 101 and Article 55(2). The standing of procuratorates authorized by Article 101 is still based on state-ownership and collective-ownership, which is in nature different from EPIL. But before EPIL, this provision was rarely referred to in judicial practice. In the revision of the *Criminal Procedure Law* in 2018, some have asked for this revise this provision but eventually failed.[80] A similar problem also exists with regard to the relationship between EPIL and EEDCL.

5.5.1.3 Specific Requirements for the Pre-Trial Notification Procedure

The law and judicial interpretation documents do not specify on how the procuratorates should issue such a notification. However, several points are still worth mentioning for judges to consider in judicial practice.

First of all, it is the manner of the notification. Usually, notifications are published in national newspapers of legal categories, such as the Procuratorate Daily and Legal Daily. Occasionally, they are published in local legal newspapers, such as Shandong Legal Daily.[81] Generally speaking, it is acceptable to make notifications through official newspapers that are relevant to the geographical location or issues in dispute of the case. What is more controversial, is the practice of making a notification through newspapers that are not relevant to the case at all. For example, in the *People's Procuratorate of Qianxi v Moujia Li* (2020), a notification was made on the China Labour Security Daily, while the issue in dispute was with regard to the illegal

---

[78] Ibid 87; Liu (2019b), p. 77; Zhou (2018), p. 3.

[79] *People's Procuratorate of Wuqia v Qingquan Zhao*, Intermediate Court of Kizilsu Kirghiz Autonomous Perfecture, Criminal Division, Appeal, No. 52, 2019 (in Chinese).

[80] Bai (2018).

[81] *People's Procuratorate of Rizhao v Bin Li*, Intermediate Court of Rizhao, Civil Division, First Instance, No. 344, 2019 (in Chinese).

purchase, transportation and selling of endangered wild animals.[82] Such notification should be considered inappropriate, and the standard of relevance should be applied here. In practice, many have chosen to notify on the official website of the Procuratorate Daily.[83] Given the popularity of e-government and internet media in China, this form of notification could be preferred as an efficient and convenient approach.

Moreover, the notification is made to eligible plaintiffs of EPIL in general and the willingness to sue should be made explicitly within 30 days. During the piloting stages, the notification was made individually to eligible plaintiffs within the jurisdiction of the procuratorates and the notified parties should provide a written reply on whether they are willing to sue within 30 days.[84] This process, however, is a waste of resources and is thought to be inefficient. In addition, the restriction of the notification to eligible plaintiffs within the jurisdiction of the procuratorates impairs the standing to sue of NGOs outside the jurisdiction.

Finally, the notification should not attach other conditions. In practice, if the notified NGOs revealed their willingness to sue, the procuratorates could require NGOs to bring the EPIL within certain time limits or make additional requirements regarding their ability to sue.[85] The top-down internal performance evaluation pressure and the pursuit of litigation efficiency could result in the reluctance of the procuratorates to hand over the opportunity to sue to NGOs.[86] In fact, NGOs rarely exercise their priority to sue through the notification process.[87]

5.5.2 Administrative Authorities

Although Article 55(1) allows 'relevant organizations and authorities' to bring EPIL, it also requires that it should be 'prescribed by law'. Namely, without explicit

[82] *People's Procuratorate of Qianxi v Moujia Li*, Criminal Division, First Instance, No. 24, 2020 (in Chinese).

[83] See Justice Web, http://www.jcrb.com/gg2015/; *People's Procuratorate of Daan v Yan Li*, People's Court of Daan, Criminal Division, First Instance, No. 88, 2020 (in Chinese).

[84] *Implementation Measures for Pilots on People's Procuratorates Initiating Public Interest Litigation*, Procuratorial Committee of the Supreme People's Procuratorate, 16 December 2015, art 13 (in Chinese); *People's Procuratorate of Liaocheng v Rongtai Lu*, Intermediate Court of Zibo, Civil Division, First Instance, No. 57, 2017 (in Chinese).

[85] Ge, above note 36.

[86] Ibid.

[87] Exceptionally, the NGO did confirm to initiate the case as the plaintiff. But the NGO and the Procuratorate have closely cooperated on that case before the notification was actually issued. See Yuanyuan Chen, *Procuratorate of Dongguan Assisted an NGO to Bring Public Interest Litigation* (8 August 2020) <http://sthjj.gz.gov.cn/hbyd/zjhjb/content/post_6491326.html> (in Chinese).

authorisation, administrative authorities are not allowed to bring EPIL.[88] For example, in the *Environmental Protection Bureau of Fumian District, Yulin City v Quzhou Xuntong Logistics Co., Ltd. (2018)*, the court in the appeal ruled that the plaintiff was not the 'relevant organizations and authorities prescribed by law'; therefore, they did not have the standing to bring EPIL.[89] But before the revised *Civil Procedure Law* came into effect in 2013, judicial practice had recognized the standing of administrative authorities to bring EPIL within their administrative jurisdiction, such as the *Local People's Government of Yexie Town, Songjiang District, Shanghai v Rongxiang Jiang and Shengzhen Dong (2012)*. The polluted river was located within the jurisdiction of the plaintiff. The court therefore concluded that the plaintiff had the power to restore the environment and was entitled to sue.[90] Despite the farfetched legal reasoning, this case was later selected by the SPC as one of the nine typical environmental cases in China.[91] This issue was further complicated when the EEDCL was introduced.[92] In this chapter, we will not elaborate on this issue; instead, we will add relevant discussion to the analysis of Article 89 of the Marine EPL.

Article 89 of the Marine EPL is often understood as the only statutory authorisation thusfar for administrative authorities to bring EPIL. This provision, however, has already existed since 1999, much earlier than Article 55 and Article 58. Controversies abound, however, with the application of Article 89, Article 55 and Article 58, particularly as to whether NGOs can bring an EPIL to protect the marine environment.

In practice, the attempts by NGOs to bring such EPIL cases have been rejected by the courts in most circumstances. For example, in *Dalian Environmental Protection Volunteers Association v PetroChina Fuel Oil Co., Ltd.* et al. *(2015)*, the court ruled that the Marine EPL as a special law should be applied prior to the EPL as a general law; therefore, the standing can only be granted to administrative authorities in charge of marine environment protection, and the NGOs in this case were not qualified plaintiffs.[93] This ruling is also supported by the SPC, which stated that

---

[88] Wang (2012), p. 105.

[89] *Environmental Protection Bureau of Fumian District, Yulin City v Quzhou Xuntong Logistics Co., Ltd.*, Intermediate Court of Yulin, Civil Division, Appeal, No. 929, 2018 (in Chinese).

[90] *Environmental Protection Bureau of Jiangyin v Wenfeng Wang* et al., Basic Court of Jiangyin, Civil Division, First Instance, No.3, 4 December 2013 (in Chinese); *Local People's Government of Yexie Town, Songjiang District, Shanghai v Rongxiang Jiang and Shengzhen Dong*, Basic Court of Songjiang, Civil Division, First Instance, No.4022, 28 June 2012 (in Chinese).

[91] *The SPC Announces Nine Typical Environmental Cases* (2014).

[92] The nature of EEDCL is highly controversial. Representative theories include national interests litigation, private litigation and public interests litigation. See Li (2019), p. 55.

[93] Ibid; *Dalian Environmental Protection Volunteers Association v PetroChina Fuel Oil Co., Ltd.* et al., Dalian Maritime Court, Registration Division, First Instance, No. 5, 17 June 2015 (in Chinese); *Chongqing Liangjiang Voluntary Service Center and Guangdong Environmental Protection Foundation v Guangdong Century Tsingshan Nickel Industry Co., Ltd.* et al., Intermediate Court of Maoming, Civil Division, First Instance, No.122, 2016 (in Chinese).

EPIL brought by administrative authorities to protect the marine environment is a special kind of EPIL and different from the one brought by NGOs.[94]

Occasionally, the courts bypass the problem of allowing NGOs to bring EPIL that is related to marine environment protection. For example, in *Chongqing Liangjiang Voluntary Service Center and Guangdong Environmental Protection Foundation v Guangdong Century Tsingshan Nickel Industry Co., Ltd.* et al. *(2016)*, the defendants were accused of dumping slag on local mudflats, wetlands and mangrove forests. The court recognized the standing of the plaintiffs in the sense that the defendant not only polluted the marine environment, but also damaged land ecosystems.[95]

Similar to NGOs, the standing of procuratorates has also been challenged in practice, such as the *Guangzhou People's Procuratorate v Weilai Li (*2017*)*. The court, however, recognized the standing of the plaintiff based on Article 55(2) of the *Civil Procedure Law*. The plaintiff also complied with the pre-trial notification requirement and stated that no eligible administrative authorities and NGOs are willing to sue.[96] This could be considered as an affirmative decision on the application of Article 55 in the EPIL to protect marine resources.

However, the SPC attempts to settle the dispute in a different way. In the Judicial Interpretation on EPIL to Protect Marine Natural Resources and Environment issued in 2022, Article 3 clarifies that procuratorates can also bring EPIL to protect marine natural resources in case the administrative authorities in charge of marine environment protection failed to do so after receiving notification from the procuratorates.[97] NGOs, however, are not mentioned by the judicial interpretation at all. On the one hand, this interpretation seems to recognize that Article 55 can also be applied in EPIL to protect the marine environment. Otherwise procuratorates should not be allowed to sue. On the other hand, NGOs are still precluded from doing so, which is hard to justify from a doctrinal perspective after the standing of procuratorates is recognized.

In fact, if Article 89 were in fact a kind of EPIL, the alleged conflict between this Article and Article 55 does not really exist. Although Marine EPL is a special law compared to the EPL, Article 58 itself is a special requirement on the standing of NGOs in EPIL. Article 58 of the law clearly states that it applies to the protection of environmental public interests. Marine environmental protection is clearly subjected to this provision.

---

[94]Opinions on Strengthening Judicial Work to Protect Environment and Natural Resources and Safeguard the Construction of Ecological Civilization, [2014] No. 11 of Supreme People's Court (in Chinese).

[95]*Chongqing Liangjiang Voluntary Service Center and Guangdong Environmental Protection Foundation v Guangdong Century Tsingshan Nickel Industry Co., Ltd.* et al., Intermediate Court of Maoming, Civil Division, First Instance, No.122, 2016 (in Chinese).

[96]*Guangzhou People's Procuratorate v Weilai Li*, Guangzhou Maritime Court, Civil Division, First Instance, No. 431, 2017 (in Chinese).

[97]*Judicial Interpretation on EPIL to Protect Marine Natural Resources and Environment*, No. 15 [2022] of the Supreme People's Court, 15 May 2022 (in Chinese).

Moreover, Article 89 has already existed since 1999 (Article 90 at the time), while Article 58 was adopted in the 2010s. If they were all considered special rules, the new ones are superior to the old ones. Even if Article 55 and Article 58 were treated as general rules and Article 89 was considered as a special rule, this dilemma of law application should be dealt with by the Standing Committee of the National People's Congress according to Article 94 of the Legislative Law. The judiciary cannot decide on this issue by itself.[98]

A more fundamental question to ask, however, is whether Article 89 is indeed a kind of EPIL. As stated in the law, standing is granted to administrative authorities in charge of marine environment protection to seek compensation from polluters and destroyers on behalf of the State. In nature, administrative authorities' standing to sue is based on state-ownership of marine natural resources, which is in line with the traditional rules concerning the standing.[99] In addition, administrative authorities are suing on behalf of the State only, which indicates that the administrative authorities are more likely litigation counsels rather than plaintiffs. In practice, however, they are considered as plaintiffs, which further increases the ambiguity of Article 89.

Recent years have also witnessed the development of EEDCL, which further complicates the situation. Notably, the State Council issued the Reform Plan for the Compensation of Environmental Damages in 2017 to allow certain administrative authorities to bring EEDCL against infringers to seek compensation for environmental damages within their administrative jurisdictions. A special judicial interpretation document (for trial implementation) was then issued by the SPC in 2019 to elaborate on procedural arrangements.[100] In particular, Article 1 of the Judicial Interpretation on EEDCL (For Trial Implementation) (2019) authorizes the following administrative authorities to sue in case they failed to reach an agreement with polluters: the provincial and municipal governments, institutions and departments designated by them, and departments authorized by the State Council to exercise state ownership.[101]

This provision reflects a mixed feature of both private and public law. On the one hand, one could argue that the departments authorized by the State Council to exercise state ownership seem to gain their standing based on national property rights. On the other hand, the provincial and municipal governments seem to gain their standing based on their administrative jurisdiction. As long as a pollution activity occurred within their jurisdiction, they have the right to bring EEDCL, even if it affects private or collective property rights. This interpretation, in fact, reveals the essence of the EEDCL. Namely, it is a quasi-punitive measure against

---

[98] Wu (2015), pp. 40–42.

[99] A legal person of public law can bring civil cases if its interests or rights are harmed. See Zhou (2014), p. 90.

[100] *Judicial Interpretation on EEDCL (For Trial Implementation)*, No. 8 [2019] of the Supreme People's Court, 20 May 2019, art 1 (in Chinese).

[101] Ibid.

polluters in the disguise of a tort claim.[102] While administrative and criminal punishments are subject to a range of strict substantive and procedural restrictions, this quasi-punitive measure is free from the above restraints.

In many ways, Article 89 resembles more like EEDCL than EPIL. The SPC has already dealt with the relationship between EEDCL and EPIL. Article 16 of the *Judicial Interpretation on EEDCL* (For Trial Implementation) (2019) states that if an EPIL case was brought against the same activities during the trial of the EEDCL case, the same court should accept the EPIL case if the plaintiffs are eligible. Article 17 further states that the EPIL case should be suspended pending the result of the EEDCL case, and the court should continue to decide on claims that have been covered by EPIL but unresolved by the EEDCL decision. It is obvious that the standing of NGOs is not affected under such circumstances.[103]

Compared to the preclusion of NGOs from bringing marine EPIL at all, the above interpretation provides a more balanced way to protect the procedural rights of NGOs and save judicial resources. In fact, NGOs have already attempted to justify their standing by arguing the different nature of Article 89 compared to EPIL.[104] But it has not been accepted by the court in practice.

This could be understood, however, from a realistic perspective. Marine environmental pollution is usually very complicated, which means high litigation costs and a high demand for scientific and legal professional knowledge.[105] Given the status quo of civil society development in China, the professionalism of NGOs could be doubted to bring a marine EPIL. Similar arguments can be made for the standing of procuratorates as well.[106] It is therefore suggested that both NGOs and procuratorates should not be allowed to bring marine EPIL following the above logic. In addition, marine pollution could become an international dispute, which is considered sensitive for NGOs to participate in.[107] Having recognized the above genuine concerns, however, the exclusion of NGOs to sue still falls short of convincing doctrinal explanations.

[102] In fact, there are disputes on whether administrative or criminal fines should be deducted when calculating environmental damages in EEDCL cases. See *People's Procuratorate of Tongren v Guizhou Yuping Xiangsheng Chemical Co., Ltd.*, Intermediate People's Court of Zunyi, Civil Division, First Instance, No. 520, 30 January 2018 (in Chinese).

[103] *Judicial Interpretation on EEDCL (For Trial Implementation)*, No. 8 [2019] of the Supreme People's Court, 20 May 2019, arts 16, 18 (in Chinese).

[104] Friends of Nature used the same argument to ask the court to recognize its standing to sue. *Friends of Nature v Rongcheng Weibo fishery Co., Ltd.*, High Court of Shandong, Civil Division, Appeal, No. 1334, 2018 (in Chinese).

[105] Yang (2021), pp. 120–129.

[106] Ibid 127.

[107] Ibid 130.

5.6 Concluding Remarks

Private enforcement in the form of EPIL was envisioned to complement public enforcement against the polluters and to increase the influence of NGOs. To these ends, the standing to sue in EPIL is expanded beyond individual rights. However, certain restrictions were also imposed to ensure professionalism and prevent vexatious litigation and the intrusion of individual rights in the name of public interests.

As revealed in the above analysis, the Chinese courts have generally taken a relaxed position on the standing requirements of NGOs in terms of EPIL, especially with regard to the five-year duration standard and the relevance test. In some respects, Article 58 of the EPL is even more relaxed that the counterpart in Germany. For example, the standing of NGOs in EPIL is not restricted to those with legal personalities.

The real problem that affects the performance of NGOs, lies in the fledging status of civil society development in China and the urgency to promote a court-centred environmental movement in China. Instead of gradually enhancing an enabling environment for the civil society, China is eager to open a door for judges to exercise their power broadly to ensure the realization of a pro-environmental policy. As a result, procuratorates and administrative agencies have been encouraged to bring EPIL and EEDCL. Although the role of procuratorates is envisioned to be complementary, NGOs rarely exercise their priority to sue through the notification process. NGOs are still precluded from bringing marine EPIL into judicial practice, while administrative authorities and procuratorates are allowed to do so. The EPIL and EEDCL cases brought by administrative authorities are indeed quasi-punitive measures against polluters in the disguise of tort claims. This eventually turns the private enforcement of environmental law in the form of EPIL into another kind of public enforcement which is more powerful but less restricted than traditional administrative approaches. Special attention should be given to prevent the intrusion of individual rights in the name of public interests.

With procuratorates and administrative authorities increasingly involved in holding polluters accountable for environmental pollution, the public attention is likely to shift from the responsibility and accountability of the government itself. Different from procuratorates, NGOs in China have not yet been allowed to bring EPIL to challenge administrative decisions. A breakthrough in this direction should be the focus of future legal reforms on the standing of NGOs in EPIL.

Acknowledgement This work was supported by The National Social Science Fund of China, grant number 21BFX188, 2021.

References

All-China Environmental Protection Federation (2008) Report on the Status of Development of Environmental Social Organizations in China. https://www.docin.com/p-2133623362.html. Accessed 18 May 2022

Bai L (2018) Legislation on Public Interest Litigation with Chinese Features in the New Era. The Procuratorate Daily, 8 March 2018 (in Chinese)

Boyer B, Meidinger E (1985) Privatizing regulatory enforcement: a preliminary assessment of citizen suits under federal environmental laws. Buffalo Law Rev 34(834):836–837

Charters of Henan Center to Promote Corporate Social Responsibility. http://www.hncsr.org/QiYeZeRen_News_Type_7.html. arts 3 and 9. Accessed 18 May 2022

Chen Y (2020) Procuratorate of Dongguan Assisted an NGO to Bring Public Interest Litigation http://sthjj.gz.gov.cn/hbyd/zjhjb/content/post_6491326.html. Accessed 18 May 2022

China Biodiversity Conservation and Green Development Foundation (2018) Introduction of CBCGDE. http://www.cbcgdf.org/Resume!jianjie.action. Accessed 18 May 2022

Coffee JC Jr (1983) Rescuing the private attorney general: why the model of the lawyer as bounty hunter is not working. Maryland Law Rev 42(2):215–288

Coffee JC Jr (1986) Understanding the plaintiff's attorney: the implications of economic theory for private enforcement of law through class and derivative actions. Columbia Law Rev 86(4): 669–727

Degenhart C (1981) Kernenergierechlt: Schiwerpunkte, Entscheididtngsstrukturen, Entwicklungslinien, Heymann

Feng G (2017) A Practical Perspective on Environmental Public Interest Litigation. Speech delivered at Environmental and Resources Law Institute, 3 June 2017 (in Chinese)

Greve MS (1989) The non-reformation of administrative law: standing to sue and public interest litigation in West German environmental law. Cornell Int Law J 2(197):201

Henry HP (2001) A shift in citizen suit standing Doctrine: *Friends of the Earth, Inc. v. Laidlaw Environmental Services*. Ecol Law Q 28(2):233–252

Huang X (ed) (2011) Report on Social Organization in China (2010–2011). Social Sciences Academic Press (China), p 5 (in Chinese)

Huang X (ed) (2020) Report on Social Organization in China (2020). Social Sciences Academic Press (China)

Judicial Interpretation on EPIL to Protect Marine Natural Resources and Environment, No. 15 [2022] of the SPC, 15 May 2022 (in Chinese).

Kelsen H (1945) General Theory of Law and State (trans: Wedberg A). Harvard University Press.

Li D (ed) (2017) Review of public interest litigation in environmental protection (2016). Law Press China

Li H (2019) Research on the nature of EEDCL and other relevant issues: analysis from the perspective of EPIL. Adm Law Rev 4:55. (in Chinese)

Lihong Lin (2006) The definition and scope of public interest litigation Chinese J Law 6:148, 149. (in Chinese).

Liu J (2019a) The predicaments of the civil public interest litigation attached to criminal cases and their solutions. Polit Sci Law 10:84–86. (in Chinese)

Liu Y (2019b) Research on the synergy problem of civil public interest litigation attached to criminal cases. Crim Sci 5:77. (in Chinese)

Ministry of Civil Affairs, Hongyu Liu (2019) The Member of the National Committee of CPPCC: Build Legal Capacity for NGOs and Increase the Capacity to Sue. http://mzzt.mca.gov.cn/article/zt_2019lh/dbwyhmz/zxwy/201903/20190300015780.shtml. Accessed 18 May 2022

National People's Congress, *Second Reading Draft for the Amendment of Environmental Protection Law* (17 July 2013). http://www.npc.gov.cn/npc/xinwen/lfgz/flca/2013-07/17/content_1801189.htm. Accessed 18 May 2022

Peng D (2013) Third Reading of Environmental Protection Law: Profound and In-depth Changes. http://www.npc.gov.cn/npc/zgrdzz/2013-12/12/content_1816460.htm. Accessed 18 May 2022

Rabkin JA (1998) The secret life of the private attorney general. Law Contemp Probl 61(1): 179–183

Rehbinder E (2014) Collective Court Actions for Protecting the Environment in the EU and Germany. Speech delivered at the Counsellors' Office of the Shanghai Municipal People's Government, 30 October 2014

Smith T (2018) Power Surge: China's New National Supervisory Commission. https://www.thechinastory.org/yearbooks/yearbook-2018-power/forum-absolute-power-corrupts-absolutely/power-surge-chinas-new-national-supervisory-commission/. Accessed 18 May 2022

Southern Weekly (2007) Ten Years of Congjie Liang and Ten Years of Friends of Nature. Available at: http://phtv.ifeng.com/hotspot/river/green/200711/1117_2348_301229.shtml. Accessed 18 May 2022

SPC (2019) Press Conference on the Report of Development of Environmental Judicature in China (the Green Book) and Typical Cases. https://www.court.gov.cn/zixun-xiangqing-145072.html. Accessed 18 May 2022

SPC (2020) Adjudgment of Environmental Cases in China (2019) https://www.court.gov.cn/zixun-xiangqing-228341.html. (in Chinese)

Supreme People's Procuratorate (2016) The Work Report of the Supreme People's Procuratorate (2015) https://www.spp.gov.cn/spp/gzbg/201603/t20160321_114723.shtml. Accessed 18 May 2022

Supreme People's Procuratorate (2022) The Work Report of the Supreme People's Procuratorate (2022) https://www.spp.gov.cn/spp/gzbg/202203/t20220315_549267.shtml. Accessed 18 May 2022

The Paper (2021) Two Environmental Volunteers Arrested for Blackmailing, China Biodiversity Conservation and Green Development Foundation Will Verify Later. https://www.sohu.com/a/458742001_260616. Accessed 18 May 2022

The SPC Announces Nine Typical Environmental Cases (2014). Available at: http://www.chinacourt.org/article/detail/2014/07/id/1329697.shtml. Accessed 18 May 2022

The SPC Rules on Guiding Cases, No.51 [2010] of the SPC, 26 November 2010

The State Council Information Office (2019a) SPC Press Conference on Adjudgment of Environmental Cases, The Development of Judiciary and Typical Cases (2 http://www.scio.gov.cn/xwfbh/gfgjxwfbh/xwfbh/44193/Document/1691730/1691730.htm Accessed 18 May 2022

The State Council Information Office (2019b) SPC Press Conference on Adjudgment of Environmental Cases, The Development of Judiciary and Typical Cases http://www.scio.gov.cn/xwfbh/gfgjxwfbh/xwfbh/44193/Document/1691730/1691730.htm. Accessed 18 May 2022

ThePaper (2016) Final Decision Confirmed the Environmental NGO's Standing in Public Interest Litigation against Tengger Desert Pollution. http://www.thepaper.cn/newsDetail_forward_1427140 (in Chinese) Accessed 18 May 2022

Wang C (2006) Law-making functions of the Chinese courts: judicial activism in a country of rapid social changes. Front Law China 3:1

Wang S (ed) (2012) Understanding the civil procedure law of People's Republic of China. Law Press China

Wu J (2015) The procedural design of EPIL. ECUPL J 6:40. (in Chinese)

Wübbeke J (2014) The Three-year Battle for China's New Environmental Law. https://www.chinadialogue.net/article/show/single/en/6938-The-three-year-battle-for-China-s-new-environmental-law. Accessed 18 May 2022

Xin C (2014) Understanding the environmental protection law of People's Republic of China. Law Press China

Yang C (2012) Several Issues on Environmental Protection and Environmental Legal System. http://www.npc.gov.cn/npc/c541/201211/63ba26f719454bf1a673ac1762850c85.shtml. Accessed 18 May 2022

Yang H (2021) 'The Plaintiffs' standing to sue in marine EPIL. Stud Law Bus 3:120–129. (in Chinese)

Zhang J (2019) The standing to sue in marine EPIL: a study of the rongcheng Weibo Case. J Zhejiang Ocean Univ (Human Sci). 36(6):5. (in Chinese)
Zhang Z (2017) Observation Report on Environmental Public Interest Litigation Filed by Procuratorates in Piloting Areas (Speech delivered at Symposium on the Theory and Practice of Environmental Public Interest Litigation in China, Zhejiang University, 10 June 2017) (in Chinese)
Zhou C (2014) The procedural design and function of civil public interest litigation. North Leg Sci 5:90. (in Chinese)
Zhou W (2018) People's Procuratorate of Lichuan v Mingan Wu: civil public interest litigation attached to criminal case of food safety. Chinese Procurat 3:14. (in Chinese)
Zhu Q (2016) Civil law, 2nd edn. Peking University Press

Chapter 6
Environmental Public Interest Litigation and the Application of Civil Code

Gu Gong

Abstract China's environmental public interest litigation has long been plagued by the lack of substantive legal basis. For this, the Civil Code passed in May 2020 added two ecological damage liability provisions to make up. These two clauses are respectively aimed at "restoration" and "compensation", which together constitute a complete ecological environment damage compensation system, and provide a clear substantive law basis and specific rule guidance for environmental public interest litigation, which is of positive significance. However, these two provisions are still in principle and general. There are still many uncertainties in the legal attribute, the application of punitive damages, the definition of the concept of ecological environmental damage, the proof of causality and other specific aspects, which still need special legislation to improve the situation.

Keywords Environmental public interest litigation (EPIL) · The civil code · The liability for ecological damage · Substantive legal basis · China

6.1 Introduction

In China, public interest litigation is divided into "civil public interest litigation"and "administrative public interest litigation". The former refers to the litigation in which the plaintiff, as the representative of the public interest, files and applies for relevant civil liability against the environmental infringer to remedy the damage to the ecological environment caused by its action. This litigation dominates the environmental public interest litigation, and is the type of environmental public litigation which will be discussed in this chapter.

The emphasis on the "civil" attribute of litigation is a prominent feature of China's environmental public interest litigation. This not only means that it takes the order of "civil public interest litigation going first and then administrative public

G. Gong (✉)
Peking University Law School, Beijing, China

X. Wang et al. (eds.), *Environmental Public Interest Litigation in China*,
https://doi.org/10.1007/978-3-031-26526-6_6

interest litigation"in legislation.[1] The administrative public interest litigation, which is more in line with the public attribute of public interest protection and which has been widely adopted in many countries, is only open to procuratorial organs so far.[2] More importantly, the whole public interest litigation features a strong civil law thinking from the nature orientation to the system design, as it is mainly based on the rules of civil law and implemented in accordance with the relevant procedures of civil cases. In this context, it is not surprising that China's Civil Code, officially adopted in May 2020 and positioned as the "Encyclopedia of citizens' life",[3] specifically stipulates the "legal liability clause for compensation for ecological environmental damages" (hereinafter referred to as the "ecological damage liability clause") applicable to environmental public interest litigation. Therefore, it has become one of the highlights of the "Chinese characteristics, practical characteristics and characteristics of the times" in the Chinese civil code,[4] and has an important impact on environmental public interest litigation.

The provisions of the Civil Code reflect the importance and support of the state in environmental public interest litigation. However, civil law is a private law norm with the goal of private interest relief, which is not naturally consistent with the goal of public interest protection which is of great importance in public interest litigation. The practice of environmental public interest litigation is complex, and the required system rules cannot be fully provided by the abstract provisions of the Civil Code. Therefore, there are still many questions in the "ecological damage liability clause" of the Civil Code in theory and practice, which need to be discussed and resolved. What are the provisions of China's Civil Code on public interest litigation; what impact does it have on the practice of public interest litigation; what are the controversial issues, and how to deal with and improve them? This chapter will analyze potential solutions to each of these questions.

6.2 Origin and Content of Ecological Damage Liability Clause in the Civil Code

6.2.1 Legislative Reasons

For a long time, Chinese scholars and lawyers have regarded environmental public interest litigation only as a "litigation system", focusing on the "litigation" rules, and the relevant legislation has also focused on the same. A representative view holds that the reason why the environment is not fully protected is that the traditional procedural law does not provide an institutional mechanism for pursuing ecological

[1] Xi (2016), p. 114.

[2] Mingyuan (2016), p. 64.

[3] Zhongbin (2020), p. 50.

[4] Liming (2020), p. 13.

damage through the application of civil liability. "For civil damage to the public environment, although individuals can bring civil litigation of private interests for the direct damage they have suffered, they cannot bring civil litigation for social interest; without special legal authorization, the administrative department generally has no right to directly adjudicate civil compensation responsibilities, nor can they bring civil litigation to the court as a direct victim. Based on the principle of "no complaint from the people, no intervention from the officials", the judicial organs usually do not take the initiative to intervene in civil cases. "In response to the damage to public interests and the absence of the plaintiff, the public interest litigation system comes into being."[5]

Environmental civil public interest litigation is positioned as an infringement litigation with the expansion of plaintiff's qualification. When the public environment is damaged, the plaintiff, as the "representative" of the damaged public interest, requests the infringer to bear civil liability, and seeks full relief on behalf of the environment by pursuing civil liability. Therefore, China's environmental public interest litigation legislation has remained as a procedural law for a long time. The focus of amendments has been on relaxing the plaintiff's qualification to bring an action and reducing the litigation threshold. The specific steps are to amend the civil procedure law first and add a broad authorization to allow "organs and relevant organisations specified by law" to sue, so as to pave the way for relaxing the plaintiff's qualification in the field of environmental protection;[6] make specific provisions for public interest litigation in specific fields through specific legislative provisions such as the environmental protection law, and clarify the scope and conditions of social organizations or state organs with public interest litigation rights. In terms of the substantive law, there are no new legal provisions. Actions for ecological environment damage can be only taken under the traditional "environmental tort damage". Applying the traditional environmental tort rules, the available remedies for environmental damage are traditional civil liabilities such as "restoration", and "compensation for losses". Some flexibility in the implementation of environmental public interest litigation has been achieved through judicial interpretation. For example, Article 20 of the Interpretation on Several Issues concerning the Application of Law in the Trial of Environmental Civil Public Interest Litigation Cases adopted by the Supreme People's Court (SPC) in 2014 stipulates that "if the plaintiff requests restoration, the People's Court may judge the defendant according to the law to restore the ecological environment to the state and function before the damage. If it cannot be completely repaired, alternative restoration methods may be allowed". The Court is allowed to judge that the defendant shall bear the "ecological environment restoration costs" including "the costs of formulating and implementing the restoration plan and the costs of monitoring and supervision". The judicial interpretation also gives the Court and the power to make decisions ex

[5] Tao (2013), p. 101.

[6] Article 55 of Civil Procedure Law (1991, amended 2007, 2012 and 2017).

officio in determining litigation claims, evidence collection, counterclaim, order mediation, reconciliation, withdrawal and so on.[7]

However, public interests and private interests are different, and so are the corresponding liability rules. Given the obvious differences in nature and content between public interest relief and private interest relief, the above-mentioned comprehensive application of civil law rules to deal with environmental public interest issues has been questioned in theory and difficulties have been encountered in practice. For example, some scholars hold that "restoring the ecological environment"and "restoring the original state" in the civil law are quite different in terms of relief objects, standards and methods. They impose different types of responsibilities and should be distinguished and stipulated separately. They also believe that "restoring the ecological environment" is the main object of environmental public interest relief, and it should be stipulated in "special environmental legislation".[8]

Some scholars believe that the ecological environment is a public good, and environmental restoration and compensation is a kind of public responsibility undertaken by individuals for the society, which must be undertaken in accordance with the legal rules. It will raise the issue of legitimacy if environmental restoration and compensation are directly imposed in accordance with the tort law, since there is no explicit provision of the substantive law in this regard. There are many problems such as insufficient legitimacy, abstract judgment of public interests divorced from the provisions of the public law, and treatment of the ecological environment as "private property" by doing so.[9]

Taking the perspective of division of power, some scholars believe that the court handles environmental public interest cases as "civil cases", which goes beyond the scope of judicial power and damages the reasonable division of labor and power balance between judicial power and administrative power,[10] resulting in the fact that the gains outweigh the losses from the perspective of the rule of law.[11] If judges can decide what is public interest independently without the legislation, they will invade the territory of legislation and break the separation of powers. It is dangerous to the rule of law. Some scholars think that the legal liability of the defendant in environmental civil public interest litigation is essentially administrative liability, but as the action of civil liability is taken together with the administrative liability caused by the same illegal act, it results in duplication of liability.[12]

Although there are still disputes on the details, most people agree that environmental public interest litigation is different from private interest litigation, and the

[7] Article 9, 14, 15, 17, 25, 26, 27 of SPC Interpretation on Several Issues Concerning the Application of Law in Environmental Civil Public Interest Litigation (2015).

[8] Zhongmei and Haiyang (2017), p. 125.

[9] Gong (2019), p. 136.

[10] Mingyuan (2016), p. 94.

[11] Xi (2021), p. 1.

[12] Xuede (2019), p. 106.

legal liability in environmental public interest is different from the traditional tort liability based on private interest relief. Therefore, there is an urgent need for the improvement of the environmental public interest litigation system to establish independent and suitable substantive law norms, to avoid the mechanical application of traditional private interest infringement rules and share the same norms with private interest relief. To this end, logically, there are two options. First, formulate special f environmental public interest litigation legislation in the nature of public law, create specific rules according to the practical needs of environmental public interest relief, and systematically construct the corresponding system. Secondly, formulate special provisions on the legal liability of public interest litigation in the civil code, and establish the norms of public interest liability which are different from the traditional tort liability provisions for the purpose of private interest relief. Environmental law scholars mostly support the former path.[13]

Some civil law scholars also believe that "the environmental civil public interest litigation system and the environmental and ecological tort system are two different systems, and the corresponding environmental civil public interest litigation liability and environmental and ecological tort liability are also two different responsibilities", so therefore the Civil Code should not provide for this.[14] However, under the modern civil law view that "the Civil Code is the basic private law, with the protection of private rights as its core, this does not mean that it does not involve the protection of public interest at all".[15] The important position of ecological environment damage compensation as one of the core measures for the construction of China's ecological civilization system, has meant that legislators have chosen the latter path, in Chap. 7 "Liability for Environmental Pollution and Ecological Damage" of Compilation VII "Tort Liability" of the civil code. Two liability provisions for "ecological environmental damage" other than traditional civil rights and interests are created, namely article 1234 and article 1235, which have become special provisions of environmental public interest litigation. For reference, they are summarized below:

6.2.2 *Interpretation of the Content*

Article 1234 and article 1235 of the Civil Code are connected and coordinated. The former aims at "restoration" and the latter at "compensation", which together constitute a complete system of "compensation for ecological environment damage".

---

[13]Xi (2016), p. 101; Mingyuan (2016), p. 49; Gong (2019), p. 127; Zhongmei and Haiyang (2020), p. 118.

[14]Xinbao (2017), p. 68.

[15]Liming (2021), p. 224.

6.2.2.1 Restoration Clause

Article 1234 of the Civil Code stipulates that if the ecological environment is damaged in violation of state regulations and the ecological environment can be restored, the organ specified by the state or the organization specified by law has the right to request the infringer to bear the responsibility for repair within a reasonable period of time. If the infringer fails to restore within the time limit, the organ prescribed by the state or the organization prescribed by law may have it restored by itself or entrust others, and the expenses incurred shall be borne by the infringer.

This article focuses on the responsibility of "restoration" of the ecological environment, so it is also known as the "restoration clause". There are two applicable conditions: first, the act is illegal and causes environmental damage. The "violation of state regulations" here means "violation of law" in a broad sense. Legislators deliberately use the expression "violation of state regulations" rather than "legal regulations" in order to expand the scope of the "law" violated. Therefore, the violation of regulations, rules, judicial interpretations and environmental protection provisions in national policy documents outside the law constitutes a violation of this article and may lead to liability to restore environmental damage. Second, there is a requirement that the ecological environment "can be restored".

The body taking the action must be "an organ prescribed by the state or an organization prescribed by law". This provision is based on the notion of "compensation obligee", which has been stipulated by the Reform Plan of Ecological Environment Damage Compensation System jointly issued by the CPC Central Committee and the State Council, and the "plaintiff of public interest litigation", which has been stipulated in the Civil Procedure Law and the Environmental Protection Law. Among them, the "organs prescribed by the state" are provincial, municipal and prefecture level governments and their designated "departments or institutions". These are departments entrusted by the State Council or provincial governments to exercise the duties or ownership of public-owned natural resource assets, and the procuratorial organs in accordance with article 55 of the Civil Procedure Law. The "organization prescribed by law" refers to environmental protection organizations that meet the provisions of Article 58 of the Environmental Protection Law.

Article 1234 designs a "two-step" responsibility system. In case of compensable scenarios, the claimer shall first request the infringer repair the damage within a certain time limit. When it fails to be restored within that time limit, the claimer may restore it by himself or entrust others to do it, and the infringer shall bear the expenses.

6.2.2.2 Compensation Clause

Article 1235 of the Civil Code stipulates that if violation of state regulations causes damage to the ecological environment, the state organ or the organization prescribed by law has the right to request the infringer to compensate for the following losses and expenses:

(1) Losses caused by the dysfunction from the damage of ecological environment to the completion of restoration.
(2) Losses caused by permanent damage to ecological environment function.
(3) Expenses for investigation, identification and assessment of ecological environment damage.
(4) Expenses for removing pollution and restoring ecological environment.
(5) Reasonable expenses incurred to prevent the occurrence and expansion of damage.

This article focuses on the responsibility of "restoration" of the ecological environment, so it is also known as the "compensation clause". It is similar to the "restoration clause" in terms of application and claimers, but only the five types of specific compensation items specified above can be claimed.

6.2.3 Progress and Significance

The ecological damage liability clause of the Civil Code clearly stipulates the legal liability rules for environmental public interest relief for the first time at the legal level, which reflects the confirmation and standardization of the provisions of relevant national policy documents and the exploration of the court, and adopts and responds to some views of the academic community. It has at least the following significance for the practice of environmental public interest litigation in China.

First, establish the substantive legal basis and enhance the legitimacy of the system. The ecological damage liability clause "solves the problem of insufficient substantive law basis of public interest litigation",[16] and conforms to the original legislative intention and social expectation of providing a clear legal basis for the trial of environmental public interest litigation cases.[17]

Second, for environmental public interest relief, special provisions different from private interest relief are formulated to avoid the confusion of indiscriminate and mixed application of public and private responsibilities.

Third, emphasize legitimacy control and make the rules more reasonable. Both articles take "violation of state regulations" as the precondition and add "illegality" in the constituent elements. This is significantly different from "no fault" and "not necessary for illegality" of environmental private interest tort liability, which is more in line with the public attribute of legal liability of public interest litigation. It also limits the scope and number of cases that can be brought.

[16] Kaiyuan (2020), p. 7.
[17] Wei (2020), p. 461.

Fourth, stress the priority of restoration and strengthen the recovery orientation. The Civil Code distinguishes “restoration liability” from “compensation liability”, which requires that priority should be given to restoration if it can be done. It states that it is preferable for the infringer to pay for the restoration, with third parties carrying out the restoration only if this does not take place. This is more reasonable and efficient.

Fifth, standardize and unify the current public interest litigation mechanism. These two provisions put the reform of the eco-environmental damage compensation system led by local governments or environmental regulators and the environmental civil public interest litigation filed by environmental protection organisations and procuratorial organs in the same framework for unified provisions. This avoids the confusion of the former two functioning separately. The enumeration of five types of “compensation items” in article 1235 also effectively solves the disadvantages of allowing the plaintiff to put forward in the name of traditional “civil liability” different expressions, chaotic types and different rules.

In short, although there are only two articles, the above provisions of the Civil Code have established a specific, clear, unified and standardized environmental public interest relief responsibility system, which can significantly promote and enhance the legitimacy and rationality of China’s environmental public interest litigation system.

However, environmental public interest litigation has a special nature, unique content and many rules. While the two provisions on ecological damage liability in the Civil Code are very simple, there are no clear provisions on many matters that must be faced in the process of system implementation, leaving many problems to be studied theoretically and explored in practice.

6.3 The First Focal Question: Public Law Norms or Private Law Norms

The academic and practical circles have reached a consensus on the particularity of ecological damage relief which is different from traditional infringement. Experts from the legislative department have pointed out that “compensation for ecological environment damage cannot be equated with general environmental infringement”, holding that they are different in value orientation, illegality, damage composition, the proof of causality and so on.[18] The interpretation book compiled by the Professional Committee of Civil Trial Theory of China Trial Theory Research Association also believes that the Civil Code “distinguishes the two liability modes of ecological environment restoration and civil law restitution, which are of practical significance”.[19] The representative view of academic circles is that articles 1229 to 1233

[18] Wei (2020), pp. 461–462.

[19] Civil Trial Theory Professional Committee of China Trial Theory Society (2020), p. 253.

of the Civil Code stipulate the tort liability for infringement of private interests, and articles 1234 to 1235 concerning infringement of public interests.[20] Experts from the court system also hold this view and believe that the coexistence of the two types of liability means that the Civil Code " classifies the environmental tort liability into environmental private interest tort liability and environmental public interest tort liability".[21]

It is agreed by all parties that the two should be treated differently, and different rules should be applied on many specific issues. However, is the difference between the two essential and fundamental? More specifically, is the "environmental public interest tort liability"in the Civil Code still the same private law liability as the traditional tort liability? Is the ecological damage liability clause still a private law norm with the same nature as the traditional tort norm? All parties were vague and did not give a definite answer. According to the existing research, there are three views:

The first is the civil law norm as a special provision of the general tort rules. This view defines the liability for ecological damage as civil tort, and the relevant provisions are positioned as the traditional civil law norms. They are fully understood and applied in accordance with the concepts and principles of private law, and the general tort rules are automatically applied in other aspects except those clearly stipulated by the law.[22] On the surface, this understanding is most consistent with the formal logic. It looks like naturally that the legal norms in the Civil Code should belong to the "civil law norms", but it is the most infeasible in practice. Because the basic principles and basic rules of ecological damage relief and traditional tort liability are fundamentally different, there are essential differences. As far as the implementation mechanism is concerned, what will be formed according to this idea will be "pure civil litigation" parallel to the existing public interest litigation system, which is what occurs in practice.

The second is the "special tort norms" different from the ordinary tort rules. This view regards ecological damage compensation as a special infringement—"public interest infringement", which is different from the traditional infringement, and regards the relevant legal provisions as a special "environmental tort liability clause for the public interest with ecological and environmental damage".[23]

Therefore, the relevant liability features both tort liability and public interest purpose. It not only follows the basic principles and rules of tort law, but also flexibly adapts to the actual needs of environmental public interest protection practice. In terms of implementation, it can seamlessly connect environmental civil public interest litigation and ecological environment damage compensation, which is the mainstream understanding widely recognized at present. Its deficiency is that it brings "public interest relief" and "public interest responsibility" into the civil law

---

20 Liming (2021), p. 219.

21 Xuguang (2020), p. 25.

22 Shaokun and Yudong (2020), p. 153.

23 Xuguang (2020), p. 24.

system positioned in private law, which challenges the division of public and private law and leads to the confusion of the civil law system. In terms of the provisions of the substantive law, this understanding also conflicts with the basic provisions clearly stipulated in the Civil Code, such as "this compilation adjusts the civil relationship arising from infringement of civil rights and interests"[24] and tort liability is the relief for the damage of "other people's civil rights and interests".[25]

The third is the "public law norms" as the basis of the public law responsibility system. This view holds that ecological damage compensation belongs to public legal liability and relevant legal norms belong to public legal norms. It is only stipulated in the Civil Code as a special legislative case for trade-off reasons such as legislative convenience and reference to the application of relevant tort law norms. Therefore, although these two provisions are indeed the "provisions of the Civil Code" stipulated in the Civil Code, they are not "Civil Code norms" in nature. They should be understood and applied in accordance with the relevant principles of public law, and the specific rules should be improved from the perspective of the construction of the public law system. In the future, it is still necessary to formulate special legislation with the nature of public law.[26] This understanding is in line with the basic principle of the division of public and private law and the trend of special legislation on ecological damage compensation and public interest litigation, but it still needs to be accepted and recognized by more people. After all, interpreting a legal norm stipulated in the Civil Code as a public law norm goes beyond the traditional legal cognition.

6.4 The Second Focal Question: The Application of Punitive Damages

Similar to the liability for ecological damage, Chapter 7 of Compilation VII of the Civil Code also added article 1232 "punitive liability" as one of the innovative provisions for implementing Xi Jinping's Thought on Ecological Civilization,[27] which has been applied in some cases.[28] Article 98 of the Public Interest Litigation Treatment Rules of the People's Procuratorate published by the Supreme People's Procuratorate in June 2021 also clearly stipulates that "if the defendant intentionally pollutes the environment and destroys the ecology in violation of the provisions of the law, resulting in serious consequences, the people's procuratorate may file claims such as punitive damages". In January 2022, the SPC issued the Interpretation on the Application of Punitive Compensation in the Trial of Ecological Environment

[24] Article 1164 of Civil Code.

[25] Articles 1165, 1166 of Civil Code.

[26] Gong (2019), p.128; Zhongmei and Haiyang (2020), p. 126; Youhai and Qian (2019), p. 69.

[27] Chen (2020).

[28] Intermediate People's Court of Shenzhen City (2021).

Infringement Dispute Cases. This specified that Article 12 allows "the organ specified by the state or the organization specified by law, as the representative of the infringed, to request that the infringer be ordered to bear the responsibility of punitive damages". However, whether punitive damages should be used in environmental public interest litigation is still controversial.

Opponents believe that as far as the purpose of liability is concerned, punitive damages play the function of punishment and containment by compensating "victims", which is different from the purpose of ecological damage which is to compensate the "environment". In terms of the expression of the article, the compensation obligee stipulated in the article is the one "infringed", which refers to the specific victims whose civil rights and interests are infringed. This is different from the plaintiff in public interest litigation. From the perspective of system, the position of punitive compensation lies before the ecological damage liability clause and between the ordinary environmental tort clauses, which shows that it is only for the relief of private interest damage. Judging from the effect, it is also not appropriate for the plaintiff of public interest litigation to obtain this part of the compensation. For enterprises, the combination of punitive compensation, environmental restoration and ecological compensation will excessively increase the burden of enterprises and violate the principle of proportion.[29] From the perspective of institutional necessity, the existing systems of administrative fines and criminal forfeitures in the field of environment can serve the function of punishment, containment and deterrence,[30] while the liability for ecological damage itself has the nature of punishment,[31] Hence there is no need to use punitive compensation.

Supporters believe that environmental violations often cause "duality" of private interest damage and public interest damage at the same time. When punitive compensation can be used to remedy the former, in the case of exclusion of the application of the latter, it will lead to the unreasonable situation that the relief of "ecological environment"is lower than that of "people". According to the analysis of specific cases, it is difficult to realize full compensation and relief if punitive compensation is not applicable in the complex situation of environmental public interest litigation, when the liability for compensation is too light due to the difficulty in identifying ecological damage, or one-sided due to the concealment of social and public spiritual damage.[32] Some scholars provide arguments from the perspective of "private law enforcement", arguing that administrative and criminal liability alone cannot achieve the purpose of punishing those who deliberately damage the environment.[33] Some scholars even think that punitive compensation is more consistent

[29]Liming (2021), pp. 221, 223.

[30]Xuemin (2020), p. 68.

[31]Yongfei (2021), p. 30; Dan (2020), p. 248.

[32]Youhai (2022), pp. 29–32.

[33]Xiaoran (2018), p. 52; Weixing and Yulin (2021), p. 94.

with the system of environmental public interest litigation in terms of nature and value.[34]

6.5 Other Controversial Issues

6.5.1 The Definition of the Concept of "Ecological Environmental Damage"

Although "ecological environmental damage" is the core concept of the ecological damage liability clause, this term is not defined, so is left to the discretion of the court—resulting in uncertainty as to its meaning. In practice, it is mostly based on the concepts defined in the Reform Plan of Eco-environmental Damage Compensation System and the General Outline of Technical Guidelines for Eco-environmental Damage Identification and Assessment formulated by the Ministry of Ecology and Environment.(MEE) However, the former is national policy and the latter are technical documents, which are not explicitly authorized by law. Therefore, strictly speaking, it is not the corresponding interpretation of the provisions of the Civil Code. Moreover, the "eco-environmental damage" in these two documents refers to all environmental changes that have adverse effects, without specifying any requirements on results and degree. Therefore, it is obviously too broad and needs to be refined and limited.

6.5.2 Rules of the Proof of Causality

Article 1230 of the Civil Code sets a special provision of "inversion of the burden of proof" for the causality proof of traditional environmental tort liability, but the clause on ecological damage liability does not mention causality. It is in question whether the above provisions can be copied. If the ecological damage liability is regarded as a special type of general environmental tort, and it is considered that the traditional rules can be applied except those specially stipulated in the special provisions, the former can be copied in terms of causality. However, from the perspective of the essential difference between private interest liability and public interest liability, as well as the difference in the status of the parties in ordinary tort litigation and environmental public interest litigation, it is another issue. After all, most of the victims in environmental infringement cases are usually individuals, whose personarum and property have been seriously damaged, and their lives and health have been threatened. They often have fewer financial means and status than the infringers, so they need preferential protection. However, the plaintiff in

[34] Shuyi and Lin (2017), p. 64.

environmental public interest litigation is either a state organ or an environmental protection organization with administrative power or professional ability. Both parties will often have the same financial and professional means at their disposal. If the inversion of evidence is also implemented, it may be unfair. However, environmental damage is not easy to prove. Even in public interest litigation, the rules of proof should be moderately relaxed. How to balance the situation remains to be stipulated later.

6.5.3 Judgment Standard of "Can Be Repaired"

The fact that the damaged environment "can be restored" is a special element for the application of restoration rules. This seems simple, but it actually gives rise to many problems. As far as "restoration" is concerned, there are two concepts of "environmental restoration"and "ecological restoration" in the General Outline aforementioned, and the latter has different types such as "basic restoration", "compensatory restoration" and "supplementary restoration". The specific application needs to be clarified by legislation and varies according to the situation. "Can" mainly refers to the feasibility at the technical level, but it should also include the economic level.[35] Compared with restoration, environmental restoration can have less economic constraints, but this is mainly aimed at the ecological benefits brought by restoration and the market value of the restored natural elements, which does not mean that the cost is completely ignored, and it is still necessary to have a positive "cost-benefit". Specific rules also need to be clarified. For the ecological environment of "flowing water purifying itself", time is also an important factor to judge whether it can be restored. The "reasonable period" stipulated in article 1234, needs to be clarified with a comprehensive judgment from all aspects of the situation.

6.5.4 Ranking Relationship Between Different Subjects

The "or" between "the organ prescribed by the state" and "the organization prescribed by law" means that in the case of litigable circumstances, only one can by chosen from the two types of plaintiffs and they cannot act together, let alone repeat the prosecution, which is correct and necessary. However, in practice, we will face the problem of how to determine the order of different subjects. For this, the ecological compensation clause itself does not touch upon the issue and needs to be clarified by legislation. In particular, it is worth noting that the so-called "organs stipulated by the state" here include not only procuratorial organs, but also local governments and regulatory departments that have obtained the right to claim

[35] Xiaogang (2021), p. 1.

according to the Reform Plan of Ecological Environment Damage Compensation System. According to the requirements of the Civil Procedure law, the civil public interest litigation right of procuratorial organs lies behind social organizations, while as per the Several Provisions of the SPC on Hearing Cases of Compensation for Ecological Environment Damage (Trial), the litigation right of local governments and regulatory departments takes precedence over social organizations.[36] It is unclear whether the introduction of the ecological compensation liability clause of the Civil Code will bring impact and change to this complex ranking relationship.

6.5.5 *Applicable Relationship Between Compensation Items*

Strictly speaking, article 1235 "compensation clause" only stipulates the "scope of compensation", not a complete "compensation rule". This is because it mainly lists the specific compensation items "entitled to claim compensation from the infringer" and does not clarify the applicable relationship and rules between them. Given the complexity and diversity of the actual situation and the inevitable mutually exclusive relationship between some items (such as "ecological environment restoration cost" and "permanent loss of ecological environment function"), this clause can only be considered as confirming all possible compensation items in ecological damage compensation cases, to provide a practical legal basis for specific practice. However, what and how to use in specific cases still needs to be handled differently according to the type of case and specific circumstances.

6.5.6 *Relationship with Other Forms of "Civil Liabilities"*

The situation of environmental cases is complex and the demand for environmental protection is diverse. In the past practice of environmental public interest litigation, besides the liability for compensation, the order in relation to "ceasing the infringing act" with preventive function and the "apology" with moral care have also been widely applied.[37] However, the ecological damage liability clause only specifies two liability modes: restoration and compensation. If so, is it possible to apply other liabilities? If the answer is no, it is unfavorable to environmental public interest relief. If the answer is yes, what is the substantive legal basis of relevant liabilities?

---

[36] Article 55 of Civil Procedure Law (1991, amended 2007, 2012 and 2017); Article 17 of SPC Several Provisions on the Trial of Cases on Ecological Damage Compensation (for Trial Implementation) (2019, amended 2020).

[37] Hui (2014), p. 58; Gong (2016), p. 16; Zhanwen (2019), p. 119.

6.5.7 Relationship with Other Public Law Responsibilities and Mechanisms

The legal liability of civil public interest litigation is similar to relevant public law liability in content and function. For example, much environmental legislation in China stipulates the "administrative restoration responsibility" with the basic content of "order to restore the original state" and "performance on behalf of others".[38]

Can ecological damage liability be used together with it? In other words, when the environmental infringer has been subject to administrative sanctions and punishment, can it be sued by an environmental public interest lawsuit and be held liable for ecological damage? If so, the party risks of being hold liable twice for the same action'.[39] If not, it seems to be inconsistent with previous practice and is not conducive to the practice of environmental public interest litigation, because most of China's environmental public interest litigation is the "Prosecution" of cases that have been sanctioned by public law.[40]

In addition, some scholars believe that criminal forfeiture, administrative fines and punitive compensation for ecological environment damage should be allowed to be deducted from each other, because they are essentially the same. They are economic punishment measures imposed on serious violations and should be allowed to be offset from each other.[41] This understanding is very sensible, and the practical treatment is also reasonable. However, if we think that the legal liability of environmental public interest litigation is "civil liability" which is very different from administrative liability and criminal liability, this understanding is difficult to stand in theory and will encounter legal obstacles, because Article 187 of the Civil Code clearly stipulates that "If a civil subject should bear civil liability, administrative liability and criminal liability for the same act, the bearing of administrative liability or criminal liability does not affect the bearing of civil liability; if the property of the civil subject is insufficient to pay, priority shall be given to bear civil liability".

6.6 Conclusion

Generally speaking, the ecological damage liability clause of the Civil Code provides a substantive legal basis for China's environmental civil public interest litigation, which is of positive significance to enhance the legitimacy of the system and the rationality of the rules. However, in terms of the content of the clauses, its

[38] Zhiping (2020), p. 185.

[39] Xuede (2019), p. 106.

[40] Gong (2016), p. 26.

[41] Youhai (2022), p. 32.

function is still relatively limited, and many problems requirement legislative amendment. Although the ecological damage liability clause is stipulated in the Civil Code, it should not be treated as pure "civil liability" and cannot be interpreted and applied in full accordance with the principles and rules of traditional civil law. The formulation of special legislation with the nature of public law ought to be the direction for the improvement of China's environmental public interest litigation legislation.

References

Chen W (2020) Explanation on the Draft Civil Code of the People's Republic of China: Delivered at the Third Session of the 13th National People's Congress on 22 May 2020 http://www.npc.gov.cn/npc/c30834/202005/50c0b507ad32464aba87c2ea65bea00d.shtml. Accessed 10 Jan 2022
Civil Trial Theory Professional Committee of China Trial Theory Society (2020) Understanding and judicial application of tort liability in civil code. Law Press, Bei Jing
Dan L (2020) The limitation of the environmental damage punitive compensation claimers. Guangdong Soc Sci 3:46–253
Gong G (2016) Empirical analysis of environmental civil public interest litigation in China in 2015. Law Sci 9:16–33
Gong G (2019) Reflections on the positioning of the nature of environmental civil public interest litigation. Chinese J Law 3:127–147
Hui Z (2014) Assumption of responsibility in environmental public interest civil litigation. Legal Forum 6:58–64
Intermediate People's Court of Shenzhen City (2021) The first public interest lawsuit for Maozhou River environmental pollution was pronounced. Accessed 10 January 2022
Kaiyuan T (2020) Practical implementation of the "Civil Code" green clause which guided by the Xi Jinping thought of ecological civilisation. Natl Judge Coll J 23:3–7
Liming W (2020) The characteristics of China, practice and times in civil code. Law Mod 4:1–15
Liming W (2021) Highlights of environmental pollution and ecological damage liability in civil code. Soc Sci Guangdong 1:216–225
Mingyuan W (2016) On the development direction of environmental public interests litigation: analysis on administrative right and jurisdiction. China Legal Sci 1:49–68
Shaokun F, Yudong Z (2020) On the new development of tort liability norms in civil code. Law Soc Dev 4:144–161
Shuyi W, Lin L (2017) Punitive damages and its application in environmental infringement cases. Study Pract 8:64–72
Tao B (2013) New starting point of environmental public interest litigation legislation. Law Rev 1: 101–106
Wei H (2020) Interpretation of the civil code of the People's Republic of China. China Legal Publishing House, Bei Jing
Weixing W, Yulin H (2021) On the prudent application of punitive damages in the ecological environment damage compensation system. Nan Jing Soc Sci 9:91–100
Xi W (2016) On the legislation order of environmental public interest litigation. Tsinghua Univ Law J 6:101–114
Xi W (2021) Drawbacks in the system design of the environmental public interest litigation and ways of improvement. J Kunming Univ Sci Technol 1:1–8
Xiaogang W (2021) Interpretation of ecological environment restoration cost theory in Article 1235 of civil code. J Gansu Univ Polit Sci Law 1:1–10

Xiaoran Z (2018) On the construction of punitive damages in environmental civil public interest litigation. J Cent-South Univ (Soc Sci) 2:52–62
Xinbao Z (2017) Study on the legislation of tort liability part of the civil code. China Legal Sci 3: 49–70
Xuede G (2019) On the repeated liability in environmental civil public interest litigation after the sanction of public law. Adm Law Rev 5:106–118
Xuemin C (2020) Regulation of the punitive damages system in the environmental tort damage: based on the reflection of Article 1232 of the Civil Code 57. J China Univ Polit Sci Law 6:57–69
Xuguang W (2020) The rule construction and understanding of green clause in civil code. Natl Judge Coll J 23:14–29
Youhai S, Qian W (2019) Research on the Limitation of green regulation in Tort Liability Chapter of Civil Code: Disagreement on the inclusion of ecological environmental damage liability into the Civil Code from the perspective of "public-private division". J Gansu Polit Sci Law Inst 562–569
Youhai S (2022) Justification and application of punitive compensation for ecological environment damage. J China Univ Polit Sci Law 1:26–37
Zhanwen Q (2019) The application and restriction of apology in civil public interest litigation. Tribune Polit Sci Law 4:119–128
Zhiping L (2020) Research on ecological environment restoration mechanism of administrative order type. Law Rev 3:184–196
Zhongbin L (2020) Let the spirit of the civil code be rooted in everyone's heart. China's Natl People's Congr 22:50
Zhongmei L, Haiyang D (2017) Responsibility of "restoration of ecological environment": an empirical analysis. Chinese J Law 3:125–142
Zhongmei L, Haiyang D (2020) Reconstructing environmental tort relief system through "ecological recovery theory". Soc Sci China 2:118–140

Part II
Substantive and Procedural Issues About EPIL in China

Chapter 7
Procuratorates at the Crossroad: Performance, Controversies and Prospects of Procuratorial EPIL in China

Yun Ma and Wenzhen Shi

Abstract Procuratorial environmental public interest litigation (EPIL) has shown its vitality in addressing environmental concerns through the channel of PIL and gained momentum in China's national agenda-setting. Procuratorial activism is signaled and revealed by mobilizing procuratorates at all levels, activating PIL in all types of litigation, seeking breakthroughs in new fields and replacing the role of gap filling with a paramount one in practice. It is noteworthy that the efficacy of procuratorial EPIL may be discounted due to its cherry-picking feature in case screening. Caution should be placed against procuratorial activism, which may risk crowding out NGOs by a turn of policy focus to criminal prosecution with incidental civil EPIL, and intruding administrative discretion by a comparatively low threshold to initiate administrative EPIL and tightened judicial review. Procuratorial EPIL needs to be founded on a rational design of procuratorates' role in the overall EPIL system by coordinating between dual roles of EPIL litigants and the constitutionally entrenched role of legal supervision, which decides the design of the follow-up substantive and procedural details. Connection mechanisms among different types of PIL should be established to bring each into its full play and reach synergy thereof. Legal empowerment and top-down design should be adopted to reduce discrepancies and increase cohesion.

Keywords Environmental public interest litigation (EPIL) · Public interest litigation (PIL) · Procuratorial PIL · NGO · Judicial review · China

---

Y. Ma (✉)
Administrative Law Department, Law School of China University of Political Science and Law, Beijing, China
e-mail: Mayun@cupl.edu.cn

W. Shi
Law School of China University of Political Science and Law, Beijing, China

X. Wang et al. (eds.), *Environmental Public Interest Litigation in China*,
https://doi.org/10.1007/978-3-031-26526-6_7

7.1 Introduction

EPIL has gained momentum in China's national agenda-setting and has been emphasized in policy-making as an efficient tool to improve environmental law enforcement in China.[1] A complex framework of legislation and rule-making by various bodies has taken shape at the central level over the last ten years, and empirical data now demonstrates that the system is maturing in terms of overall efficiency.[2] Notwithstanding this, academic and scholarly opinions differ on nearly every issue, ranging from its very constitutional legitimacy to the design of its substantive and procedural elements, as well as future reform recommendations for EPIL.[3] Much discussion in the literature has centered on civil EPIL initiated by non-governmental organisations (NGOs) against polluters. Such discussions focus on the obstacles faced by NGOs under the scenario of public participation and the development of civil society under an authoritarian political regime.[4] It is noteworthy that, along with the shift of policy focus, EPIL brought forward by procuratorial organs (hereafter "procuratorial EPIL") has gained momentum in the nation's environmental agenda-setting, either in terms of the case numbers or policy priorities. With procuratorates nationwide mobilized and resources concentrated, procuratorial organs are now actively defending environmental public interests in China. Such procuratorial activism is not without controversy. Questions have arisen as to how the function of the state apparatus of procuratorates should be legally defined by coordinating its new role as an EPIL litigant with its constitutionally entrenched role as a legal supervisor. A further issue is how procuratorial organs fit within the overall structure of EPIL and how the different players, including NGOs, environmental authorities and judiciaries, work in concert.

With a focus on procuratorial EPIL, this chapter intends to fill the gap of existing literature by critically analyzing the theoretical foundations and the impetuses for change that underpin the promotion and institutional design of procuratorial EPIL. After outlining the legislative framework of procuratorial EPIL (Sect. 7.2) and evaluating its performance by probing into empirical data (Sect. 7.3), this chapter explains the major controversies arising from implanting procuratorial EPIL into Chinese law and analyzes the problems of its practice. Suggestions for future reform and avenues for further research are also offered.

[1] See generally Gao and Whittaker (2019); Ma (2019); Zhai and Chang (2018); Zhang and Mayer (2017).

[2] See Sects. 7.2 and 7.3 below.

[3] See Sect. 7.4 below.

[4] See generally Xie and Xu (2021); Carpenter-Gold (2015); Zander (2018).

7.2 Implanting Procuratorial PIL into Chinese Law: A Struggling Move Forward

PIL is comparatively new to China. It was initially legalized in 2012 and NGOs were authorized to bring civil PIL with the amendment of the Civil Procedure Law (CPL)—although academic propositions and policy guidelines were made before this date. However, the development of procuratorial PIL does not synchronize with the path of PIL as a whole. In contrast with its counterpart of NGOs, the procuratorial organ, conventionally in charge of criminal prosecution in criminal litigation, was seldom a player in this field until 2017. In that year, the provision of procuratorial PIL was written into statutes after a two-year pilot reform project. In reality, when the Administrative Litigation Law (ALL) was amended in 2014, the proposal to add a provision authorizing procuratorial PIL was thwarted by the Legal Affairs Office of the State Council's opposition.[5] Implanting procuratorial PIL into Chinese law is a product of compromise. On the one hand, it brings challenges to the conventional role of procuratorial organs and creates the need for transformation thereof; on the other hand, it requires adaptation of existing legal systems, especially administrative litigation rules, to make the whole system more systematic and integral. This section will identify the main controversies, concerns and comprises thereof concerning the implantation of procuratorial PIL into Chinese law after presenting the general legislative framework thereof.

7.2.1 *An Overview of the Legislative Framework of Procuratorial EPIL*

The most remarkable feature of PIL is that it makes the judicial review of activities that threaten public interests—either private behaviors, administrative actions or inaction—available by expanding the standing rule and allowing qualified plaintiffs to trigger a trial. EPIL is the most frequently practiced field and heralds the rule-and-policy-making of PIL in China. As early as 2005, the State Council officially proposed to establish EPIL to strengthen environmental protection.[6] Until 2012, when the second amendment was made to the CPL, EPIL was incorporated into the Chinese legal system (Article 55),[7] although sporadic practices of EPIL had been

[5]He (2018), p. 173.

[6]State Council, *Decision on the Implementation of the Outlook on Scientific Development and Strengthening Environmental Protection* (关于落实科学发展观加强环境保护的决定) (3 December 2005) (Article 27).

[7]The CPL of 1991 was first amended in October 2007. However, EPIL provisions were not included in this amendment.

accepted by local courts previously.[8] PIL was initially crystalized in the field of civil litigation, while subsequent administrative PIL involved a comparatively lengthier process of legalization until the 2017 amendment to the ALL.

The amended CPL confers legal standing on "legally mandated organs and relevant organisations" to file civil suits against "actions that harm the public interests such as environmental pollution and infringement of groups of consumers' interests". With a scarcity of the-then statutes authorizing "legally mandated organs" and confusion about the qualifications of "social organisations" to initiate PIL, the 2012 amendment did not result in a surge of PIL cases.[9] EPIL was revitalized in 2014 when the Environmental Protection Law (EPL) was amended to empower environmental NGOs with the standing in EPIL, subject to specific qualifications.[10] Furthermore, the scope of EPIL was also expanded from "environmental pollution" as stipulated in CPL to "environmental pollution and ecological damage" in EPL (Article 58).

Although EPIL had made its debut in the field of civil litigation since 2012, when the ALL was overhauled in 2014 (25 years after its enactment in 1989), the PIL provision was not included.[11] Procuratorial organs officially entered the field of EPIL in 2015 when the Standing Committee of National People's Congress (SCNPC) authorized the Supreme People's Procuratorate (SPP) to conduct a two-year pilot project of PIL in 13 selected provinces. After the two-year experiment, in June 2017, Congress amended both the CPL and ALL and authorized procuratorial organs nationwide to bring both civil and administrative PIL.

A rudimentary framework of procuratorial PIL was established by adding two sub-articles into the CPL and ALL [respectively Article 55(2) and Article 25(4)] by Congress. Detailed procedures were not fleshed out until the issuance of a series of interpretations and work rules on instituting and adjudicating PIL cases, respectively and jointly, by the Supreme People's Court (SPC) and SPP afterward (Table 7.1). Collectively, they form a complex framework of rules governing procuratorial PIL in

---

[8]The first environmental court was established in Guizhou province in 2007 and similar special courts were subsequently established in Jiangsu and Yunnan provinces in 2008. All of them have accepted some PIL cases brought by administrative agencies or NGOs, although without legislative support. Several courts have established PIL procedures on a trial basis. See Wang and Gao (2010), pp. 37–44; Zhang (2016), pp. 177–181.

[9]Under the-then legislative framework, there was only one statute, i.e., 1999 Maritime Environmental Protection Law, prescribing that "departments responsible for maritime environment regulation and management" may bring compensation claims against polluters that damage maritime environment, aquatic resources and maritime protected areas (Article 90(2)). As a result, the scarcity of such statutes makes authorized organs rather limited in practice. Although marked as a breakthrough of the 2012 amendment, the number of EPIL cases turned out to be zero in 2013. See Lin and Wang (2014), pp. 157–174.

[10]The qualifications of NGOs include the following: 1. registration at prescribed authority of civil affairs at the level of a city with districts or above; and 2. engagement of environment-protection-related public-interest activities for at least five consecutive years and with no record of illegal activity (Article 58).

[11]See He (2018), p. 173. For more discussions on the hesitation of legislators, see Sect. 7.2.2.

Table 7.1 The framework of legislation and rules on procuratorial EPIL in China

Stage	Effective date	Issuing body	Title
Pilot stage	Approved by Central Leading Group for Comprehensively Deepening Reform in April 2015, and later publicly released by SPP in July 2015	Central Leading Group for Comprehensively Deepening Reform	Pilot Scheme of Reform on Instituting PIL by Procuratorial Organs
	July 2015	SCNPC	Decision of the SCNPC on Authorizing the SPP to Conduct PIL Pilot Project in Selected Regions
	December 2015	SPP	Implementation Rules of People's Procuratorates on the Pilot Work of Initiating PIL
	March 2016	SPC	Implementation Measures for Pilot Work of People's Courts in Hearing Public Interest Litigation Cases Filed by People's Procuratorates
	April 2017	SPC	Work Rules of the SPC on Adjudicating EPIL Cases (for trial implementation)
Legalization	June 2017	SCNPC	Amendment to CPL and ALL
Post-legalization rule-making	March 2018 (amended in December 2020)	SPC and SPP	Interpretation of the SPC and SPP on Several Issues Regarding the Application of Law in Procuratorial PIL Cases
	March 2018	SPP	Guidelines for Handling Civil public interest Litigation Cases by Procuratorial Organs
	June 2019 (amended in December 2020)	SPC	Several Provisions of the SPC on the Trial of Cases on Compensation for Damage to the Ecological Environment (for Trial Implementation)
	December 2020 (initially be effective in January 2015)	SPC	Interpretations of the SPC on Several Issues About Application of Law in Adjudicating Civil EPIL Cases (amended)
	July 2021	SPP	Rules for the Handling of Public Interest Litigation Cases by People's Procuratorates

China. Among them, the most remarkable ones are the Interpretations co-issued by SPC and SPP in 2018 (hereafter "2018 Interpretations"), which are further modified and became effective in January 2021 by further expanding the scope of procuratorial civil PIL to the infringement upon the name, portrait, reputation or honor of a hero or martyr. Although harmonization efforts were made between the two supreme judicial organs, it is not difficult to imagine a potential collision and discrepancies within a complex system of rulemaking. Furthermore, since the nature of judicial interpretation is "interpreting" the law during the process of a legal application instead of "making" the law, the legitimacy of such rules is questioned since it essentially replaces statutes enacted by representative legislatures.[12] For example, these rules largely touch upon the litigious rights and obligations of procuratorates, the burden of proof and other substantive issues which are absent in ALL and CPL. A summary of the history of procuratorial EPIL in China follows in Table 7.1.

Based on the legal framework of procuratorial PIL shown above, procuratorial EPIL mainly takes place in two forms, i.e., civil and administrative PIL. Apart from these two, the mechanism of incidental litigation may also embed PIL in other forms (Table 7.2). To cope with the intertwinement of criminal, administrative and civil liability in a particular dispute and yield a more thorough resolution thereof, prosecutors may bring administrative PIL with an incidental civil action or criminal litigation with incidental civil PIL.[13] However, the rules governing incidental EPIL are still debated and far from clear.[14]

It is worth mentioning that, along with the development of procuratorial EPIL, there is another remarkable parallel reform, named "ecological damage compensation system reform", which authorizes selected authorities, as representatives of the owners of state-owned natural resources, to bring civil litigation against polluters and the ability to claim for ecological compensation directly or in case of a failure of negotiation between the two parties.[15] Ecological damage compensation litigation stands out as an innovative type of litigation due to its adoption of the structure of "governments versus citizens" which is distinct from either the conventional structure of "citizens versus citizens" in civil litigation or the structure of "citizens versus governments" in administrative litigation. Arguably classified as part of civil EPIL,[16]

---

[12] Wang (2022), p. 168.

[13] ALL (Article 61); Criminal Procedure Law (Article101) (Criminal Procedure Law as the legal foundation of incidental civil PIL is debated, see Sect. 7.4.1.2 below).

[14] See Sect. 7.4.1.2.

[15] The mandated authorities to sue include, see generally General Office of CPC Central Committee & General Office of State Council, *Pilot Plan of Ecological Environmental Damage Compensation System Reform* (生态环境损害赔偿制度改革试点方案) (3 December 2015) and SPC, *Several Provisions of the SPC on Trying Ecological Compensation Cases (On Trial Basis)*(最高人民法院关于审理生态环境损害赔偿案件的若干规定(试行)) (effective on 5 June 2019 and modified on 23 December 2020). For more details, see Ma (2019), pp. 275–276.

[16] While other dissenting voices termed it "state-interest litigation" and placed it abreast with private-interest and public-interest litigation, see Lv (2017), most scholars insist that it should be classified as a special type of civil EPIL, see Bo (2020), p. 40; Li (2019), pp. 60–61.

Table 7.2 Overview of parties, scope and pre-litigation proceedings in different types of PIL

Type	Parties				Scope of acceptable cases by courts	Pre-litigation proceedings
	Qualified plaintiff		Supporter	Defendant		
	Primary order	Secondary order				
Civil PIL	Legally mandated organs and relevant organisations	PO	PO	Private violators	Ecological environment and resource protection; food and drug safety, etc.	Public notice of intent to bring PIL for 30 days
	PO		/	Private violators	Heroes and martyrs protection	Public notice of intent to bring PIL for 30 days, or directly consult with the opinions of the relatives of heroes and martyrs
Civil EPIL (Ecological compensation litigation)	Provincial, municipal governments and their designated organs, or organs entrusted by State Council to exercise the ownership of state-owned natural resources		PO in selected local rule-making practices	Private violators, e.g., polluters and ecological environment damagers	Ecological environmental damage	Negotiation between qualified plaintiffs and private violators
Administrative PIL	PO		/	Administrative agencies	Ecological environment and resource protection; food and drug safety; state-owned property protection; transfer of state-owned land use rights etc.	Issuance of procuratorial recommendations to agencies

(continued)

Table 7.2 (continued)

Type		Parties				Scope of acceptable cases by courts	Pre-litigation proceedings
		Qualified plaintiff		Supporter	Defendant		
		Primary order	Secondary order				
Incidental PIL	Administrative PIL with incidental civil PIL	PO		/	Administrative agencies+ private violators	Overlap between civil and administrative PIL	Overlap between civil and administrative PIL
	Criminal litigation with incidental civil PIL	PO		/	Criminal suspects + private violators	Same with civil PIL	Same with civil PIL

PO = procuratorial organs

this type of suit thus formulates the necessity of coordination within the emerging procuratorial EPIL. Local procuratorates and environmental authorities have made collective moves towards coordination and connection between the two. Joint rulemaking, joint periodical meetings and co-signing of memorandums by procuratorates and environmental authorities are more and more practiced at the local level. Various coordination mechanisms are created from these procedures, which range across the whole of pre and post litigation issues including the mandated negotiation procedure, the collection of evidence and the management of the ecological compensation fund.[17]

In terms of the rule of legal standing, the roles of procuratorial organs differ between administrative and civil PIL. In civil PIL, procuratorial organs play the role of "filling in the gap"[18] with the exception of PIL instituted for protecting heroes and martyrs.[19] In EPIL, environmental NGOs enjoy the priority of bringing PIL and procuratorial organs are allowed to sue when the former have not taken such an initiative or be "supporters of PIL" when NGOs have already sued (Article 58 of CPL). Accordingly, prescribed as the pre-litigation proceedings of civil EPIL, procuratorial organs are required to issue public notice of their intent to bring civil EPIL for at least 30 days. In contrast, procuratorial organs are the sole party that has the standing to bring administrative PIL [Article 25(4) of ALL]. Pre-litigation proceedings are prescribed as well in terms of administrative PIL, i.e., they should issue "procuratorial recommendations" to administrative agencies and urge them to perform their statutory duties if procuratorates find during the performance of their functions illegal exercise of power or inaction of relevant authorities, and PIL would be filed if authorities fail to comply with the recommendations.

7.2.2 *Transforming Procuratorial Organs: Empowerment of Standing and Reconciliation of Dual Roles*

Scholars have advocated for the adoption of PIL in administrative litigation since the 2000s.[20] However, as previously stated, procuratorial PIL was not endorsed when the ALL was amended for the first time in 2014. The main concerns of the-then legislators were two-fold: first, whether it is appropriate to endow procuratorial organs with the legal standing to bring PIL and, if so, how to position its role with

---

[17]For local rulemaking, see generally Jiangxi province (August 2021), Guangzhou city (September 2021), Zhuzhou city (December 2021) etc.; for joint meetings and co-signing of memos, see generally Chongqing city (June 2020) and Jizhou district of Hengshui city (September 2020).

[18]Lv (2018), pp. 92–106.

[19]Procuratorial organs are the only qualified plaintiffs to file PIL in the field of protection of heroes and martyrs. See Article 25 of the Law on the Protection of Heroes and Martyrs.

[20]See Yu (2001), p. 16; Wang (2002), p. 42.

NGOs and other authorized bodies in civil EPIL; and second, whether PIL matches the internal structure of ALL.

In terms of the appropriateness of procuratorial standing, the issue concerns the constitutionality of procuratorial organs in bringing PIL. According to the Chinese Constitution (Article 134) and Organic Law of People's Procuratorate (Article 2), procuratorial organs are defined as "state organs for legal supervision". Although constitutionally defined, the attribution of procuratorial power is always debatable given an adherence to the accepted doctrine of separation of powers. Different from the legislative-executive-judicial trilogy in western legal culture, China adopts the political regime of "one government, two *yuan,* one commission" under the representative NPC, in which procuratorates and courts form the aforementioned two *yuan* and the newly established National Supervisory Commission forms the third part.

Similar to their western counterparts, Chinese procuratorial organs are in charge of commencing prosecutions against criminal suspects, which could be defined as an exercise of executive power. However, they are also empowered with a range of other functions viz. the supervision of the investigation, prosecution, trial, and the enforcement of criminal, civil and administrative judgments and rulings.[21] Scholars, therefore, opine differently on the nature of procuratorial power i.e. as an exercise of executive or judicial power or as a hybrid of the two.[22] The implantation of procuratorial PIL marks a significant shift in the function of procuratorial organs, which triggered a new round of discussion on the constitutionally prescribed role as "legal supervisor" for procuratorial organs. Various reasoning is provided in literature to defend the constitutionality of the reform of procuratorial PIL, for example, by an expansive reading of "legal supervision" covering both non-binding procuratorial recommendations as previously mandated and initiation of PIL as newly instructed,[23]by resorting to a historical reading of the Constitution and identifying the "original intent" of constitutional framers towards PIL as an entrenched form of legal supervision.[24]

Besides the issue of constitutionality, scholars also debate the institutional capacities of procuratorial organs, and in particular, whether these organs should have exclusive standing in administrative PIL. Some scholars hold a positive attitude by emphasizing a procuratorial organs' advantages, in comparison with NGOs, viz. to organize and oversee an investigation including the collection of evidence and to the existence of long established individual and institutional communication mechanisms established among procuratorial organs, administrative agencies and courts during previous practices and, sufficient financial support from state revenue and

[21]See Criminal Procedure Law 2018 (Article 3, 8, 243, etc.); CPL 2017 (Article 14), ALL 2017 (Article 93).

[22]Chen (2002), pp. 3–4.

[23]Xu (2017), p. 80.

[24]Liu (2017), p. 10.

fewer risks of vexatious litigation.[25] Others challenge the compatibility between procuratorial organs' previous role and the newly mandated one of PIL for its lack of expertise and capacity to bring EPIL due to inherent scientific complexity in environmental disputes.[26] Furthermore, along with the ongoing reform of the state supervision (*jiancha*) system, the power to investigate those crimes committed by officials by taking advantage of their authorities is transferred from procuratorial organs to the newly established state supervision organs, which may weaken the former's supposed institutional advantage of evidence collection.[27]

The 2017 amendment of CPL and ALL solves the legality issue. However, there is a trickier issue ahead with regard to reconciliation between the dual roles of procuratorial organs as both plaintiffs and legal supervisors. Different roles point to different paths of both the substantive and procedural design of the system of procuratorial PIL. For example, the answers may vary significantly under different contexts with regard to whether litigant privileges should be afforded to procuratorial organs during court proceedings.

In terms of how to reconcile the dual roles and which one should be prioritized, academia holds opposing opinions. The pro-plaintiff school believes that a firewall should be established to reduce the risk of procuratorial organs interfering with adjudication in the name of legal supervision. Procuratorial power of investigation associated with its function of legal supervision should not be allowed to extend to civil PIL.[28] Their basic arguments include that procuratorial organs should not have privilege over other plaintiffs or defendants in civil EPIL and that the principle of equality in civil actions should be adhered to.[29] In contrast, the supporters of the pro-supervisor model stress the peculiarity of procuratorial organs' standing in comparison with ordinary plaintiffs.[30] Liu holds that the traditional theory of standing should evolve with the development of PIL. The standing of procuratorial organs in PIL is underpinned by *ad hoc* legal recognition rather than a right of relief based on the conventional standard of "affected interests". It is in nature a legal undertaking to participate in proceedings (*susong dandang*) arising from the role of legal supervisors possessed by procuratorates.[31]

Courts and procuratorates opine differently on the aforementioned issues as well. For example, during the pilot stage, the SPP and SPC diverged over the issue of whether procuratorates should protest against effective judgments or appeal to courts at the next higher level, with the SPP holding the first opinion while SPC the

[25] See, e.g., Liu (2017), pp. 10–11; Lu (2018), p. 33; Jiang (2015), p. 22; Wu (2018), p. 79; Zander (2018), p. 634.

[26] Lv (2017), p. 249.

[27] Zhu (2018), p. 3.

[28] Zhan and Shi (2020), p. 76.

[29] See, e.g., Zhan and Shi (2020), pp. 75–76; Lv (2016), p. 250.

[30] See, e.g., Liang (2019), p. 132; Xu (2017), pp. 79–80.

[31] Liu (2017), p. 13.

second.[32] These controversies were partially clarified by the 2018 Interpretations. It seems that the privileges of procuratorial organs are abridged by stating that "courts should issue a notice of appearance instead of summons to procuratorial organs" (Article 8), and "procuratorial organs should appeal to the courts at the next higher level" (Article 10). However, a reluctant attitude in recognizing procuratorial organs as normal plaintiffs could still be observed. The status of procuratorial organs is legally defined as "public-interest litigants" (*gongyi susong ren*) (Article 4), which is specifically chosen to showcase procuratorial organs' difference from "plaintiff" as used in normal civil litigation. However, the connotations of such a definition are far from clear. The dual role of the procuratorial organ awaits further clarification in both theory and practice.

7.2.3 Challenges on the Administrative Litigation Structure by Procuratorial PIL

In terms of the second concern, the challenge of implanting procuratorial EPIL is also reflected in the mismatch between the structure of PIL and conventional administrative litigation in China, which are respectively featured with objective and subjective litigation.

The distinction between subjective and objective litigation (*contentieux objectif* vs. *contentieux subjectif*) can be traced to Leon Duguit in 1911, later developed by German and Japanese scholars, and now has become a key concept in the administrative law of continental law countries.[33] Chinese academia normally deems German law as a typical example of subjective litigation with a key feature of protecting individual rights (subjective public rights) while the French system of objective litigation has the starting point of guaranteeing public interests (objective legal order).[34] Although a cutting-off line is hard to draw, scholars summarize several indicators to distinguish the two models, including the purpose of litigation, standing, cause of action, the scope of judicial review, categories of judgments, effects of judgments, etc.[35] With regard to the choice of Chinese administrative litigation, although debated,[36] Chinese law has been greatly influenced by the

[32] See SPP, *Implementation Measures on the Pilot Project of Procuratorial PIL* (人民检察院提起公益诉讼试点工作实施办法) (16 December 2015) (Article 25); SPC, *Implementation Measures on the Pilot Project of People's Courts in Adjudicating Procuratorial PIL Cases* (人民法院审理人民检察院提起公益诉讼案件试点工作实施办法) (1 March 2016) (Articles 10 & 23).

[33] Brown et al. (1998).

[34] Xue and Yang (2013), p. 29.

[35] Xue and Yang (2013), pp. 33–34.

[36] Xue and Yang argue that Chinese administrative law has presented a mixed feature, termed "inner cracking", with some indicators showing a subjective feature, for example, standing rules, and others showing objectively, such as the scope of judicial review. Such distortion results in the

German model and has followed the subjective approach,[37] which means that, for instance, the scope of standing is limited to those citizens who have suffered legal wrongs. Article 25(1) of ALL states that plaintiffs that are "subject to or otherwise have an interest in an administrative action" may have the standing to sue, which established the so-called "interest (*lihai guanxi*) standard" of standing. In contrast, administrative PIL brought by procuratorial organs presents clear characteristics of objective litigation.[38] The following issue arises: whether an expansive reading of the "interest standard" should be adopted to endorse procuratorial organs' or even private individuals' standing in the future. Chen argues against this proposal because the expansive reading of "interests" or "injury" may dilute and hollow out the well-established "interest standard". Procuratorial organs' standing should not be unduly broad because it is *ad hoc* and exceptional in nature.[39]

Besides the challenges brought to the theoretical model of standing, the adoption of PIL sharpens the distinction between the two models of subjective and objective litigation. It is argued that the implantation of PIL with a distinctive objective feature into the previously subjective-oriented ALL may destroy the integrity of the whole statute,[40] which is especially reflected in incompatibility between PIL and existing procedural rules, such as pre-litigation proceedings, the burden of proof, evidential rules and appeal rules.[41] Discussions on whether a separate PIL law is a better choice than inserting a PIL provision into ALL endure from the initial stage of amending ALL until now. It is asserted that ALL is not equipped with a systematic framework and rules suitable for procuratorial PIL and a separate set of rule-making for PIL, either as an independent chapter within ALL or a separate procuratorial PIL law, should be developed.[42]

7.3 Evaluation of the Performance of Procuratorial EPIL Based on Empirical Evidence

After an overview of the legislative framework of procuratorial PIL by reviewing the process of its implementation, this section will examine and evaluate the performance of procuratorial PIL in general and EPIL in particular based on empirical

ineffectiveness of administrative litigation to respond to right claims of individual citizens and to guarantee the objective public law order. *See* Xue and Yang (2013), p. 37.

[37] See Yu (2001), p. 16.

[38] Liu (2018b), pp. 43–47.

[39] Chen (2017), pp. 1214–1231.

[40] See Li (2017), pp. 52–54.

[41] For example, how the burden of proof should be distributed in PIL is broadly discussed by scholars. See generally Wang (2017), p. 91.

[42] See, e.g., Li (2017), p. 52; Wang (2022), pp. 174–182; Gong (2021), p. 70.

evidence, followed by a critical inspection of the real efficacy of procuratorial EPIL in practice.

7.3.1 *An Overview of the Performance of Procuratorial PIL*

7.3.1.1 The Surging Number of Procuratorial PIL Cases: The Effect of Mobilizing Procuratorates

The pilot project of procuratorial PIL is a monumental shift in policy and is a national priority. Its practices have been labelled a success by Chinese authorities, as demonstrated by the figure of procuratorial performance. The success of procuratorial PIL, at least in the numerical sense, is largely due to the model of political control and pressure-based propelling that it has followed.[43] Such a model has quickly mobilized the vast resources of more than 3600 procuratorial organs in China.[44]

During two years of the pilot program, i.e., from 1 July 2015 to 30 June 2017, 9053 cases were accepted by all procuratorates in pilot areas in which 1150 cases of PILs were brought to courts with 437 thereof concluded in the form of judgements. Once legalized in 2017, the number of procuratorial PIL cases has skyrocketed to 10,925 in the latter half of 2017,[45] which is 1.2 times the total number in the past two years. The very year after the pilot period has witnessed another 10 fold increase in the number, and this number has climbed to 169,790 in 2021, with an average annual increase of 14.5% in comparison with 2018 (Fig. 7.1). The number of PIL cases concluded by courts has seen a similar surge. From 2018 to 2020, the numbers of PIL cases concluded by courts at all levels increased steadily from 1526 in 2018, 3238 in 2019 to 5976 in 2020, with the increase rate creeping up to almost 100% annually.[46]

7.3.1.2 The Expanding Province of Procuratorial PIL: Seeking Breakthroughs in New Fields

EPIL has made up a large proportion of the total PIL cases handled by procuratorial organs. During the pilot stage, EPIL was set as one of the four enumerated fields of the PIL pilot project,[47] and the number of EPIL cases accounted for 72.10% of the

---

[43] Lu (2018), pp. 25–26.

[44] Xie and Xu (2021), p. 449.

[45] Data are from SPP (2018).

[46] Data are from SPP (2019a), SPP (2020), SPP (2021).

[47] The other three include protection of state-owned property, transfer of state-owned land use right and food and drug safety.

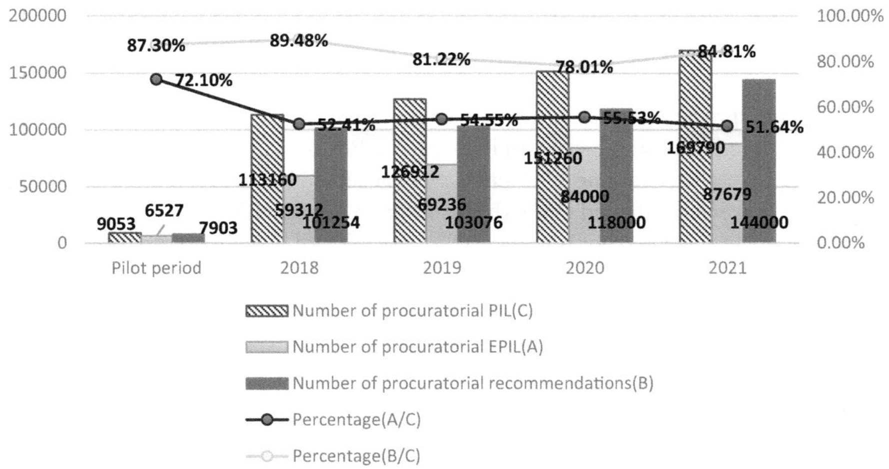

Fig. 7.1 PIL and EPIL cases brought and recommendations issued by procuratorial organs from 2015 to 2021 (Data during the pilot period (2015.06–2017.06) is from SPP (2017), in 2018 is from SPP (2019a), in 2019 from SPP (2020), in 2020 from SPP (2021) and in 2021 from SPP (2022). Due to differences in benchmark-setting, there are minor discrepancies among the statistics from different sources and in different statistical periods)

total (Fig. 7.1). After its legalization in 2017, the scope of procuratorial PIL has witnessed a gradual expansion since procuratorial organs are legally authorized to seek more breakthroughs in other fields, such as protection of heroes and martyrs, private information, minors' interests and military servicemen's interests.[48] As a result, approximately 27,000 PIL cases are handled by procuratorates in newly pioneered areas such as the protection of cultural relics and the rights and interests of disadvantaged groups in 2020, which accounts for 3.4 times the number in 2019, and the number climbed to 45,263 in 2021.[49] Compared with the dominant role of EPIL cases during the pilot stage, the proportion of EPIL cases in terms of the total number of PIL cases has witnessed a gradual decline afterwards. Nonetheless, the number of EPIL cases still accounts for more than half of the total number of PIL cases from 2018 to 2021 (Fig. 7.1).

[48] See generally Law on Protection of Heroes and Martyrs (2018, Article 25), Law on Protection of Minors (2020, Article 106), Law on Protection of the Status, Rights and Interests of Military Servicemen (2021, Article 62), Law on Protection of Personal Information (2021, Article 70).

[49] Data are from SPP (2021), SPP (2022).

7.3.1.3 An Overwhelming Percentage of Pre-litigation Proceedings: Filtrating PIL Disputes

Pre-litigation proceedings are broadly used in administrative PIL, and they prove to be an effective filtration mechanism in resolving PIL disputes (Fig. 7.1). During the two-year pilot project, among 9053 cases were accepted by all procuratorates in pilot areas, 7903 were conducted through pre-litigation proceedings (87.3%) in which 5162 administrative agencies have corrected their illegal behaviors after reception of procuratorial recommendations accounting for 77.14%.[50] The role of procuratorial recommendations became even more dominant after the pilot period. For example, in 2018, among 101,254 procuratorial recommendations issued to agencies, 97.2% were adopted by the addressees. Similarly, the number was as high as 87.5% in 2019 and 93.8% in 2020.[51] Seen from the high rate of agencies' self-correction, pre-litigation proceedings show the effect of increasing the initiative of administrative agencies to actively correct their illegal behaviors despite the softness of non-binding recommendations. This is partly guaranteed by the deterrence effects of potential litigation brought by procuratorial organs that may follow the recommendations thereof.[52] This phenomenon also indicates that the efficacy of PIL largely depends on the quality of communication and settlement between procuratorial organs and administrative agencies, with comparatively less involvement of the courts. In addition to legally mandated forms of recommendations, during the pilot term, many local procuratorial organs have innovated other mechanisms to increase the efficacies of pre-litigation proceedings, such as "appointment and conversation (*yuetan*) with authority leaders" and pre-litigation round-table meetings.[53] Due to their advantage of saving costs and judicial resources, the usage of pre-litigation proceedings has therefore been advocated in academia and further strengthened by the SPP.[54]

7.3.1.4 The Imbalance Among Different Types of Procuratorial PIL: Signaling Procuratorial Activism

In terms of the internal composition of procuratorial PIL cases, there is an apparent imbalance with administrative PIL greatly outnumbering civil PIL. As shown in Table 7.3 above, the number of administrative PIL cases brought by procuratorial organs is more than 30 times that one of civil PIL cases during the pilot period. After

---

50 Data are from SPP (2017).

51 Data are from SPP (2019a), SPP (2020), SPP (2021).

52 Liu (2018b), pp. 39–46.

53 Lu (2018), p. 27.

54 For instance, Zhang Jun, the chief procurator of SPP, points out that it is the best and the most ideal way to protect public interests during the pre-litigation stage when he delivered the SPP report on PIL to SCNPC in 2019. See SPP (2019b).

Table 7.3 Different types of PIL and EPIL cases (2015 to 2020) (The data during the pilot period is collected from SPP (2017) and SPC (2017), in 2018 from SPP (2019a) and SPC (2019), in 2019 from SPP (2020) and SPC (2020), and in 2020 from SPP (2021) and SPC (2021). Note: the number of 150 EPIL cases brought by NGOs during the pilot period shown in the table is calculated from the effective date of EPL in January 2015, instead of June 2015 as the starting month of the pilot of procuratorial PIL. Due to differences in benchmark-setting, there are minor discrepancies among the statistics from different sources and in different statistic periods)

	Civil PIL			Administrative PIL			
		EPIL cases					
	Total (%)	Brought by NGOs	Brought by Ps	Total (%)	EPIL cases brought by Ps to courts	Criminal litigation with incidental civil EPIL	Environmental damage compensation cases adjudicated by courts
Pilot period	321 (3.5%)	*150**	94	8705 (96.2%)	730	25	/
2018	4393 (3.9%)	65	113	108,767 (96.1%)	376	1248	8
2019	7125 (5.6%)	179	312	119,787 (94.4%)	355	1642	36
2020	≈14,000 (9.3%)	/	/	≈137,000 (90.6%)		/	62

Ps = procuratorates

the legalization in 2017, the percentage of administrative PIL cases in the total soared to 96.1% in 2018 and remained above 90% in the following years. This can partly be attributed to the narrow scope of standing in administrative PIL and the supplementary role played by procuratorial organs in civil PIL. Such an imbalance in the numbers of civil and administrative PIL cases also indicates an apparent imbalance between procuratorial organs and NGOs.

Imbalance also exists between EPIL cases brought by procuratorial organs and NGOs with the former greatly outnumbering the latter. This is true not only in terms of the total number of both civil and administrative EPIL cases but is also the case in terms of civil EPIL cases only. For example, in 2018, 113 civil EPIL cases were brought by procuratorial organs, while only 65 were brought by NGOs. Another 376 administrative EPIL cases were brought by procuratorial organs in the same year. A similar imbalance continues today.

It is noteworthy that there is now a high rate of criminal litigation with incidental civil EPIL, especially after the legalization in 2017. As shown in Table 7.3, in 2018, procuratorial organs brought 1248 criminal cases with incidental civil EPIL which accounts for 71.8% in comparison with 376 administrative EPIL cases (21.6%) and 113 civil EPIL cases (6.5%). This shows a sharp turn of procuratorial EPIL with a new focus on criminal litigation with incidental civil EPIL. The result as such is that the room for NGOs is further restricted or even crowded out by procuratorial activism, which will be discussed later.

7.3.1.5 Performance of Procuratorates Before Courts: Compelling Judicial Deference

In administrative PIL, the dominant type of procuratorial claim is to appeal to courts ordering the defendants to perform their duties. In contrast, the least type is to revoke their administrative actions. By December 2016, among 318 environment-related bills of complaints made by procuratorates, there were only 5 with claims of revocation while 308 were claims to perform duties.[55] In another empirical study, among 540 administrative EPIL cases during the period of 2017 to 2020, the combined claims of "confirmation of illegality of administrative action" plus "order of duty performance" account for 71.48 while all other types of claims account for the remaining 28.52.[56] This reflects procuratorial deference to administrative authorities and its intent to avoid aggravating the burden on citizens if previous administrative actions were to be revoked and a new one made.[57]

---

[55] See Liu (2018b), pp. 46–47.

[56] Tian and Xu (2021), p. 18.

[57] Liu (2018b), p. 47.

Procuratorates' performance before courts is remarkable. It is noteworthy that during the pilot stage, procuratorial organs never lost a case.[58] Although subsequent results show an increase in the number of cases lost, the winning rate is still stunningly high. Among 947 claims made by procuratorates in the aforementioned 540 cases from 2017 to 2020, 824 claims (accounting for 87.01%) were affirmed by courts.[59] Another study shows that among 155 cases adjudicated in 2019, there were only 4 cases where the plaintiffs' claims were dismissed and 2 where the disputed administrative actions were revoked and the defendants were ordered to remake their administrative action.[60] More than 94.5% of procuratorial administrative EPIL ends at the first instance, and the winning rate of procuratorial organs in the first instance soars to 97.4% and 88.9% in the second instance.[61] This seems to reflect judicial deference to the claims made by procuratorial organs. This trend is of increasing concern to scholars.[62]

7.3.2 *Efficacy of Procuratorial PIL Inspected and Revisited*

Notwithstanding the seeming success of procuratorial PIL in practice, there are concerns about its efficacy in reality. The foremost concern is the selective feature of procuratorial PIL in terms of the particular fields of cases. During the pilot stage, similar to other pilot reforms in China, procuratorial PIL cases are screened, distributed with a quota to the pilot provinces that are cautiously chosen, and approved in a level-by-level structure.[63] This allowed the procuratorial agency to cherry-pick amongst particular cases. This ability has stretched into the post-pilot stage and shows a close connection between particular case clusters and existing political priorities. Wildlife protection is an example. Based on a group of 108 PIL cases brought by procuratorial organs with regard to wildlife protection from 2018 to 2021, the number of cases adjudicated by courts varies significantly in different years, with 3 in 2018, 21 in 2019, 63 in 2020 and 8 in 2021.[64] It is not difficult to discover that the surge of such cases in 2019 and 2020 was driven by the ban on the consumption of wild animals in response to the outbreak of COVID-19, which led to unprecedented social attention on wildlife protection. It is noted that the key purpose

[58] During the pilot stage, among 458 cases concluded by courts at all levels, there are 437 cases in which procuratorial claims are upheld. Among the remaining 21 cases, 15 were withdrawn by procuratorates due to rectification of errors or performance of duties by defendants, and the remaining 6 cases were dismissed due to resolution of disputes through mediation. It is fair to say that procuratorial organs actually never lost a case during the pilot stage. See SPP (2017).

[59] See Tian and Xu (2021), p. 18.

[60] Xia and Shang (2021), pp. 29–41.

[61] Xia and Shang (2021), pp. 29–41.

[62] See Tian and Xu (2021), p. 18.

[63] Lu (2018), p. 26.

[64] Data are collected through a search in the database of PKULAW by March 2022.

of PIL is not only to compensate for the infringement of public interests but also to emphasize the importance of certain political decisions and other institutional determinations.

Some scholars express their concerns of the phenomenon of "bad money drives out good", which means that procuratorates are more inclined to work on easy cases with "symbolic effect" while leaving those "hard" ones behind, which may relate to issues which are more crucial in terms of the scope of environmental damage or the public interest.[65] Taking EPIL as an example, based on 207 EPIL cases publicly released by SPP as "model cases", Qin shows that the largest percentage thereof relates to cases concerning damage to farmland and forestland caused by the illegal occupation, with 84 cases in total accounting for 40.6%.[66] In the cases stated above, most violators are disadvantaged farmers and herdsmen who reclaim and cultivate farmland and forestland as a means of livelihood. Apparently, the degree of environmental damage in such cases is comparatively lower than that in notorious water and soil pollution incidents. There is less resistance faced by procuratorial organs should they choose these easier cases. The remarkable performance of procuratorial PIL could be documented as such. In this sense, the real efficacy of procuratorial EPIL in vindicating environmental interests is questionable.

A similar characteristic of cherry-picking is also observed in other fields of procuratorial PIL. For example, procuratorial organs are inclined to sue against agencies on their inactions to recover payment of land-transferring fees while seldom touching upon issues of illegal acquisition of land ownership and illegal construction thereon, which may easily, and has, resulted in violent confrontations.[67]

7.4 Controversies, Reflections and Prospects of Procuratorial PIL

After outlining the legislative framework and evaluating the performance of procuratorial EPIL in China, this section intends to identify major controversies debated in academia and obstacles that are confronted with in practice together with prospects for future reforms. The following aspects will be touched upon, i.e., how to place procuratorial organs in civil EPIL especially considering the necessity of coordination with NGOs, how the relationship between procuratorial organs and judiciary is reflected in administrative EPIL, the potential impacts on administrative enforcement and how to formulate synergy across different types of PIL.

---

[65] Jiang et al. (2020), p. 29.
[66] Qin (2019), p. 91.
[67] Qin (2019), p. 96.

7.4.1 Procuratorial Organs' Role in Civil EPIL Revisited: A Turn for Procuratorial Activism or Restraint?

Procuratorial organs play an increasingly important role in civil EPIL, as evidenced by both the overwhelming percentage of civil cases brought by procuratorates over those brought by NGOs and the burgeoning criminal litigation with incidental civil EPIL. The role of NGOs as environmental interest defenders is therefore largely restricted or even crowded out.

7.4.1.1 "No More Filling-the-Gap": Procuratorial Organs as the Bellwether of EPIL

Since NGOs are barred from bringing administrative EPIL according to current law, although expanding the legal standing of NGOs to keep pace with civil PIL is heard from time to time,[68] it is the field of civil EPIL that NGOs enjoy priority. As previously noted, it is legally prescribed that procuratorial organs play a filling-the-gap role in civil EPIL, which means that NGOs enjoy priority to bring EPIL. However, this does not necessarily mean that the relationship between NGOs and procuratorial organs is "dominant-supplementary" as anticipated. The reality is quite the reverse.

As previously presented, the EPIL cases brought by procuratorial organs greatly outnumber the ones brought by NGOs in terms of both the total number as a whole and civil EPIL cases alone. As a result, the room left for NGOs is compressed and the role thereof is restricted. Procuratorial organs are deemed and have already become the bellwether of EPIL.[69] Furthermore, criminal litigation with incidental civil EPIL is gaining momentum and now accounts for an increasingly large percentage of the total workload of procuratorial organs. Such a high number shows that the designated "supplementary role" of procuratorial organs in the civil EPIL is no longer the case. The room for NGOs is further compressed because once the potential EPIL claim proposed by NGOs could be attached to ongoing criminal prosecution proceedings, there is a great chance that the incidental litigation will be brought forward by procuratorial organs.

It is acknowledged that the predicaments of NGOs in delivering better performance in EPIL are not solely caused by the intrusion of procuratorial organs. The restrictive factors that hamper the environmental performance of NGOs are already well documented in the literature, such as high litigation costs due to the insufficiency of the legal aid mechanism, strict regulations on fundraising, a considerably high threshold in terms of qualifications of standing set in law, insufficiency in the disclosure of public information and barriers to accessing information, economic

[68] Zhai and Chang (2018), p. 384; Gong (2021), p. 69.

[69] Weng and Zhou (2017), pp. 218–219.

slowdown and a cautious approach taken by courts in handling EPIL cases.[70] For example, out of 700 potential NGOs that are qualified to bring civil EPIL by 2014, it is roughly calculated that less than 30 thereof are eligible and willing to make such a move.[71] Although the number and composition of NGOs as EPIL plaintiffs has gradually increased and is becoming more diversified,[72] NGOs that have both the willingness and capacity to bring EPIL are still rare in practice. A hostile political atmosphere towards grass-root NGOs and civil society even worsens the plight of NGOs. Considering the risks of vexatious litigation and *xinfang* (letters and visits, a petition system in China) that may follow the expansion of standing to NGOs, the state is prone to choose quasi-official organisations and procuratorial organs to spur environmental EPIL, so as to ensure that EPIL would remain under control by the state.[73] NGOs are thus deemed a foe to be taken precautions against instead of friends to cooperate with in good faith.

Procuratorial organs' active stance is largely due to their impetuses to seek breakthroughs and enhance the political standing of procuratorial organs as a whole. As previously mentioned, procuratorial PIL reform comes along with another significant state reform, which is the state supervision system reform launched in 2018. The traditional primary function of procuratorial organs in anti-corruption campaigns and the incidental power to investigate those crimes committed by officials by taking advantage of their authorities has been transferred to the newly established state supervision organs.

In this context, procuratorial organs, as part of the state apparatuses, are tasked with making PIL a new point of strength. Procuratorial organs are now politically mobilized with resources concentrated and agenda-setting prioritized, which turned out to be very effective.

7.4.1.2 Seeking Out a Niche for Procuratorial Organs: A Debatable Path Forward

With regard to the appropriate role of procuratorial organs in civil EPIL, academic opinions are divided on this issue. Some applaud procuratorial organs' active stance and argue for facilitating more involvement of procuratorial organs in civil EPIL, for example, by quashing the "filling-the-gap" rule and its replacement by a "first-come-first-served" rule.[74] In theory, it is asserted that the nature of civil EPIL is to "subrogate administrative enforcement". Since neither NGOs nor procuratorial organs directly possess regulatory power *per se*, the priority of standing should be provided to the one better equipped with necessary capacities and resources, i.e.,

[70] Zander (2018), pp. 615–628.

[71] He (2015).

[72] Zhang and Mayer (2017), pp. 213–216.

[73] Lu (2018), p. 34.

[74] Gong (2019), p. 138; Gong (2018), p. 32.

procuratorial organs in this regard.[75] Arguments are also made based on reflections on the practice of civil EPIL. It is contended that in terms of pre-litigation proceedings, in a few cases, NGOs respond and join in after the expiry date of the public notice made by procuratorial organs, which results in a significant waste of procuratorial resources.[76]

In contrast, another school argues against the expansive role of procuratorial organs in civil EPIL and is concerned by the lack of balance between procuratorial agencies and civil plaintiffs. A policy shift to prioritize procuratorial EPIL is termed the "nationalization of environmental interests".[77] It is claimed that the dominant role of procuratorial organs in civil EPIL highlights the path-dependency of solely relying on the state to deal with environmental problems. The participatory approach in environmental governance by involving and incentivizing all relevant stakeholders is thus distorted.[78] It is argued that the scope and coverage of pre-litigation public notice by procuratorial organs should be expanded to facilitate more NGOs and relevant bodies to be well informed and thus to sue or participate.[79] Holding a more revolutionary view, Lin explicitly questions the legitimacy of civil PIL brought by procuratorial organs and argues for taking it as the last resort to public interest defense.[80]

Similar debates continue with the reform of criminal litigation with incidental civil EPIL with concerns about its derogation of the right to sue by NGOs. As previously stated, a public notice of intent to sue is prescribed as mandatory pre-litigation proceedings to be accomplished by procuratorial organs in civil EPIL. However, research in 2019 showed that most criminal litigation with incidental civil PIL (accounting for 96% of the total) does not have a pre-litigation announcement.[81] Jiang Bixin, the-then vice president of SPC, publicly stressed the NGOs' litigant rights and warned against any violation of NGOs' prioritized rights to sue by procuratorial organs' initiation of incidental civil EPIL.[82] Such a gap between law in book and law in action became a joint concern of SPP and SPC and finally led to a joint reply confirming the necessity of pre-litigation notice in proceedings of incidental litigation and the absence of such a procedure is deemed a procedural flaw which should be corrected by court order.[83]

---

[75] Gong (2021), p. 68; Gong (2019), p. 138.

[76] Gong (2018), p. 32.

[77] See e.g., Liu (2019a), p. 128; Chen and Zhou (2019), pp. 71–72.

[78] Liu (2021), p. 33.

[79] Liu (2021), p. 33.

[80] Lin (2018), pp. 62–64.

[81] Xie (2019), p. 98.

[82] Jiang (2019), p. 11.

[83] See SPC & SPP, *Reply of the SPC and the SPP on Whether the Procedure of Pre-litigation Public Notice Should be Performed When People's Procuratorates Initiate Criminal Litigation with Incidental Civil PIL* (最高人民法院 最高人民检察院 关于人民检察院提起刑事附带民事公益诉讼应否履行诉前公告程序问题的批复), *Fashi* No. 18 (2019).

The debates surrounding pre-litigation procedures mentioned above boil down to how to position the role of procuratorates in civil EPIL. Liu holds a positive attitude by asserting that the supplementing role of procuratorates should be abandoned and the priority in initiating incidental civil PIL should be provided to procuratorial organs once the disputes drop within the legally authorized scope of civil PIL brought by procuratorial organs.[84] Consequently, she contends that the pre-litigation proceedings of public notice required in normal civil EPIL should be exempted when procuratorial organs bring an incidental civil EPIL.[85] Opposing opinions are heard by questioning the legality of incidental civil EPIL. It is argued that the Criminal Procedure Law could not be deemed the legal foundation thereof, which renders amendment of the law a necessity.[86] Since both criminal prosecution and civil PIL aim to protect public interests, the two types of suits overlap with each other in addressing public interest concerns. Furthermore, the existing mechanism of criminal prosecution with incidental civil litigation could already satisfy the needs of redressing the loss of state-owned or collectively owned public property which makes incidental civil PIL redundant.[87]

7.4.2 *Tandem Between Procuratorate and Judiciary in Administrative EPIL: Crowding Out Administrative Enforcement*

This section will discuss the role and performance of procuratorates in administrative EPIL with a focus on the relationship between procuratorates and judiciaries and the impacts thereof on administrative enforcement of law.

7.4.2.1 Monitoring Administration Through Procuratorial EPIL: Risks of Inappropriate Intrusion

Agencies are conventionally profiled as environmental public interest defenders through environmental law enforcement activities. Considering the well-documented situation of poor enforcement of environmental law in China, procuratorial EPIL acts as a useful monitoring tool for governments to solve the long-perceived problem of insufficient environmental law enforcement. A new layer of supervision of environmental agencies' performance of statutory duties has been

[84] Liu (2019b), p. 83.

[85] It should not be required to submit evidence that the civil PIL pre-litigation announcement procedure has been completed when filing a criminal litigation with incidental civil EPIL. See Liu (2019b), p. 83.

[86] Liu (2018a), p. 87; Gong (2019), p. 136.

[87] Cheng (2018), p. 119.

formed. In contrast, procuratorial supervision of administrative power was previously reflected in procuratorial organs' authorized power to protest against (*kangsu*) judgments of administrative litigation and to correct administrative agencies' illegal behaviors via non-binding procuratorial recommendations (Article 93 of ALL).

However, it is also well recognized that the administrative authorities enjoy broad discretionary power in enforcing the law which should be deferred to by both the judiciaries and procuratorates. This phenomenon is even more salient in the field of environmental management. Li notes that the dilemma between the prevalence of unlawful behaviors and limited resources of law enforcement forces agencies to make decisions on resource allocation to realize the maximization of law enforcement benefits.[88] This dilemma is most evident in environmental enforcement activities where the scarcity of resources is salient, where stringent regulatory tools may cause antagonism from the counterpart and disparity among different regions results in discrepancies in defining fulfillment of statutory duties.[89] In this sense, the tandem between procuratorates and judiciaries formulated in the process of EPIL may bring inappropriate constraints on or even crowd out administrative power.

Potential constraints on administrative power exist in the stages of commencing proceedings against administrative agencies by procuratorial organs and the subsequent judicial review. By exploring how these proceedings are commenced and processed, the dynamics of interaction among procuratorial, administrative and judicial powers and the constraints stated above could thus be better revealed. Considering the compulsory nature of pre-litigation proceedings, the issues above could be further broken down into the following: first, when should procuratorial organs issue procuratorial recommendations to agencies (hereafter "Recommendation Standard")? Second, when should they file PIL claims after the issuance of recommendations (hereafter "PIL Standard")? Third, how do courts react to procuratorial organs' findings? (hereafter "Judicial Review Standard")

7.4.2.2 A Low Threshold of PIL and Tightened Judicial Control in Law and Practice

The multilayered system of standards in setting the threshold of PIL gradually takes its shape via continuous practices of law-and-rule making. As shown in Table 7.4, the Recommendation Standard and the PIL Standard initially formulated in the Pilot Scheme (2015) are further clarified by SPP and SPC in the rules they issued shortly afterwards (with slight differences in phrasing), and later legalized in 2017 in the amended ALL.[90] The abstract expression of "non-performance of statutory duties

[88] Li (2017), p. 59.

[89] Lu (2018), pp. 30–31.

[90] There are minor changes between the initial expression of the Recommendation Standard and the fixed one in 2017 ALL. The term "state and social public interests" stated in previous 2015 plan and rules were changed to "state or public interests" in 2017 ALL.

Table 7.4 Choices of recommendation standards and PIL standards in law-and-rule-making of procuratorial PIL

Stage	Title (effective date)	Recommendation standard	PIL standard
Pilot stage (2015–2017)	Pilot Scheme of Reform on Instituting PIL by Procuratorial Organs (2015.07.02)	Competent authorities illegally exercise their powers or their behaviors constitute inactions that cause infringement upon state and public interests	Authorities refuse to correct their illegal behaviors or do not fulfil their statutory responsibilities that continue to infringe upon state and public interests
	Implementation Rules of People's Procuratorates on the Pilot Work of Initiating PIL (2015.12.24)	*Ibid* (Article 32)	*Ibid* (Article 41)
	Implementation Rules of People's Courts on the Pilot Work of Adjudicating PIL Cases Initiated by People's Procuratorates (2016.03.01)	/	Competent authorities or other legally authorized bodies illegally exercise their powers or their behaviors constitute inactions that cause infringement upon state and public interests (Article 11)
	Work Rules of the SPC on Adjudicating EPIL Cases (for trial implementation) (2017.04.01)	/	Administrative actions made during the pre-litigation stage by agencies are insufficient to protect public interests, or the status of violation of public interests continues albeit actions taken by administrative agencies would have sufficed to protect public interests (Article 53)
Legalization	2017 ALL (2017.07.01)	Competent authorities illegally exercise their powers or their behaviors constitute inactions that cause infringement upon state and public interests [Article 25(4)]	Agencies do not fulfil their duties and responsibilities in accordance with law (Article 25(4))
Post-legalization stage (2017 till current)	Interpretation of the SPC and SPP on Several Issues Regarding the Application of Law in Procuratorial PIL Cases (2018.03.02)	*Ibid* (Article 21)	Agencies do not fulfill their duties and responsibilities in accordance with law (Article 21); Procuratorial organs should provide evidence showing that agencies do not fulfil their duties and responsibilities in accordance with law or

(continued)

Table 7.4 (continued)

Stage	Title (effective date)	Recommendation standard	PIL standard
			correct their illegal behaviors after pre-litigation proceedings (Article 22)
	Rules for the Handling of Public Interest Litigation Cases by People's Procuratorates (2021.07.01)	Administrative agencies illegally perform their duties and responsibilities that cause infringement upon the state or public interests (Article 75)	Administrative agencies do not legally perform their duties and responsibilities after issuance of procuratorial recommendations; state or public interests are under the status of infringement (Article 81) Including the following circumstances that agencies: 1. do not reply to procuratorial recommendations after the expiry of the deadline, nor take effective rectification measures; 2. have made plans of rectification measures, but do not substantively enforce them; 3. do not take any or only partially take rectification measures, although responses to recommendations are made in time; 4. still should continue to perform their duties and responsibilities in accordance with the law, although violators have been held criminally liable or relevant cases have been transferred to judicial organs in charge of criminal cases; 5. the corrective action plan is difficult to implement on schedule because of an objective obstacle, but the corrective action fails to be

(continued)

Table 7.4 (continued)

Stage	Title (effective date)	Recommendation standard	PIL standard
			resumed in a timely manner after removal of the objective obstacle; 6. take rectification measures that are against laws and regulations; or 7. there are other circumstances under which agencies do not perform their duties and responsibilities in accordance with law (Article 82)

and responsibilities" adopted in the 2017 ALL causes considerable confusion in practice.

The primary confusion lies in the requirements of the degree of public interest restoration before PIL claims are initiated before courts. Generally speaking, two types of standards are adopted, i.e., "administrative process standard" (i.e., any dereliction of duty during the whole process that damages public interests constitute non-performance) and "consequentialist standard" (i.e., the consequences of continuity of citizens' non-compliance with law constitutes non-performance). Choices of the two above-mentioned standards are reflected in both rule-making and judicial practices.

First, as shown in Table 7.4, in the Work Rules issued by the SPC in 2017, it is clear that the SPC adopts the "consequentialist" standard by stating that courts should accept the EPIL claims made by procuratorial organs if the status of violation of public interests continues albeit actions taken by administrative agencies in pre-litigation proceedings would have sufficed to protect public interests (Article 53). In contrast, an apparent turn to "administrative process standard" is seen in the 2021 Procuratorial Work Rules. For example, the circumstance that the "administrative agencies have comprehensively taken corrective measures to perform their duties in accordance with law" is prescribed as one of the circumstances to end a case (Article 71). Seven circumstances under which "non-performance of statutory duties" may be upheld are enumerated in Article 82, which shows that a pro-process standard is gradually being established. However, as an internal working rule, this document has no binding legal force.

Second, turning to judicial practices, it is found that in terms of judicial review of "non-performance of statutory duties by administrative agencies", a harder look is taken in the judicial review of PIL in comparison with administrative litigation cases in general. With regard to the former one, the standards of statutory duties in defining non-performance are set as high as "particularized instead of general, actual instead of conjectural".[91] In contrast, in PIL cases, considerably stricter requirements of performance of duties are placed upon agencies, especially *ex-post* monitoring duties.[92] Empirical studies show that "statutory" duties are generally grounded in both statutes and administrative penalty decisions made by agencies in which the latter category accounts for around 15% of the total.[93] To be specific, should the contents of the penalty decisions not be fulfilled due to inaction of administrative agencies, procuratorial EPIL may be initiated based on non-performance of "statutory" duties. For example, in People's Procuratorate of Miyun District of Beijing versus Miyun Landscaping Bureau (2017),[94] the defendant was sued because the species of trees planted by the violator did not match that ordered by the defendant.

The issue of whether *mens rea* of competent authorities is a compulsory part in constituting their "non-performance of statutory duties" is frequently debated in the courts. Since procuratorial organs are prone to take a "tunneled" view on the existence of unrestored ecological damage *per se*, environmental authorities claim that they should be exempted should no fault be observed. For example, in *Dunhua Municipal Procuratorate v Dunhua Land Resources Bureau* (2018),[95] the defendant claims that it has submitted the application for adjusting integrated land use planning to the Ministry of Land and Resources for approval; therefore, the failure to obtain approval for the land use development should not be attributed to it. In *People's Procuratorate of Pinggu District of Beijing v Pinggu Landscaping Bureau* (2017),[96] the defendant argued that it was not the appropriate season to re-plant trees; therefore, its inaction should not be deemed "non-performance of duties".

Along with diversification of cases and the deepening of the reform, procuratorial organs began to lose their cases although it is a very small percentage of cases. One typical case in which the defendant was successful was *Huichun Municipal Procurotarate v Huichun Land Resources Bureau* (2018).[97] The defendant was accused of dilatory conduct in relation to its correction of misbehaviors of deforestation and mining in core zones of Manchurian Tiger Nature Reserve by the violator. The defendant claimed that the case could not be concluded within a short period due to the difficulties arising from the death of the violator, the difficulty of evidence collection and other objective factors. Both courts at the first and second instance found in favor of the defendant in that it showed an intention and had already taken steps to perform its duties.

Scholars' opinions vary in relation to the choice of judicial review standard. Some argue for a hard look at administrative performance in EPIL cases and therefore

[91] Zhang (2010), p. 18.

[92] Lu (2018), p. 30.

[93] Liu and Zhang (2020), p. 70.

[94] (2017) *Jing* 0118 *Xingchu* No. 60, District People's Court of Miyun District, Beijing.

[95] (2018) *Ji* 2401 *Xingchu* No. 42, Intermediate People's Court of Yanji City, Jilin Province.

[96] (2017) *Jing* 0117 *Xingchu* No. 20, District People's Court of Pinggu District, Beijing.

[97] (2018) *Ji* 24 *Xingzhong* No. 104, Intermediate People's Court of Yanbian Korean Autonomous Prefecture, Jilin Province.

contend that the consequentialist standard should be upheld.[98] Others argue for the opposite. Liu urges for a restrictive interpretation of "non-performance of duties" and opposes the strict consequentialist standard of initiating procuratorial EPIL adopted in practice. He claims that resorting to administrative agencies to fully realize the restoration of ecological damage is futile and beyond what the law requires.[99] Li argues that a broad reading of administrative inaction may result in an inappropriate intervention on administrative discretion. PIL claims should only be brought forward once illegal actions performed by authorities have caused serious damage to the state and social public interests.[100] Lin and Liu contend that administrative agencies should exhaust legally prescribed regulatory measures to curb and punish illegal behaviors and urge violators to take science-based and rational measures to restore the damaged environment, regardless of the actual results of the restoration, and they should be exempt from being sued in PIL.[101] There are also scholars seeking a middle ground between different standards. Wang argues for a hybrid standard. He proposes that both standards have advantages and disadvantages. For example, the process standard has advantages of quantification and ease of monitoring but cannot fully realize the compensation purpose of EPIL. The consequentialist standard has the risk of inappropriately aggravating the burden of administrative agencies. He asserts that it is up to the courts to select the most appropriate standard in a case-by-case scenario.[102]

Together with the call for loosening restraints on administrative discretion, reform plans on procedural designs are also recommended. For example, the deadline for environmental agencies to perform their duties after receiving procuratorial organs is set at two months (Article 21 of 2018 Interpretations). Suggestions are made to replace the fixed period with more flexible deadlines in accordance with the differences in difficulties and durations required by ecological restoration.[103]

7.4.3 Connecting Civil and Administrative EPIL: Seeking Synergy

Since environmental and resource protection is the typical overlapping field that a procuratorial organ can bring both civil and administrative PIL, the question arises as to how to choose between the two. For example, facing polluting enterprises, procuratorial organs may sue against enterprises urging them to cease infringement, remove the obstruction, eliminate danger, restore to the original status and

[98] Huang (2018), p. 112.
[99] Liu (2021), p. 34.
[100] Li (2017), p. 59.
[101] Lin (2022), p. 80; Liu and Zhang (2020), p. 71.
[102] Wang (2020), p. 141.
[103] Liu (2021), p. 34.

compensate for losses, or they may sue against environmental authorities urging them to order polluting enterprises to cease infringement, abate pollution or pay relevant charges, etc. Overlap exists in both the contents and effects of the two types of litigation, so do potential conflicts. Synergy between the two is therefore needed.

Academic opinions are varied. Some argue for the priority of initiating civil EPIL over administrative EPIL because the primary aim of litigation is to obtain compensation,[104] while others claim for the priority of administrative EPIL to achieve better performance of administrative authorities[105] or contend a middle ground by emphasizing their respective values and facilitating connecting mechanisms between the two.[106] The deeper divergence among different opinions lies in their attitude toward civil PIL *per se*, as elaborated below. It is noteworthy that although the questioning of the function of civil EPIL is not particularly due to the involvement of procuratorial organs as such, the increasingly strong role of procuratorial organs therein, as mentioned earlier, should not be overlooked.

Although the development of civil EPIL has led to the reform of EPIL in China, more and more critical thinking about its crowding-out effect on administrative enforcement and a restrictive call towards civil EPIL are being voiced in academia.[107] In essence, civil EPIL replaces administrative enforcement when polluting enterprises' non-compliance with environmental laws and standards occurs while environmental authorities suffer from institutional inertia. This results in the adjudication function of the court being replaced by a function to oversee the enforcement of environmental law, which intrudes upon and bypasses the administrative enforcement of environmental laws.[108] Compared with NGOs and procuratorates, environmental authorities are better equipped with expertise and resources to cope with environmental issues, especially considering the inherent features of complexities, dynamics and uncertainties embodied in environmental issues. However, in civil EPIL cases, especially when highly technology-laden environmental issues are of concern, administrative authorities are called upon by the judiciary to assist them by providing technological support. The role of the environmental authority is thus curtailed and the priority of developing administrative EPIL over civil EPIL is proposed. Some even call for revoking authorization on NGOs and procuratorates to initiate civil EPIL and abolishing this system.[109]

Considering that violations could be ceased via stricter administrative enforcement of environmental law while leaving environmental damage unrecovered by resorting to administrative responsibility only, scholars are arguing that the connection mechanism between civil and administrative EPIL may be a better and more rational choice for now. In terms of the connection mechanism, Gong proposes that

---

[104] Ma (2010), p. 96.

[105] Zhu (2015), p. 113; Lin (2022), pp. 78–79; Wang (2016a), p. 64.

[106] Gong (2021), p. 69; Gong (2018), pp. 30–34.

[107] See generally Wang (2016b); Wang (2016a); Ma (2019); Gong (2021); Li and Wu (2016).

[108] Wang (2016a), p. 55.

[109] Wang (2016a).

administrative EPIL with incidental civil EPIL could be a choice and should be activated in practice. Initiation of such incidental litigation should be limited to the circumstances under which any single type of EPIL cannot not fully address and compensate for the damage.[110] With regard to the procedural design of civil EPIL, although it is prescribed that courts should notify relevant environmental authorities within ten days after accepting civil PIL claims as a coordinating mechanism,[111] it is argued that a pre-litigation requirement of exhaustion of administrative remedies should be adopted in civil EPIL.[112] Zhang and Mayer propose that allowing NGOs to notify environmental authorities of their intention to file PIL, leaving the possibility for the latter to pre-empt such action, may produce more efficiency.[113]

Besides the aforementioned reform plans, in order to ease the crowding-out effect of civil EPIL on administrative enforcement, some scholars point to the newly established ecological compensation litigation system as an alternative to civil EPIL. In Bo's words, the sequences among different mechanisms in vindicating environmental interests should be set as follows. Administrative authorities should first aim at fully enforcing environmental laws and then bring ecological compensation litigation against polluters once ecological damage occurs. Administrative EPIL may step in by procuratorial organs and other potential plaintiffs, including NGOs and individuals, once non-performance of statutory duties or delay in bringing ecological compensation litigation occurs.[114] Overall, with considerable varieties of available mechanisms in the Chinese legal system, how synergy may be achieved among them still awaits further research and practice.

7.5 Conclusions

Although with a lengthier legalization process in contrast with the legal empowerment of NGOs, procuratorial EPIL has gradually gained momentum in the country's environmental agenda. Promoted as a model of political mobilization, procuratorial EPIL has made significant progress. There is no doubt that the newcomer of procuratorates into the field of EPIL has created another layer of checks and balances and the active stance of procuratorates in expanding the sphere of EPIL has shown its deterrence effects on potential violations. However, concerns remain about the symbolic effects of procuratorial EPIL with an apparent feature of "cherry-picking" revealed in practices of case screening. The extent to which environmental benefits are obtained *de facto* through the co-efforts of procuratorates and judiciaries in

[110] Gong (2018), p. 30.

[111] 2015 Interpretations (Article 12); Interpretation of the SPC on the Application of CPL (Article 286).

[112] Wang (2016a), p. 55.

[113] Zhang and Mayer (2017), p. 225.

[114] Bo (2020), p. 42.

which huge fines by way of ecological compensation are ordered by courts is questionable.

Moreover, since procuratorial EPIL is deemed the bellwether of the country's environmental course, the phenomenon of "nationalization of environmental interests" is reflected in the top-down design of procuratorial EPIL. Responding to the call of designers, procuratorial activism is reflected in various aspects, either by activating PIL in all types of litigation or by testing water in newly pioneered areas. Procuratorial activism has shown its side effects of crowding out both NGOs as competing plaintiffs and administrative enforcement as the candidates for procuratorial inspection. For the former, the policy shift towards criminal prosecution with incidental civil EPIL further deteriorates the already dwarfed role of NGOs. For the latter, the tandem between a low threshold of procuratorial PIL and a tightened judicial standard in reviewing authorities' actions and inaction may inappropriately intrude with administrative discretion.

Compared with the vigorous development of procuratorial EPIL in practice, split opinions on the legitimacy and prospects of future reforms thereof spread among scholars. It is fair to say that there is seldom any reform that has aroused such polarized scholarly opinions as it is in the field of procuratorial PIL, ranging from the legitimacy of criminal litigation with incidental civil EPIL to the order of standing in bringing PIL by different plaintiffs. The polarization of scholarly reform plans is large because the function of procuratorates and the appropriate interaction with other players is far from clear at both levels of legislation and legal practice.

It is never too late to emphasize that in order to bring procuratorial EPIL into full play, a niche in the overall EPIL system needs to be found for procuratorates. In this sense, the key to future reform does not lie in details of procedural design, but in delimitating the functional design of procuratorial EPIL in particular and EPIL at large. Promising changes have been observed from the one-step-by-one-step adjustments in rule-making made by both SPP and SPC. Nevertheless, more coordination efforts are of necessity among the two top designers. Furthermore, empowerment at the level of national legislation with binding legal effects instead of judicial rule-making practices is a choice worthy of consideration in order to reduce discrepancies and increase cohesion.

References

Bo X (2020) From the ternary mode to the binary mode--on the reconstruction of litigation system for environmental public interest (三元模式归于二元模式——论环境公益救济诉讼体系之重构). J China Univ Geosc (Soc Sci Ed) (中国地质大学学报(社会科学版)) 20:34–47. https://doi.org/10.16493/j.cnki.42-1627/c.2020.04.004

Brown L-N, Bell J-S, Galabert J-M (1998) French administrative law, 5th edn. OUP, Oxford

Carpenter-Gold D (2015) Castles made of sand: public-interest litigation and China's new environmental protection law. Harv Environ Law Rev 39:241–274

Chen W (2002) Reflection and reconstruction of procuratorial power in China--analysis with public prosecution as the core (我国检察权的反思与重构——以公诉权为核心的分析). Chin J Law (法学研究) 2:3–19
Chen P (2017) Multi-layered construction of standing in administrative litigation (行政诉讼原告资格的多层次构造). Peking Univ Law J (中外法学) 29:1214–1232
Chen H, Zhou H (2019) Reflection on the nationalization of public interest litigation (公益诉讼"国家化"的反思). North Leg Sci (北方法学) 13:70–79. https://doi.org/10.13893/j.cnki.bffx.2019.06.007
Cheng L (2018) The negation of criminal incidental civil public interest litigation (刑事附带民事公益之否定). North Leg Sci (北方法学) 12:117–124. https://doi.org/10.13893/j.cnki.bffx.2018.06.011
Gao Q, Whittaker S (2019) Standing to sue beyond individual rights: who should be eligible to bring environmental interest litigation in China. Transit Environ Law 8(2):327–347
Gong G (2018) An analysis of the joint mechanism of the public prosecutor's "two litigation" in prosecutorial organization—based on the improvement of "interpretation of prosecutor's public interest litigation" (检察公益"两诉"衔接机制探析——以"检察公益诉讼解释"的完善为切入). J Zhejiang Gongshang Univ (浙江工商大学学报) 5:27–34. https://doi.org/10.14134/j.cnki.cn33-1337/c.2018.05.003
Gong G (2019) Reflections on the positioning of the nature of environmental civil public interest litigation (环境民事公益诉讼性质定位省思). Chin J Law (法学研究) 41:127–147
Gong G (2021) The nature and legislative perfection of public interest litigation (公益诉讼的属性及立法完善). J Natl Prosecutors Coll (国家检察官学院学报) 29:55–71
He Z (2015) Will there be vexatious litigation in EPIL?(环境公益诉讼会出现滥诉现象吗?) China Environmental Daily (中国环境报) (4 February 2015)
He H (2018) How much progress can legislation bring? The 2014 Amendment of the Administrative Litigation Law of PRC. Univ Pa Asian Law Rev 13:137–173
Huang H (2018) The analysis of essential standards of judicial adjudication in environmental administrative public interest litigation instituted by procuratorial authorities (检察机关提起环境行政公益诉讼的司法裁判标准研究). Law Sci Mag (法学杂志) 39:107–113. https://doi.org/10.16092/j.cnki.1001-618x.2018.08.012
Jiang T (2015) The system of administrative PIL brought by the procuratorial organs: thinking of a Chinese problem (检察机关提起行政公益诉讼制度:一个中国问题的思考). Tribune Polit Sci Law (政法论坛) 33:19–29
Jiang B (2019) Practical development and institutional consummation of Chinese environmental public interest litigation (中国环境公益诉讼的实践发展及制度完善). J Law Appl (法律适用) 1:5–12
Jiang M, Yuan Y, Mou Q (2020) From zero to one and from one to fine: the difficulties of environmental prosecution public interest litigation and the solution--taking criminal incidental civil environmental prosecution public interest litigation cases of C city as an empirical research paradigm (从无到有与从有到精:环境检察公益诉讼的困局与破局——以C市刑事附带民事环境检察公益诉讼案件为实证研究范式). J Law Appl (法律适用) 18:24–32
Li H (2017) Approach to legalization of procuratorial administrative PIL (检察机关提起行政公益诉讼的法治化路径). Adm Law Rev (行政法学研究) 5:52–62
Li H (2019) A study on the essence of ecological damage compensation lawsuit and related issues: an analysis from the perspective of environmental civil public interest lawsuit (生态损害赔偿诉讼的本质及相关问题研究). Adm Law Rev (行政法学研究) 4:55–66
Li Y, Wu K (2016) Role of procuratorates in the environmental public interest litigation (论检察机关在环境公益诉讼中的角色与定位——兼评最高人民检察院《检察机关提起公益诉讼改革试点方案》). J Renmin Univ China (中国人民大学学报) 30:2–13
Liang H (2019) Procuratorial public interest litigation: logic, significance, omissions and improvement (检察公益诉讼:逻辑,意义,缺漏及改良). J Anhui Norm Univ (Hum Soc Sci) (安徽师范大学学报(人文社会科学版)) 47:126–134. https://doi.org/10.14182/j.cnki.j.anu.2019.03.016

Lin L (2018) On the space for system of procuratorial organs initiating civil public interest litigation (论检察机关提起民事公益诉讼的制度空间). Adm Law Rev (行政法学研究) 6:55–66
Lin L (2022) A further analysis on the institutional space of civil public interest litigation initiated by the procuratorate--comparing with the scope of administrative public interest litigation (检察机关提起民事公益诉讼之制度空间再探——兼与行政公益诉讼范围比较). Adm Law Rev (行政法学研究) 2:77–92
Lin Y, Wang X (2014) Back to the start: EPIL in 2013 (2013环境公益诉讼回到原点). In: Friends of nature, green book on environment: annual report on environment development of China (2014) (环境绿皮书:中国环境发展报告(2014)). Social Sciences Academic Press, Beijing, pp 157–174
Liu Y (2017) Judicial practice and theoretical exploration of procuratorial PIL (检察公益诉讼的司法实践与理论探索). J Natl Prosecutors Coll (国家检察官学院学报) 25:3–18+170
Liu J (2018a) The predicaments of the civil public interest litigation attached to criminal cases and their solutions (刑事附带民事公益诉讼的困局与出路). Polit Sci Law (政治与法律) 10:84–94
Liu Y (2018b) Establishing the mechanism of objective litigation in administrative PIL (构建行政公益诉讼的客观诉讼机制). Chin J Law (法学研究) 3:39–50
Liu Q (2019a) On non-conventionality of environmental public interest litigation (论环境公益诉讼的非传统性). Sci Law (J Northwest Univ Polit Sci Law) (法律科学(西北政法大学学报)) 37:123–132. https://doi.org/10.16290/j.cnki.1674-5205.2019.01.011
Liu Y (2019b) Study on synergy in criminal incidental civil public interest litigation (刑事附带民事公益诉讼的协同问题研究). Crim Sci (中国刑事法杂志) 5:77–91. https://doi.org/10.19430/j.cnki.3891.2019.05.006
Liu J (2021) On the function orientation and procedure optimization of environmental public interest litigation filed by the procuratorial authorities (论检察环境公益诉讼的职能定位及程序优化). J China Univ Geosci (Soc Sci Ed) (中国地质大学学报(社会科学版)) 21:28–40. https://doi.org/10.16493/j.cnki.42-1627/c.2021.04.004
Liu W, Zhang F (2020) An empirical study on the illegal inaction of the administrative organ and its judgment in the environmental administrative public interest litigation (环境行政公益诉讼中行政主管机关不作为违法及其裁判的实证研究). J Soochow Univ (Law Ed) (苏州大学学报(法学版)) 7:68–77. https://doi.org/10.19563/j.cnki.sdfx.2020.02.007
Lu C (2018) From judicial process to organization incentive: Chinese pilot of administrative PIL (从司法过程到组织激励:行政公益诉讼的中国试验). Stud Law Bus (法商研究) 5:25–35
Lv Z (2016) Judicial rationality on environmental protection should not be overridden by the pricey compensation (环境司法理性不能止于"天价"赔偿). China Leg Sci (中国法学) 3:244–264
Lv Z (2017) Providing legalized plan for ecological damage compensation system (为生态损害赔偿制度提供法治化方案), Guangming Daily (22 December 2017)
Lv W (2018) Study on the system for initiation of public interest litigation by procuratorates in China. China Leg Sci 6:92–106
Ma M (2010) On the prosecution of administrative public interest litigation (检察机关提起行政公益诉讼制度研究). J CUPL (中国政法大学学报) 6:92–99+159
Ma Y (2019) Vindicating environmental public interests in China: a balanced approach to institutional interaction in public interest litigation system. ELR 21(4):269–291. https://doi.org/10.1177/1461452919881342
Qin H (2019) An empirical study on the practice of administrative public interest litigation (检察机关提起行政公益诉讼的实证考察). Adm Law Rev (行政法学研究) 3:87–100
SPC (2017) White Paper on Environmental and Resource Adjudication in China 2016-2017. http://www.cbcgdf.org/NewsShow/4856/2610.html. Accessed 25 Feb 2022
SPC (2019) White Paper on Environmental and Resource Adjudication in China 2017-2018. http://news.cnr.cn/dj/20190302/t20190302_524527648.shtml. Accessed 25 Feb 2022
SPC (2020) White Paper on Environmental and Resource Adjudication in China 2019. https://www.court.gov.cn/zixun-xiangqing-228341.html. Accessed 25 Feb 2022

SPC (2021) White Paper on Environmental and Resource Adjudication in China 2020. https://www.court.gov.cn/zixun-xiangqing-307471.html. Accessed 25 Feb 2022
SPP (2017) The transcript of the press conference of SPP in "To Comprehensively Enforce Procuratorial PIL". http://www.scio.gov.cn/xwfbh/qyxwfbh/Document/1557150/1557150.htm. Accessed 25 Feb 2022
SPP (2018) Work Report of the Supreme People's Procuratorate (2017). https://www.spp.gov.cn/spp/gzbg/201803/t20180325_372171.shtml. Accessed 25 Feb 2022
SPP (2019a) Work Report of the Supreme People's Procuratorate (2018). https://www.spp.gov.cn/spp/gzbg/201903/t20190319_412293.shtml. Accessed 25 Feb 2022
SPP (2019b) Work Report of the Supreme People's Procuratorate on the Development of Public Interest Litigation Procuratorial Work at the Second Session of the Thirteenth National People's Congress. https://www.spp.gov.cn/spp/tt/201910/t20191024_435925.shtml. Accessed 25 Feb 2022
SPP (2020) Work Report of the Supreme People's Procuratorate (2019). https://www.spp.gov.cn/spp/gzbg/202006/t20200601_463798.shtml. Accessed 25 Feb 2022
SPP (2021) Work Report of the Supreme People's Procuratorate (2020). https://www.spp.gov.cn/spp/gzbg/202103/t20210315_512731.shtml. Accessed 25 Feb 2022
SPP (2022) Work Report of the Supreme People's Procuratorate (2021). https://mp.weixin.qq.com/s/k3rMQwYhm12yfQc97XqVNA. Accessed 13 Mar 2022
Tian Y, Xu J (2021) A study on the precision of claim in environmental administrative public interest litigation -- an empirical analysis based on 540 judicial documents (环境行政公益诉讼的诉讼请求精准化研究——基于540份裁判文书的实证分析). J Nanjing Tech Univ (Soc Sci Ed) (南京工业大学学报(社会科学版)) 20:17–30+111
Wang T (2002) On administrative PIL (论行政公益诉讼). Chin J Law (法学研究) 5:42–53
Wang M (2016a) On the development direction of environmental public interests litigation: analysis on administrative right and jurisdiction (论我国环境公益诉讼的发展方向:基于行政权与司法权关系理论的分析). China Leg Sci (中国法学) 1:49–68. https://doi.org/10.14111/j.cnki.zgfx.2016.01.003
Wang X (2016b) On the legislative order of environmental public interest litigation system (论环境公益诉讼制度的立法顺序). Tsinghua Univ Law J (清华法学) 10:101–114
Wang D (2017) The burden of proof in administrative PIL (检察机关提起行政公益诉讼的举证责任). Public Adm Law (行政与法) 10:91–98
Wang Q (2020) The assessment standard for administrative omissions in environmental administrative public interest litigation (环境行政公益诉讼中行政不作为的审查基准). Tsinghua Univ Law J (清华法学) 14:129–142
Wang C (2022) On developing a special legal system for public interest administrative litigation (独立行政公益诉讼法律规范体系之构建). Peking Univ Law J (中外法学) 34:165–182
Wang A, Gao J (2010) Environmental courts and the development of environmental public interest litigation in China. J Court Innov 3:37–44
Weng X, Zhou X (2017) A study on the phenomenon of "executing the main and the punishment" in the pilot program of public interest litigation--also on the different responsibilities of the procuratorial organs in the two types of cases (公益诉讼试点中的"行主民辅"现象研究——兼论检察机关在两种案件类型中的不同担当). Soc Sci Front (社会科学战线) 11:218–229
Wu S (2018) Theoretical analysis of EPIL brought by procuratorial organs: from the perspective of law and economics (检察机关作为原告提起环境公益诉讼的理论分析——以法经济学角度切入). Fujian Law (福建法学) 3:74–79
Xia Y, Shang J (2021) Review and perfection of the adjudication mode of environmental administrative public interest litigation (环境行政公益诉讼判决方式的检视及其完善). J Nanjing Tech Univ (Soc Sci Ed) (南京工业大学学报(社会科学版)) 3:29–41
Xie X (2019) Criminal incidental civil public interest litigation: institutional innovation and practice breakthrough (刑事附带民事公益诉讼:制度创新与实践突围——以207份裁判文书为样本). Crim Sci (中国刑事法杂志) 5:92–111

Xie L, Xu L (2021) Environmental public interest litigation in China: a critical examination. Transnatl Environ Law 10:441–465
Xu Q (2017) Functional design and institutional construction of administrative PIL brought by procuratorial organs (检察机关提起行政公益诉讼的职能定位与制度构建). Adm Law Rev (行政法学研究) 5:77–86
Xue G, Yang X (2013) On China's administrative litigation structure: "subjective litigation" or "objective litigation"?(论我国行政诉讼构造: "主观诉讼"抑或"客观诉讼"?) Adm Law Rev (行政法学研究) 4:29–37+78
Yu A (2001) Research on administrative PIL and objective litigation (行政诉讼的公益诉讼和客观诉讼问题). Law Sci (法学) 5:16–17+37
Zander S (2018) How effective a weapon is the new EPL in China's war against pollution: the past triumphs and future challenges of environmental public interest litigation. N Y Univ J Int Law Polit 50:605–636
Zhai T, Chang Y (2018) Standing of environmental public-interest litigants in China: evolution, obstacles and solutions. J Environ Law 30:369–397. https://doi.org/10.1093/jel/eqy011
Zhan S, Shi Y (2020) On procuratorial organs' litigation implementing power and role orientation (论检察机关的诉讼实施权和角色定位). Study Pract (学习与实践) 7:70–79. https://doi.org/10.19624/j.cnki.cn42-1005/c.2020.07.009
Zhang Z (2010) Administrative inaction in judicial judgement (司法判决中的行政不作为). Chin J Law (法学研究) 32:18–22+24–29+31–33
Zhang Z (2016) Empirical assessment of the specialization of environmental justice: environmental trial organ and mechanism-centric (环境司法专门化发展的实证检视:以环境审判机构和环境审判机制为中心). China Leg Sci (中国法学) 6:177–181
Zhang Q, Mayer B (2017) Public interest environmental litigation in China. Chin J Environ Law 1(2):202–228. https://doi.org/10.1163/24686042-12340013
Zhu Q (2015) On administrative public interest suit brought by procuratorial organs: characteristic, mode and procedure (论检察机关提起行政公益诉讼: 特征,模式与程序). Law Sci Mag (法学杂志) 36:112–118. https://doi.org/10.16092/j.cnki.1001-618x.2015.04.014
Zhu X (2018) Consolidation and development of procuratorate system after national supervision system reform (国家监察体制改革后检察制度的巩固与发展). Chin J Law (法学研究) 40:3–19

Chapter 8
Litigation for Ecological and Environmental Damage Compensation in China: EPIL with Typical Chinese Characteristics

Wenjun Luo

Abstract Litigation for ecological and environmental damage compensation is an important achievement of China's environmental justice reform in recent years, and an important component of China's ecological civilization system reform. As a special civil environmental public interest litigation (EPIL) system, it has a unique operating mechanism. In terms of the liability mode, priority should be given to ecological environment restoration and compensation for ecological environment damage. In the connection with China's civil EPIL, the principle of "special before general" should be followed, and the factors of unification of adjudication standards and conservation of judicial resources should also be considered. The development of the ecological and environmental damage compensation litigation (EEDCL) system is deeply influenced and shaped by China's effort to reform the environmental rule of law, which more directly reflects the progress made in environmental legislation and justice in China in recent years. Of course, this system needs to be tested and improved in practice.

Keywords Ecological and environmental damage compensation litigation · EPIL · Burden of proof · No-fault liability

8.1 Introduction

Litigation for ecological and environmental damage compensation (EEDCL hereinafter) is a significant outcome of China's environmental law reform in recent years and an important part of China's ecological civilization system reform. It is "a new litigation system and litigation category based on the requirements of comprehensive deepening reform in China and with distinct Chinese characteristics and innovative

W. Luo (✉)
Law School, Hubei University of Economics, Hubei, China
e-mail: luowenjun@hbue.edu.cn

X. Wang et al. (eds.), *Environmental Public Interest Litigation in China*,
https://doi.org/10.1007/978-3-031-26526-6_8

means".[1] EEDCL can be described as a kind of civil judicial activity where the concerned authority—i.e., a government at the provincial, municipal, or prefecture-level or the department or agency entrusted with the responsibility to exercise ownership of all natural resource assets—files a lawsuit against a natural person, a legal person, or an unincorporated organization that has caused ecological and environmental damage. In China, the compensation litigation system for ecological and environmental damage is a special civil environmental public interest litigation system with a unique operating mechanism.

This chapter analyzes this system in detail from four perspectives: (1) the creation and development of the compensation litigation system for ecological and environmental damage; (2) the nature and characteristics of the EEDCL; (3) the components and operating mechanism of the compensation litigation system for ecological and environmental damage; and (4) the coordination between the EEDCL and environmental public interest litigation (EPIL). Finally, a brief conclusion is drawn. The compensation litigation system for ecological and environmental damage is an important legal system that is rooted in China's political and legal foundations. The system is a response to China's special ecological and environmental issues, and it has typical Chinese characteristics. The analysis herein would help develop an overall understanding of the creation, development, and operation of this system.

8.2 Creation and Development of the Litigation System for Ecological and Environmental Damage Compensation

8.2.1 Creation of the Litigation System for Ecological and Environmental Damage Compensation

The liability system of ecological and environmental damage compensation is an important outcome of China's reform and development of an ecological civilization. In 2012, at the 18th National Congress, the Communist Party of China (CPC) delivered a report[2] which laid out a top-level design and overall arrangement for developing an ecological civilization in China. The report states "strengthening environmental supervision and improving the accountability system for ecological and environmental protection and the compensation system for environmental damage". For the first time, the idea of establishing an ecological and environmental damage compensation as an important part of the development of an ecological

---

[1]Chen (2018), pp. 20–27.

[2]Hu Jintao, We will unswervingly advance along the path of socialism with Chinese characteristics and strive to complete the building of a moderately prosperous society in all respects - Report to the 18th National Congress of the Communist Party of China. 2012-11-19, http://www.beijingreview.com.cn/18da/txt/2012-11/19/content_502297.htm.

civilization emerged in China.[3] On May 24, 2013, CPC's General Secretary Xi Jinping, while attending the 6th collective study of the Political Bureau of 18th CPC Central Committee, pointed out that "ecological and environmental issues in China have reached a very serious level and we must take the most stringent measures. Ecological and environmental protection must rely on systems and the rule of law. Only by implementing the most stringent regimes and rule of law can we provide reliable support for the development of ecological civilization."[4] Xi's remarks have played an important guiding role in accelerating China's exploration and establishment of an ecological civilization system and a compensation system for ecological and environmental damage. In November 2013, the 3rd Plenary Session of the 18th CPC Central Committee adopted the *Decision of the Central Committee of the Communist Party of China on Some Major Issues Concerning Comprehensively Deepening the Reform*, which expressly stipulated that "[A]ny person responsible for ecological and environmental damage shall be strictly subject to the compensation system."

The legal liability for ecological and environmental damage has also been confirmed by China's basic environmental legislation. In April 2014, the 8th meeting of the Standing Committee of the 12th National People's Congress revised and adopted the *Environmental Protection Law of the People's Republic of China*, which came into force on January 1, 2015, and established the legal principle of "liability for damage."[5] In addition, Article 64 of the *Environmental Protection Law* stipulated that "Where any damage is caused by environmental pollution or ecological disruption, the tortfeasor shall assume a tort liability in accordance with the relevant provisions of the *Tort Law of the People's Republic of China*." This provision suggested that China has recognized the legal liability for ecological and environmental damage in its basic environmental law.

In April 2015, the *CPC Central Committee and State Council Opinion on Accelerating the Construction of Ecological Civilization* stated that "[W]e should step up efforts to establish an operating mechanism of ecological damage compensation and, among others, an independent and impartial ecological and environmental damage assessment system." This further provided a policy basis for the implementation of the compensation system for ecological and environmental damage. In September of the same year, the *CPC Central Committee and State Council Integrated Reform Plan for Promoting Ecological Civilization* proposed that the compensation system for ecological and environmental damage be strictly implemented and systematically put forward the idea of improving the legal system and accountability for compensation for ecological and environmental damage; that is, "improving the legal system, evaluation methods, and implementation

[3] Xinhua News Agency (2012).

[4] Wang et al. (2021).

[5] The principle of liability for damage refers to that when the actor causes damage due to the behavior of polluting the environment and destroying the ecology, the actor shall bear legal responsibility according to law.

mechanisms of compensation for environmental damage to severely punish those who violate environmental protection laws and regulations according to law, determine the amount of compensation by those who cause ecological and environmental damage to the extent of the damage according to law, and to investigate the criminal responsibility of those who cause serious consequences according to law."

In December 2015, the General Office of the CPC Central Committee and the General Office of the State Council issued the *Pilot Plan for the Reform of the Compensation System for Ecological and Environmental Damage* (hereinafter referred to as the *Pilot Plan*), which stated that Jilin Province, Jiangsu Province, Shandong Province, Hunan Province, Chongqing City, Guizhou City, and Yunnan Province were selected to run a pilot program of ecological and environmental damage compensation between 2015 and 2017. The *Pilot Plan* defined the concept of "ecological and environmental damage" for the first time, and specified the general and particular requirements that should be covered by the mechanism, which include, for example, the objectives, pilot principles, scope of application, subjects of responsibility, subjects of claim, scope of compensation, litigation for damage compensation, the corresponding appraisal and assessment management and technical system, and funding and the operating mechanism. This plan marked the formal establishment and pilot operation of a compensation litigation system for ecological and environmental damage.

8.2.2 Development of EEDCL

From 2015 to 2017, the *Pilot Plan*, which was implemented in seven provinces and cities, showed that it was feasible and necessary to normalize and generalize the system of compensation liability for ecological and environmental damage in China. As a result, in 2017, the CPC Central Committee and the State Council jointly issued a *Reform Plan on the Compensation Regime for Eco-environmental Damage* (hereinafter referred to as the *Plan*), stipulating that the compensation system for ecological and environmental damage should be piloted nationwide from January 1, 2018. That meant a nationwide ecological and environmental damage compensation system, characterized by clear responsibilities, smooth approaches, technical specifications, strong guarantees, in-place compensation, and effective environmental restoration, was to be established across China.

Based on the 2015 *Pilot Plan*, the *Plan* expanded the concept of "ecological and environmental damage" by explicitly adding "forest" to the listed environmental factors. The *Plan* was also optimized based on experiences drawn from the *Pilot Plan*. That is, it made more scientific and systematic provisions in terms of the working principles, scope of application, subjects of responsibility, subjects of claim, and other aspects that have been defined by the *Pilot Plan*. Overall, the *Plan* provided more details which are required for the nationwide implementation of the compensation system for ecological and environmental damage. The condition for initiating EEDCL was changed from selective negotiation or litigation, as stipulated in the *Pilot Plan,* to prerequisite mediation as stipulated in the *Plan*; that is, the obligee and other designated departments or agencies could now file a civil action for compensation for ecological and environmental damage but only if there was no agreement reached through mediation. The *Plan* also proposed some improvements to the rules of compensation litigation, and it encouraged statutory authorities and qualified social organizations to initiate EEDCL. It proposed that the Supreme People's Court (SPC), in consultation with relevant departments, would formulates guidelines according to the actual situation to clarify the coordination between the compensation system for ecological and environmental damage and EPIL and other matters.

The introduction of the *Plan* is a milestone that marks the formal establishment of a litigation system for ecological and environmental damage compensation in China. It is worth noting that the *Civil Procedure Law of the People's Republic of China* has no clear provisions for the EEDCL. In practice, the trial of ecological and environmental damage compensation cases referred to the *Civil Procedure Law* in most cases, and the *Reform Plan* also clearly stipulated that the compensation obligee should file a civil action for ecological and environmental damage compensation. Therefore, it has now been laid down that the *Civil Procedure Law* establishes the procedural basis for EEDCL.

In May 2019, the judicial committee of the SPC passed the *Several Provisions of the Supreme People's Court on Hearing Cases of Compensation for Ecological and Environmental Damage (Trial)* (hereinafter referred to as the *Several Provisions (Trial)*), which came into force on June 5, 2019. The *Several Provisions (Trial)* were formulated in accordance with the *Environmental Protection Law of the People's Republic of China*, the *Civil Procedure Law of the People's Republic of China* and other relevant laws and in the light of the actual trials. These provisions provide an important basis for local courts to properly hear ecological and environmental damage compensation cases and investigate the compensation liability of those who caused ecological and environmental damage according to law. In 2019, courts nationwide accepted 49 cases and concluded 36 cases of EEDCL, a year-on-year increase of 145% and 350%, respectively, including 28 cases of judicial confirmation of ecological and environmental damage compensation (with 23 cases concluded) and 21 cases of EEDCL (with 13 cases concluded).[6]

On May 28, 2020, the 3rd Session of the 13th National People's Congress adopted the *Civil Code of the People's Republic of China*, which came into force on January 1, 2021. This was a significant event in the history of China's legal infrastructure. The *Civil Code* plays a fundamental role in China's legal system and is the basic law of the market economy. It adjusts and maintains social and economic order and protects the personal rights, property rights, and other legitimate rights and interests of civil subjects. Articles 1234 and 1235 of the *Civil Code (Tort Liability)* provide for the right to claim compensation for ecological and environmental

[6]The SPC (2020), p. 2.

damage in terms of substantive law,[7] which is an authoritative substantive legal basis for EEDCL.

To support the implementation of the *Civil Code*, on December 23, 2020, the judicial committee of the SPC adopted 27 revised decisions on civil judicial interpretations, four of which are directly in relation to EEDCL, including the *Several Provisions of the Supreme People's Court on Hearing Cases of Compensation for Ecological and Environmental Damage (Trial),* the *SPC Interpretation on Several Issues Concerning the Application of Law in Environmental Civil Public Interest Litigation,* and the *SPC Interpretation on the Application of Law in Adjudicating Environmental Tort Cases*. On December 30, 2020, the SPC released the Revised *"Provisions on Causes of Action in Civil Cases"*,[8] which described "EEDCL" as a new cause of action and added it to the revised Section 11 "Causes of Action In Special Procedural Cases". At the same time, Subsection 52 "public interest litigation" specified "civil public interest litigation for ecological and environmental protection" as a separate cause of action.

8.3 Nature and Characteristics of the EEDCL

The EEDCL is different from other litigation types related to ecological and environmental disputes and has the following features.

[7] Article 1234 of the *Civil Code of the People's Republic of China* provides that: 'Where a tortfeasor causes damage to the ecology and environment in violation of the State regulations and restoration is possible, the State authorized agencies or the organizations authorized by law have the right to request the tortfeasor to bear the responsibility for restoration within a reasonable period of time. Where the tortfeasor fails to restore it within the time limit, the State authorized agencies or the organizations authorized by law may initiate the restoration on its own or entrust it with others, provided that any expenses thus incurred shall be borne by the tortfeasor'; Article 1235 provides that: 'Where ecological and environmental damage is caused in violation of the State regulations, the State authorized agencies or the organizations authorized by law have the right to request the tortfeasor to compensate the following losses and expenses:

(1) losses caused by loss of service function from the time the ecology and environment are damaged to the time the restoration is completed;
(2) losses caused by permanent damage to ecological and environmental functions;
(3) expenses of investigation, appraisal and assessment of ecological and environmental damage;
(4) expenses of pollution cleanup and ecological and environmental restoration; and
(5) other reasonable expenses incurred to prevent the occurrence or aggravation of the damage.'

[8] Notice of the Decision on Amending the Provisions on the Causes of Civil Cases Issued by the Supreme People's Court, https://www.court.gov.cn/fabu-xiangqing-282031.html, 2020-12-30.

8.3.1 A Special Type of Environmental Civil Public Interest Litigation

EEDCL can be defined as a tort litigation that expands the plaintiff's standing, where in the case of ecological and environmental damage, a specific subject files a lawsuit as a "representative" to investigate the civil liability of the perpetrator, and the public interest is advanced through the compensation role and ecological restoration if practicable[9]

Hence EEDCL in China is a type of civil litigation and it applies civil procedures. First, the *Reform Plan* stipulates that the compensation obligee and its designated departments or agencies should file a civil action for compensation for ecological and environmental damage in time, which makes it clear that the EEDCL has the nature of civil litigation. Second, the *Civil Code (Tort Liability)* has laid a civil substantive basis for the EEDCL and third the SPC's *Several Provisions (Trial)* forms the procedural basis of EEDCL, which clearly defines such litigation as a civil litigation. In addition, in the SPC's *Provisions on Causes of Action in Civil Cases*, EEDCL is defined as a special civil public interest litigation. Therefore, as the policy and legal bases suggest, EEDCL involves substantive laws and procedural laws, and it is a litigation system that is rooted in civil procedures.

On the other hand, EEDCL is a kind of special environmental civil public interest litigation. Article 55 of the *Civil Procedure Law* and the *SPC Interpretation on Several Issues Concerning the Application of Law in Environmental Civil Public Interest Litigation* stipulate that the general environmental civil public interest litigation system mainly applies to litigation brought by statutory authorities, social organizations, and people's procuratorates for damage to the public interest caused by environmental pollution and ecological disruption. In environmental civil public interest litigation (EPIL), the environmental public interest affected can include personal rights, property rights, and environmental rights and interests of unspecified subjects. However, EEDCL only applies to "ecological and environmental damage", that is, adverse changes in environmental and biological factors and the degradation of ecosystem functions. Personal injury and property loss are not covered.

In addition, EEDCL focuses on ecological values, a kind of pure environmental public interest. As a result, the SPC has, by promulgating the *Several Provisions (Trial)*, created a special procedural basis for such litigation. EEDCL has different provisions from general environmental civil public interest litigation in terms of litigation, jurisdiction, prosecution conditions, prohibition of counterclaims, and rules of evidence. EEDCL embodies the conception of modern environmental governance, namely that the environment is valuable and those who cause damage shall be held accountable and whoever causes the pollution is the party to be held responsible by. Since civil tort litigation aims at damage compensation, ecological

[9] Gong (2019), pp. 127–147.

damage compensation can be regarded as the main form of action and even the core function of civil public interest litigation.

8.3.2 Negotiation as a Prerequisite for EEDCL in China

According to the *Reform Plan*, negotiation with the obligor ranks first for the obligee to claim compensation for ecological and environmental damage, and it is a procedural prerequisite to filing a lawsuit. Article 1 of the *Several Provisions* (Trial) explicitly stipulates that the compensation obligee "may file EEDCL as the plaintiff if no agreement is reached through negotiation or negotiation is impossible." This clearly suggests that negotiation is the premise and foundation of litigation, and its effectiveness and results will affect the process and effectiveness of the subsequent litigation. Therefore, litigation supports the negotiation of lawsuits on compensation for ecological and environmental damage and the enforcement of the court decision. In this sense, EEDCL can be regarded as a continuation of negotiation.

From the perspective of efficiency, taking negotiation as a prerequisite procedure to file EEDCL is conducive for administrative subjects to give full play to administrative advantages and effect timely ecological restoration. Plaintiffs in EEDCL are mostly governments, government departments, and other administrative organizations. They generally have the power of environmental administration and, therefore, through administrative law enforcement, they can order ecological and environmental polluters to fulfil their obligations and to prevent and repair ecological and environmental damage.[10]

As a flexible law enforcement action, negotiation is a breakthrough and an improvement on traditional administrative-coercive means in social management. It has the features of non-compulsion, flexibility, and selective acceptability and is a process that considers "communication" as the premise and "discourse" as its core. If the compensation obligee and obligator can reach an agreement through negotiation, then there is no need to initiate a proceeding. This not only saves litigation costs but also facilitates the starting of ecological and environmental restoration and governance at the earliest opportunity possible, which undoubtedly complies with the principle of efficiency to the maximum extent. If no agreement is reached through negotiation or the obligator fails to fulfil obligations after an agreement is reached, the initiation of proceedings provides a procedural guarantee for compensation, promotes the realization of the value of procedural justice, and conforms to the spirit of the modern rule of law.

[10]Wang (2019), pp. 11–22.

8.3.3 Governments' Top Priority to Claim in Compensation Litigation

In modern society, ecological and environmental protection is an important function of governments. The nature of public goods of ecological and environmental quality can easily result in "market failure," Therefore, it is necessary for governments that represent public interest to step in and help. Article 26 of the *Constitution of the People's Republic of China* clearly provides that "the state protects and improves the living and ecological environment, and prevents and controls pollution and other public hazards." Article 89(6) of the *Constitution* revised in 2018 adds that "leading and managing the development of ecological civilization" is an aspect of State power . This means that governments at all levels and their departments under the leadership of the State enjoy the power of and assume corresponding responsibilities for the development of ecological civilization. Governments have a special advantage in realizing the national environmental protection goal because of their professionalism and initiatives. This explains why they are tasked with "leading and managing the development of ecological civilization."[11]

"Leading by government and priority in administration" is an important feature of the compensation system for ecological and environmental damage.[12] To realize the goal of the system of compensation liability for ecological and environmental damage, governments (and their designated departments and agencies), as the compensation obligees, negotiate with the compensation obligators first and, if negotiation fails or is impossible, additionally pursue litigation as a judicial means to secure relief to ecological and environmental damage. This arrangement is in line with the principle of administrative priority and administrative efficiency. The *Reform Plan* and *Several Provisions (Trial)* are based on this logic. Although, in theory, social organizations and people's procuratorates, as defenders of environmental public interest, are also qualified as the subjects of EEDCL, China's current policies and laws confirm that governments have top priority in the right of claim in such compensation litigation. This means that governments have the standing to file EEDCL first. Environmental protection organizations and people's procuratorates may file EPIL in accordance with relevant laws and regulations only when governments do not file a lawsuit or the damage is beyond the scope of damage claimed by governments.

[11] Chen (2018), pp. 20–27.

[12] Wang (2019), pp. 11–22.

8.4 Components and Operating Mechanisms of EEDCL

8.4.1 Plaintiffs in EEDCL

Several Provisions (Trial) is the main procedural basis for EEDCL. According to Article 1 of the *Several Provisions (Trial)*, plaintiffs who have the right to file EEDCL include: (1) provincial people's governments; (2) relevant departments and agencies designated by provincial people's governments; (3) municipal and prefecture-level people's governments, including people's governments of cities divided into districts, autonomous prefectures, leagues, regions, prefecture-level cities without districts, and districts and counties of municipalities directly under the central government; (4) relevant departments and agencies designated by municipal and prefecture-level people's governments; and (5) departments entrusted by the State Council to exercise the ownership of all natural resource assets of the whole people. However, it is worth noting that there are other plaintiffs who also have the right to file EEDCL as specified in Articles 1234 and 1235 of the *Civil Code.* Those include organs prescribed by the state or organizations prescribed by law. Arguably, this scope is inconsistent with the scope of plaintiffs stated in the *Several Provisions (Trial)* and, therefore, its specific application needs be defined by relevant legislative or judicial interpretation.

8.4.2 Defendants in EEDCL

According to the *Reform Plan*, the *Civil Code,* and the *Several Provisions (Trial)*, natural persons, legal persons, or unincorporated organizations that cause ecological and environmental damage may become defendants in EEDCL.[13]

[13](1) The Reform Plan IV (2) identifies the obligor of compensation. Units or individuals who violate laws and regulations and cause ecological and environmental damage shall bear the liability of compensation for ecological and environmental damage and make full compensation;

(2) Article 2 of the Civil Code stipulates the Civil Law shall adjust the personal and property relations among natural persons, legal persons and non-legal organizations of equal subjects;

(3) Article 1of the "Several provisions (trial)" stipulates one of the following situations, or the government at the provincial or municipal level and specified in the relevant departments, agencies, or entrusted by the State Council, exercise the national department of all natural resources assets ownership with the ecological environmental damage caused by natural persons, legal persons or other organizations failing to reach an agreement through consultation or unable to negotiate, Can act as a plaintiff to file a lawsuit for ecological and environmental damage:. . . . . .

8.4.3 Conditions for Initiating EEDCL

According to the *Several Provisions (Trial)*, a plaintiff may initiate EEDCL under the following circumstances: (1) relatively major, major, or particularly major environmental emergencies; (2) environmental pollution and ecological disruption incidents in key ecological functional zones and development prohibited zones delineated in national and provincial main functional zone planning; and (3) other circumstances with serious ecological and environmental consequences. However, the *Several Provisions (Trial)* do not apply to personal injury claims, or personal and collective property loss caused by environmental pollution or ecological disruption, or marine ecological and environmental damage claims. This means that a plaintiff cannot file EEDCL under these circumstances.

8.4.4 Negotiation as a Prerequisite Procedure for EEDCL

The *Reform Plan* encourages active negotiation between the parties prior to the formal court decision. According to this principle, in the case of ecological and environmental damage, the compensation obligee should organize the investigation, appraisal, and assessment of ecological and environmental damage as well as the preparation of restoration plans and take the initiative to negotiate with the compensation obligator. Where no agreement is reached through negotiation, the compensation obligee can file a lawsuit according to law. Article 1 of the *Several Provisions (Trial)* also makes it clear that the compensation obligee can file EEDCL as the plaintiff only if no agreement is reached through negotiation (or if negotiation is impossible) with natural persons, legal persons, or unincorporated organizations that cause ecological and environmental damage. It should be noted that negotiation is a prerequisite procedure for the compensation obligee to file EEDCL.

Negotiation may have two results. One is that, if successful, an agreement on compensation for ecological and environmental damage is reached and the parties concerned can apply to a people's court for judicial confirmation. After accepting the application, the people's court should publish the agreement for a period of not less than 30 days. Once the period expires, if the people's court, after examination, considers that the agreement neither violates the mandatory provisions of laws and regulations nor harms national or public interests, it will rule that the agreement is valid. The ruling, which should be in writing, explicitly states the basic facts of the case and the content of the agreement and is made public. After the agreement comes into effect, both parties are required to perform their obligations under the agreement. If one party fails to perform their obligations, in whole or in part, the other party can apply to the people's court for enforcement. Ecological and environmental restoration (where necessary) should be organized by people's governments at the provincial, municipal, and prefecture levels with their designated departments or agencies. The other result is that negotiation fails or is impossible. Under these circumstances, the compensation obligee can file EEDCL.

8.4.5 Claims in EEDCL

According to the *Civil Code* and *Several Provisions (Trial)*, if the ecology and environment are restorable, the plaintiff has the right to request that the defendant to undertake repairs within a reasonable time. Where the defendant fails to repair within the time limit, the plaintiff may restore by themselves or others they entrust, with the expenses incurred borne by the defendant. The cost of ecological and environmental restoration covers that of the formulation and implementation of the restoration plan, monitoring and supervision during restoration, post-restoration acceptance, and the evaluation of the restoration effect.[14] In addition to the above claims, the plaintiff has the right to request the defendant to compensate for:[15]

(1) losses caused by loss of service function from the time the ecology and environment are damaged to the time the restoration is completed;
(2) losses caused by permanent damage to ecological and environmental functions;
(3) expenses of investigation, appraisal, and assessment of ecological and environmental damage;
(4) expenses of pollution cleanup and ecological and environmental restoration; and
(5) other reasonable expenses incurred to prevent the occurrence or aggravation of the damage.

---

[14](1) Article 1,234 of the Civil Code stipulates where the violation of State regulations causes damage to the ecological environment and the ecological environment can be repaired, the organ prescribed by the State or the organization prescribed by law shall have the right to request the infringer to assume the responsibility for restoration within a reasonable time limit. If the infringer fails to make repairs within the time limit, the organ prescribed by the state or the organization prescribed by law may make repairs by itself or entrust others, and the expenses required shall be borne by the infringer;

(2) Article 12 of the "Several provisions (trial)" stipulates if the damaged ecological environment can be restored, the people's court shall judge the defendant to bear the restoration liability according to law, and at the same time determine the ecological environment restoration expenses to be borne by the defendant if he fails to perform the restoration obligation.

Ecological environmental restoration costs include the costs of formulating and implementing restoration plans, monitoring and supervision costs during restoration, acceptance costs after completion of restoration, and evaluation costs after restoration effects, etc.

Where the plaintiff requests the defendant to compensate for the loss of service functions during the period from the damage to the ecological environment to the completion of the restoration, the people's court shall make a judgment according to the specific case.

Article 13, Where the damaged ecological environment cannot be repaired or cannot be completely repaired, and the plaintiff requests the defendant to compensate for the loss caused by permanent damage to ecological and environmental functions, the people's court shall make a judgment according to the specific case.

[15]Article 1235, Civil Code of the People's Republic of China.

In addition, the plaintiff may request the defendant to bear the following expenses:[16]

(1) reasonable expenses to implement contingency plans;
(2) inspection expenses for negotiation and litigation of compensation for ecological and environmental damage; and
(3) reasonable attorney fees and other reasonable litigation expenses.

8.4.6 Competent Court for EEDCL

Due to the particularity and complexity of environmental disputes posing challenges to the traditional pattern of trial, to cope with the increasingly complex environmental disputes, in recent years, China has started to vigorously strengthen environmental dispute trial specialization system reform, set up special environment courts, realize the trial mode of "three-trial-unity" or "four-trial-unity", etc. The unified trial of environmental civil, criminal and administrative cases can improve the quality and efficiency of environmental dispute trials with the specialization of the environmental justice principle. Ecological and environmental damages litigation disputes belong to a type of environmental dispute, so the provisions of "*Several Provisions (Trial)*" also comply with the logic of environmental justice specialization to arrange special jurisdiction courts and specific trial courts to conduct trials.

According to Article 3 of the *Several Provisions (Trial),* in a case of EEDCL, the jurisdiction court and Specific court of trial should be determined in accordance with the following rules:

(1) jurisdiction by forum level: a case of first instance of EEDCL shall be under the jurisdiction of the intermediate people's court or above at the place where the act of ecological and environmental damage is committed or damage occurs, or the defendant resides;
(2) designated jurisdiction: with the approval of the SPC, a higher people's court may designate an intermediate people's court within its jurisdiction to have centralized jurisdiction over a case of first instance of EEDCL;
(3) jurisdiction by order: an intermediate people's court may, if it finds it necessary, after applying to the higher people's court for approval, order a case of first instance of EEDCL under its jurisdiction to be turned over to a basic people's court that is qualified for trial; and
(4) trial by a special court: a case of EEDCL shall be tried by the environmental resources tribunal of a people's court or a designated special court.

[16] Article 14, *Several Provisions of the Supreme People's Court on Hearing Cases of Compensation for Ecological and Environmental Damage* (Trial).2020.

8.4.7 *Trial and Judgment of EEDCL*

According to Article 6 of the *Several Provisions (Trial)*, a people's court should hear and determine cases of EEDCL in the following ways.

(1) formation of a collegial panel. The people's court should form a collegial panel composed of judges and people's jurors to try cases of first instance of EEDCL.
(2) burden of proof and determination of the admissibility of evidence. The plaintiff shall bear the burden of proof on the following facts:

    (a) The defendant has committed acts of polluting the environment or damaging the ecology or is involved in other circumstances under which it should bear responsibility according to law;
    (b) Ecological and environmental damage has been caused and specific expenses are required for restoration and compensation; and

        The quantification of the value of ecological environmental damage shall follow the following principles: 1) after the occurrence of environmental pollution or ecological damage, the pollution removal expenses incurred in order to mitigate or eliminate pollution or damage to the ecological environment shall be subject to the actual incurred expenses, and the necessity and rationality of the actual incurred expenses shall be judged; 2) When the damaged ecological environment and its service functions can be restored or partially restored, the ecological environment restoration plan shall be formulated and the value of ecological environment damage shall be quantified by the use of recovery fees; 3) When the damaged ecological environment and its service functions cannot be restored, or can only be partially restored, or cannot compensate for the damage during the period, select other appropriate environmental value assessment methods to quantify the value of the ecological environmental damage of the unrestored part; 4) When the facts of environmental pollution or ecological destruction are clear, but the facts of damage are unclear or the scope and degree of ecological environmental damage cannot be determined at a reasonable cost, the virtual governance cost method is adopted to quantify the value of ecological environmental damage, and the period damage is no longer calculated.[17]
    (c) There is a correlation between the defendant's acts of polluting the environment or damaging the ecology and the ecological and environmental damage.

For the purpose of this provision, the "Certain Provisions (Trial)" is to relax the causality burden of proof for the plaintiff to claim the defendant to bear the liability for ecological and environmental damage. Different from traditional tort liability,

[17] Technical guidelines for identification and assessment of environmental damage—General principles and key components—Part 1: General (GB/T 39791.1-2020), 8.1.p8-9 issued by the Ministry of Ecology and Environment and the State Administration for Market Regulation in 2020.

which requires the plaintiff to prove that there is a strict, direct and inevitable causal relationship between the defendant's illegal act and the damage result, the ecological environmental damage compensation liability loosens the plaintiff's burden of proof and adopts the principle of "burden of proof inversion". Namely, the plaintiff need only prove the defendant environmental pollution, destruction of ecology and ecological environmental damage the relevance between the can, so if the defendant against the plaintiff argues, is the burden of proof to the defendant, it needs to prove that environmental pollution and destruction of ecology and ecological environment there is no causal relationship between damage or have legal immunity.[18] Article 7 of the Several Provisions (Trial) stipulates that "if the defendant refutes the plaintiff's claim, it shall provide evidence to prove it. The defendant shall bear the burden of proof if he claims that there are circumstances of non-liability or diminished liability as stipulated by law."

The reason why the principle of "burden of proof inversion" is adopted is that China's ecological environmental damage compensation liability adopts the principle of no-fault liability in the imputation principle. The so-called no-fault liability means that whether the actor is at fault or not, if the law stipulates that the actor should bear tort liability, the actor should bear tort liability for the damage caused by his act. The no-fault liability principle is relative to the fault liability principle. The fault liability principle is the basic imputation principle in modern tort law, which is generally applicable when the law does not clearly stipulate the application of no-fault liability. Article 1,165 and Article 1,166 of the Civil Code of China stipulate the principle of liability for fault and the conditions for the application of liability without fault.[19] The constituent elements of the fault liability principle include four elements: illegal act, damage fact, causality and subjective fault. The constitutive elements of the principle of liability without fault only include the three elements of illegal behavior, damage fact and causality, excluding the subjective fault element. Different constitutive requirements mean that the plaintiff bears different burden of proof in tort dispute litigation. Article 1229 and Article 1,230 of the Civil Code clearly stipulate the application of no-fault liability for environmental torts.[20]And Article 6 of "several provisions (trial)" further regulates the ecological environment

[18]Ibid.

[19]Article 1,165 of the Civil Code stipulates that where an actor infringes upon the civil rights and interests of another person through fault and causes damage, he shall bear tort liability.

If the actor is presumed to be at fault in accordance with the law, and he cannot prove that he is not at fault, he shall bear tort liability.

Article 1,166: Where an actor causes damage to the civil rights and interests of another person, whether or not the actor is at fault, and the law provides that he shall bear tort liability, such provisions shall prevail.

[20]Article 1229 of the Civil Code stipulates that where environmental pollution or ecological destruction causes damage to others, the infringer shall be liable for tort.

Article 1,230 of the Civil Code provides that in the event of a dispute arising from environmental pollution or ecological destruction, the actor shall bear the burden of proof in respect of the circumstances under which he is not liable or mitigates liability as prescribed by law and the absence of a causal relationship between his act and the damage.

compensation lawsuit, and the plaintiff needs only evidence to prove that the defendant has illegal behavior and damage, and the correlation between the three elements cannot prove that the defendant has subjective fault. The no-fault liability is held by the imputation principle, thus reducing the plaintiff's burden of proof. This is in line with the complexity and indirectness of disputes on compensation liability for ecological and environmental damages.

In terms of the admissibility of evidence, the *Several Provisions (Trial)* has stipulated several special rules: (1) The parties concerned are not required to prove the facts confirmed by a legally effective criminal judgment in a case of EEDCL, unless there is evidence to the contrary that is sufficient to disprove them. For facts not confirmed by a criminal judgment, if the evidence provided by the parties concerned meets the standards of proof in civil proceedings, the people's court should affirm it. (2) Incident investigation reports, inspection reports, detection reports, evaluation reports, monitoring data, etc., produced in the process of administrative law enforcement by a department responsible for it's the supervision and administration of environmental and resource protection or its entrusted agency can be used as the basis for determining the facts of the case if they are cross-examined by the parties concerned and meet the evidence standards; and (3) appraisal opinions given by an appraisal institution qualified for environmental judicial appraisal as entrusted by the parties concerned before proceeding, and inspection reports, detection reports, evaluation reports, monitoring data, etc., produced by an institution recommended by the relevant department competent in the supervision and administration of environmental and resource protection under the State Council, as entrusted by the parties concerned before proceeding, may be used as the basis for determining the facts of the case if they are cross-examined by the parties concerned and meet the evidence standards.

After trial, the people's court makes judgments mainly based on the following rules:

(1) If the defendant violates state regulations and causes ecological and environmental damage, the people's court should, in view of the plaintiff's claim and case details, reasonably direct the defendant to bear civil liabilities for restoring the ecology and environment, compensating for loss, ceasing torts, removing obstacles, eliminating hazards, and apologizing.
(2) If the damaged ecology and environment are restorable, the people's court should direct the defendant to undertake restoration according to law and, at the same time, determine the cost of restoration that the defendant should bear if they fail to fulfil the obligations of restoration. The cost of ecological and environmental restoration covers the formulation and implementation of the restoration plan, monitoring and supervision during restoration, post-restoration acceptance, and the evaluation of the restoration effect. If the plaintiff requests that the defendant provides compensation for the loss of service function from the time when the ecology and environment were damaged to the time when restoration will be completed, the people's court should make a judgment according to the case details.

(3) If the damaged ecology and environment cannot be repaired or cannot be completely repaired, and the plaintiff requests that the defendant compensate for the losses occasioned by the permanent damage to ecological and environmental functions, the people's court should make a judgment according to the case details.
(4) If the plaintiff requests that the defendant bear the following expenses, the people's court should make a judgment according to the case details: a. reasonable expenses to implement contingency plans, de-pollute, and prevent the occurrence or aggravation of damage; b. expenses of investigation, inspection, appraisal, and assessment for negotiation and litigation of compensation for ecological and environmental damage; and c. reasonable attorney fees and other reasonable litigation expenses.

Where the people's court directs the defendant to issue compensation for the loss of ecological and environmental service function and for the losses caused by permanent damage to ecological and environmental functions or to bear the cost of restoration if they fail to fulfil the obligation of ecological and environmental restoration, such compensation and cost should be paid, managed, and used in accordance with existing laws, regulations, and rules.

8.4.8 Enforcement of Judgment in EEDCL

8.4.8.1 Fulfilling Obligations in Cases of Restorable Ecology and Environment

If the court holds that the defendant should restore the ecology and environment, the defendant should restore it in a timely manner and in accordance with the judgment. The general enforcement process is as follows: (1) preparing an ecological and environmental restoration plan, organizing experts to demonstrate the plan, and after the plan is demonstrated, reporting it to the plaintiff for future reference; (2) carrying out ecological and environmental restoration as per the demonstrated plan; and (3) the plaintiff assessing the effect of the restoration.[21] If, according to the assessment, the restoration is not completed as required, the defendant should continue to restore. Organizations making the ecological and environmental restoration plans, the effect assessment reports and termination reports should fully demonstrate their scientific nature, rationality, and validity and be responsible for their authenticity and legality. The law also requires that the effect assessment report and termination report on ecological and environmental restoration should not be issued by the same organization that prepares the ecological and environmental restoration plan. If it is necessary to adjust the restoration plan in the course of the

[21] *The Handling Rules for Ecological and Environmental Damage Compensation Cases of Guizhou Province* (Trial) (Qian Huanfa (2021) No. 1).

ecological and environmental restoration, the defendant should report changes to the plaintiff in a timely manner and major changes should be reviewed.

8.4.8.2 Fulfilling Obligations in Cases of Unrestorable Ecology and Environment

The non-restorability of the ecology and the environment is both subjective and objective. Subjective non-restorability means that the ecology and environment are restorable, but the defendant does not perform their obligations of restoration in whole or in part. Objective non-restorability means that the ecological and environmental damage is too serious to be restored or completely restored, including the possibility that the ecological and environmental damage cannot be restored due to force majeure in the process of restoration.

In the event of subjective non-restorability, there are usually two ways of compelling defendants to fulfil their obligations. One is to submit to the court for enforcement. In such circumstances, the process applicable to restorability should be followed. The other is to request the defendant to pay damages for the ecological and environmental damage. If the defendant neither fulfills the obligation of restoration nor pays the damages, the plaintiff can apply to the people's court for enforcement. In practice, to urge defendants to actively fulfil their obligation of restoration, some provinces have established an enterprise environmental credit rating system, and the obligators who failed to implement valid court or administrative decisions are included in the blacklist of dishonesty. For example, Guizhou province has issued the *Measures of Guizhou Province for the Administration of Blacklist of Dishonesty in Ecological and Environmental Protection* (QHF [2021] No. 1), which stipulates that the compensation obligators who fail to implement valid court decisions on compensation for ecological and environmental damage will be listed in the blacklist of dishonesty and included in the environmental credit rating system for management. Once listed on the blacklist, the compensation obligators will be likely to receive joint discipline from over 30 of the competent administrative departments, including market access to restrict or prohibit the defendant, administrative licensing or financing behavior, stopping to enjoy preferential policies, or not approving its application for preferential policies. "Blacklist" information will also be pushed to the National Credit Information Sharing Platform (Guizhou), the National Enterprise Credit Information Publicity System (Guizhou) and the basic database of financial Credit Information of the People's Bank of China in accordance with the law. The compensation obligators will be restricted in applying for qualification assessment, project approval, land approval, policy preference and financial support in future government procurement, project bidding, credit guarantee and other activities.[22]

[22] *Measures for the Administration of Blacklist of Trust-Breaking in Ecological and Environmental Protection of Guizhou Province* (Qian Huantong (2019) No. 178).

The environmental credit rating system has greatly encouraged defendants to actively fulfil the obligation of restoration.

In the event of objective non-restorability, the defendant usually fulfils their obligations in the following ways: (1) the plaintiff urges the defendant to carry out alternative restoration in accordance with the judgment and based on the ecological and environmental damage in the region; (2) the defendant pays the compensation for ecological and environmental damage, and then the plaintiff carries out alternative restoration by themselves or third parties they entrust; and (3) where the ecological and environmental damage cannot be restored due to force majeure, the defendant or the third-party organization notifies the plaintiff within a reasonable period of time and provides a termination report on ecological and environmental restoration. After the plaintiff organizes assessment for confirmation, the termination of restoration is published in accordance with the provisions of environmental information disclosure.

8.5 Coordination Between EEDCL and EPIL

EEDCL and environmental civil public interest litigation are of the same nature and both are civil public interest litigation. The legal relationship between them can be identified as 'special and general'. When dealing with ecological and environmental damage cases, special litigation should be considered first. In the context of EEDCL and environmental civil public interest litigation, the *Several Provisions (Trial)* sets out several rules to address such priority.

8.5.1 Trial of the Two Categories of Cases by the Same Judicial Organization

Article 16 of the *Several Provisions (Trial)* stipulates that during the trial of a case of EEDCL, if the same act that damages the ecology and environment is subject to a civil public interest lawsuit and the conditions for prosecution are met, the lawsuit should be accepted by the people's court accepting the case of EEDCL and tried by the same judicial organization. This provision aims to facilitate the unification of judgment standards and save judicial resources.

8.5.2 Trial Order of the Two Categories of Cases

To avoid the relevant civil subjects being repeatedly held accountable for the same act of damaging the ecology and environment and to save judicial resources, Article

17 of the *Several Provisions (Trial)* defines the trial order of the two categories of litigation cases. That is, if EEDCL and the civil public interest litigation case were filed against the same act, the people's court accepting the case should suspend the trial of the civil public interest litigation case, complete the trial of the case of EEDCL, and then make a judgment on the claims not covered by the civil public interest litigation case.

8.5.3 Post-trial Rules

To equally protect the right of action of all kinds of subjects and ensure that ecological and environmental damage is curbed, Article 18 of the *Several Provisions (Trial)* has defined the rules of coordination under the two categories of litigation cases. One is that after the judgment of the case of EEDCL takes effect, if the organ or social organization with the right to file environmental civil public interest litigation has evidence to prove the existence of damage not found in the previous trial on the same act of damaging the ecology and environment and proceeds to file environmental civil public interest litigation, the people's court should accept the litigation. The other is that after the judgment of the environmental civil public interest litigation case takes effect, if the subject with the right to file EEDCL has evidence to prove the existence of damage not found in the previous trial on the same act of damaging the ecology and environment and proceeds to file EEDCL, the people's court should accept the litigation.

8.6 Conclusion

The compensation litigation system for ecological and environmental damage is a special environmental civil public interest litigation system with Chinese characteristics. This chapter explored the particularity of this litigation system from four perspectives: creation and development, nature and characteristics, components and operating mechanism, and its coordination with EPIL. In terms of creation and development, this system stems from the active exploration of the party and the state to promote the development of ecological civilization and from China's rich practice in building the ecological and environmental rule of law.

In terms of nature and characteristics, this litigation is a special environmental civil public interest litigation. In this system, plaintiffs, such as governments at the prefecture and municipal levels and their departments and agencies that have environmental administrative authority, have the mandate to exercise administrative public power. In designing the system, a negotiation preposing procedure is set up in accordance with the principle of administrative priority and administrative efficiency, and governments have top priority in the right of claim in compensation litigation. The condition for initiating such litigation follows the principle of "consequentialism," that is, such litigation is not against all acts damaging the ecology and environment, and it can be initiated only when a certain act damaging the ecology and environment has serious ecological and environmental consequences.

Moreover, such litigation can only claim compensation for the damage specifically occasioned to the ecology and environment (i.e., adverse changes in environmental and biological factors and degradation of ecosystem functions). It is not applicable to personal injury and property loss caused by environmental pollution or ecological disruption or to marine ecological and environmental damage. Specifically, the competent court differs from ordinary civil disputes that are generally under the jurisdiction of a basic people's court. The first instance of such litigation is directly under the jurisdiction of an intermediate people's court. For the principle of imputation, the principle of no-fault liability applies. In addition, the theory of probable causal relationships has been adopted so that a plaintiff only needs to prove the correlation between the defendant's act of polluting the environment and destroying the ecology, and the ecological and environmental damage. No direct causal relationship is needed, so this has greatly lowered the threshold of the plaintiff's burden of proof. Finally, when the form of liability is considered, there are two main forms, namely, ecological and environmental restoration per se and compensation for ecological and environmental damage.

In regard to the coordination with environmental civil public interest litigation, the principle of "special litigation first, then the general" is to be followed, and the unification of judgment standards and the saving of judicial resources are taken into consideration based on a case-by-case analysis. The creation and development of the compensation litigation system for ecological and environmental damage is deeply influenced by the comprehensive deepening reform of the environmental rule of law in contemporary China and more directly reflects China's progress in environmental legislation and justice in recent years. Nevertheless, this system still needs to be tested and perfected in practice.

References

Chen H (2018) Reflection and reconstruction of the ecological and environmental damage compensation system: from the perspective of constitutional interpretation. Oriental Law 6:20–27

Gong G (2019) Reflections on the positioning of the nature of environmental civil public interest litigation. Chin J Law 3:127–147

The SPC (2020) Environment and Resources Adjudication in China (2019). People's Court Press. September 2020, p 2

Wang X, Wei W, Liu X, Liu H (2019) Understanding and application of "several provisions (trial)" on hearing cases of compensation for ecological and environmental damage. People's Judicature 34:31–38

Wang X, Liu Y, Lin X, Make the clear waters and lush mountains benefit the people and benefit their children and grandchildren, Summary of General Secretary Xi Jinping's important discourse on ecological civilization construction, Xinhua News Agency, People's Daily, 2021-06-03, http://v.sjzntv.cn/sy/folder1733/folder1830/2021-06-03/796336.html

Chapter 9
The Role of NGOs in China's Environmental Public Interest Litigation

Mengxing Lu

Abstract As a major legal development in China's environmental law, public interest litigation (hereinafter PIL) has achieved impressive progress in boosting public participation and enhancing environmental law enforcement. This chapter aims to offer a rigorous discussion and analysis on the role of NGOs in China's EPIL. It argues that this topic deserves more attention due to the increasing importance of environmental groups in China's judicial system and the relative lack of relevant studies in this domain. More particularly, the rise and fall of NGOs in vindicating public environmental interest is worth further study. This chapter contributes to the current discussion of EPIL, by exploring the reasons behind the rise and fall of the rule of NGOs in EPIL from theoretical and empirical perspectives.

Keywords NGOs · Environmental law enforcement · Environmental public interest litigation · Environmental governance

9.1 Introduction

As the negative effects of environmental pollution and ecological damage have affected public health, economic growth, social stability and the international reputation of China, recent decades have witnessed many impressive strides toward addressing these environmental challenges.[1] After four decades of efforts,[2] China has created a comprehensive legal and institutional framework for environmental

[1] See Economy (2014), pp. 184–197.

[2] The year of 2019 marked the 40th anniversary of the implementation of the Environmental Protection Law in China, when the first nationwide legislation on environmental protection (for trial implementation) was enacted in 1979.

M. Lu (✉)
Civil Commercial and Economic Law School, China University of Political Science and Law in Beijing, Beijing, China
e-mail: mengxing.lu@cupl.edu.cn

X. Wang et al. (eds.), *Environmental Public Interest Litigation in China*,
https://doi.org/10.1007/978-3-031-26526-6_9

governance, with 35 laws, 72 administrative regulations and 17 interpretations at the central level,[3] covering a broad range of environmental issues, including pollution control, conservation and utilization of natural resources and energy, and ecological restoration.

However, rigorous efforts of the Chinese legislators have not sufficiently improved environmental conditions to meet the increasing public demand for a cleaner and healthier environment. According to the annual reports on national environmental conditions published by the Ministry of Ecology and Environment (MEE), the overall environmental conditions in China remain problematic and are even worsening in some areas. As one of most urgent environmental issues, the MEE keeps monitoring the water quality of 1937 spots across the country and the figures in general indicate an improvement in China's water quality during past years. For example, the data show that in 2020, the percentage of monitoring sections that met the classes I~III of water quality accounted for 83.4% of the total surveyed sections, growing by 8.5% compared with 2019.[4] However, it can be observed that only 7.3% of the surveyed sections met the highest standard of water quality (Class I).[5] The occurrence of acid rain is also monitored in 470 cities in China. According to the reports, there has been a remarkable growth in the percentage of cities where acid rain occurs from 19.8% in 2016[6] to 36.1% in 2017.[7]

The data above indicate that the effectiveness of China's environmental legislation and regulation has been modest at best. The crux of China's environmental policy is non-compliance and weak enforcement. This has been viewed as the biggest challenge in an extensive body of literature.[8] Many reasons have been given for the relative failure of China's environmental legislation, including the conflict between environmental protection and economic and political targets, the constraints of personnel and financial capacity to enforce environmental laws, the limited access to environmental technology, and the lack of public participation.[9] More particularly, it has been agreed in both Chinese and English literature that the lack of a robust civil society is a major problem in China's environmental governance compared to its Western counterparts.[10]

---

[3]The author counted the number of environmental laws issued by the Standing Committee of the National People's Congress, environmental regulations issued by the State Council and the Ministry of Environment and Ecology, interpretations issued by the Supreme People's courts and the Supreme People's Procuratorate.

[4]Report on National Environmental Condition (2020), pp. 18–19.

[5]According to the Environmental Standards for Surface Water (GB3838-2002) issued by the MEE, the quality of surface water is categorized into classes I to V, arranged from the highest standard to the lowest. Class I represents the highest water standard that is applied to the source of water and national nature reserve areas.

[6]Report on National Environmental Condition (2016), p. 14.

[7]Report on National Environmental Condition (2017), p. 14.

[8]See Wang (2007b), pp. 159–193; Wang (2007a), pp. 196–223; Khan and Chang (2018), p. 2547.

[9]See Wang (2007b).

[10]See Gao and Whittaker (2019), p. 342; Economy (2014), p. 190.

As a response, many efforts have been made by the Chinese leadership to boost public participation in the field of environmental protection. As one of the major legal achievements of China's environmental law in recent years, public interest litigation (hereinafter PIL) has raised public expectations as an important step in empowering civil society by providing a platform for safeguarding public environmental interests and enhancing environmental enforcement.[11] Accordingly, the theory and practice of PIL in China has attracted broad attention from both Chinese and Western scholars.

After nearly a decade of implementation, PIL has become an important part of China's environmental litigation, as China has accumulated abundant experience in exploring the role of the judicial system in the domain of environmental protection. Therefore, this chapter contributes to the current discussion concerning PIL in China by providing a holistic and integrated analysis of PIL by non-governmental organizations ("NGOs") from both normative and empirical perspectives. The key question to be addressed in this contribution is what roles NGOs play in China's environmental governance structure and in China's judicial system. The remainder of this chapter is structured as follows: Sect. 9.2 provides an overview of the legal framework for remedying environmental damage under China's authoritarian political context and depicts possible approaches that could be employed to safeguard environmental interests. Section 9.3 discusses the history of the intervention of NGOs in PIL and details particular experiments undertaken and the subsequent legislative changes. Section 9.4 provides a numerical overview of the use of PIL by NGOs. That will show that the use of PIL by NGOs first increased, and has since declined. Section 9.5 tries to address the reasons for that recent decline. Section 9.6 summaries the material that has been discussed in this chapter and suggests areas for future research.

9.2 The Legal Framework for Remedying Ecological Damage in China

In the face of increasing environmental pollution and ecological damage, the remedy for such damage has become a priority issue in China's policy agenda. According to the Reform Plan on an Ecological Environmental Damage Compensation System (hereinafter the Reform Plan),[12] ecological environmental damage[13] refers to the negative effects on environmental elements (such as air, surface water, underground

---

[11] See Gao and Whittaker (2019), p. 328.

[12] Reform Plan on an Ecological Environmental Damage Compensation System (2017).

[13] In the official document, the concept of ecological environmental damage (*shengtai huanjing sunhai*) is used. In some studies concepts like environmental damage or ecological damage are also employed. To avoid confusion, the concepts of ecological damage and environmental damage will be interchangeably used in this article. See Zhang and Mayer (2017), pp. 207–208.

water, soil and forest) and on the functions of an ecosystem, caused by environmental pollution or ecological damage incidents.[14] Pollution incidents can, together with overall ecological damage, also cause personal injury and property damage.[15] Ecological damage, by its nature, also represents damage to the overall public interest. This fact distinguishes ecological damage from the violation of private interests, such as health and property. Traditionally, China's environmental law places more emphasis on the compensation of individual loss caused by environmental pollution through the civil liability regime (more particularly tort law).[16] However, given the widespread and latent nature of environmental damage, the tort-based regime obviously falls short of remedying the damaged environment in many circumstances. In many pollution cases, there is a lack of private interest in initiating tort actions, since the personal loss is so diffuse and minimal compared with the high litigation costs, or there is a significant time gap between the commission of illicit activities and the occurrence of environmental harm.[17] Consequently, it is rare for tort litigation actions to be filed against polluters. Another notable problem of the traditional tort-based liability regime is that it is designed for individual damage compensation, not for the restoration of systemic environmental damage.[18] In certain cases, victims have been compensated on the basis of their individual or property loss, but the environmental damage is left unrestored. As a result, there is tension and even conflict between the focus on individual rights protection and the necessity of remedying environmental damage under the traditional torts-based regime.

As a response, the remedy of environmental damage has become more prominent on the policy agenda of the Chinese leadership, as the Chinese Communist Party (CPC) made a strong pledge to develop China into 'a great modern socialist country that is prosperous, strong, democratic, culturally advanced and beautiful' by the mid-twenty-first century.[19] In order to meet the growing public demand for a cleaner and healthier environment, China has made many attempts to reinforce the top-level policy design of environmental governance systems in many aspects. Although China's environmental protection continues to heavily rely on command-and-control regulations and regulatory agencies as the incarnation of public interest, defenders remain the dominant position. However, they are no longer the sole actors within China's environmental governance structure.[20] The Chinese government carried out the so-called modern environmental governance system reform, by empowering the involvement of non-state actors, and by enhancing information

[14] Reform Plan on Ecological Environmental Damage Compensation System (2017).

[15] Several concepts are interchangeably used in official documents and in the literature, such as environmental damage, ecological damage, and ecological environmental damage. To avoid potential confusion, ecological damage is used throughout this article.

[16] Lu and Faure (2020), pp. 136, 153.

[17] Faure and Partain (2019), p. 188.

[18] Goldman (2007), p. 255.

[19] Xi (2017).

[20] Guttman et al. (2018), pp. 126–135.

disclosure and transparency. In 2020, the General Office of the CPC and the General Office of the State Council jointly issued the Guiding Opinions on Establishing a Modern Environmental Governance System (hereinafter the Guidance).[21] According to this Guidance, a remarkable feature of a modern environmental governance system is the involvement of multiple actors, including governments, enterprises and the public. Due to the uncertainty and plural nature of environmental interests, a consensus has been reached in both policy and academic circles that the remediation of ecological damage cannot be achieved through a single approach.[22] This has been substantiated by China's recent legislative developments in environmental litigation.

During the past few years, a wide range of possibilities for non-state actors to vindicate environmental interests has been established through a series of legislative actions, including the amendments to the Civil Procedure Law,[23] the Administrative Procedure Law,[24] the Environmental Protection Law,[25] the enactments of the Reform Plan, the Civil Code[26] and the relevant judicial interpretations issued by the Supreme People's Court (SPC) and the Supreme People's Procuratorate (SPP).[27] Thanks to the concerted efforts of Chinese legislators, five pathways are now available for remedying ecological damage under the current legal framework (see Table 9.1): (1) administrative enforcement; (2) environmental PIL by NGOs; (3) environmental PIL by procurators; (4) criminal enforcement by procurators; and (5) ecological damage compensation litigation by the competent authorities. These five enforcement regimes can be further categorized depending on different criteria. Examples include administrative relief and judicial relief; or NGOs, procurators and public authorities (a categorization based on the party who has initiated the action).

As explicitly shown in Table 9.1, over the last decade, China has established a pluralistic legal framework for remedying ecological damage where state and non-state actors (mainly represented by NGOs), have become integral pillars of China's modern environmental governance system featuring multi-stakeholder participation.[28] This regulatory design has the advantage of providing a wide range of enforcement possibilities, thus enhancing environmental law enforcement and better safeguarding ecological interests. Particularly, the adoption of PIL is expected to provide an alternative (private) mechanism for non-state actors to enforce

[21] Guiding Opinions on Establishing Modern Environmental Governance System (2020).

[22] Hu (2016), p. 173.

[23] Civil Procedure Law (2017 amendments), Chairman Order No. 71, 1 July 2017, art. 55.

[24] Administrative Procedure Law (2017 amendments), Chairman Order No. 71, 1 July 2017, art. 25, para. 3.

[25] Environmental Protection Law (2014 amendments), Chairman Order No. 9, 1 January 2015, art. 58.

[26] Civil Code, Chairman Order No. 45, 1 January 2021, art. 1234 and 1235.

[27] See Table 9.1.

[28] General Office of the CPC Central Committee and General Office of the State Council (2020).

Table 9.1 Legal framework for remedying ecological damage

	Types of legal pathways	Enforcers	Scope of application	Legal sources
1	Administrative enforcement	Environmental protection bureau (EPBs) and other relevant agencies	Sanctioning polluters for violation of environmental regulation by using administrative instruments	EPL (2014); Specific environmental legislations at both national and local levels (soil pollution law, air pollution law, etc.)
2	Environmental PIL by NGOs	Environmental NGOs	Suing polluters for causing environmental damages through the judicial system	Civil Procedure Law (2012/2017); EPL (2014); Judicial Interpretations on environmental PIL (2015); Civil Code (2021)
3	Environmental PIL by procurators	Procurators	Civil PIL: suing polluters for causing environmental damages through the judicial system	Civil Procedure Law (2017); Judicial Interpretations on environmental PIL by procurators (2018)
			Administrative PIL: suing public authorities for illegal actions or inactions through judicial system	Administrative Procedure Law (2015); Judicial Interpretations on environmental PIL by procurators (2018)
4	Ecological damage compensation litigation by public authorities	Local governments at provincial and municipal levels	Suing polluters for causing environmental damage on behalf of state interest	Reform Plan on Ecological Environmental Damage Compensation System (2017); Judicial Interpretation on Ecological Environmental Damage Compensation Litigation (2020); Civil Code (2021)
5	Criminal enforcement by procurators (civil PIL incidental to criminal proceedings)	Procurators	Prosecuting environmental criminals while bring an incidental PIL	Criminal Law (1997) Official Reply of the SPC and the SPP on Whether the Pre-litigation Public Announcement Procedures Shall Be Performed for a Civil Public Interest Litigation Incidental to Criminal Proceedings Filed by a People's Procuratorate (2019)

environmental laws, particularly when government regulation may fail for various reasons or when the tort actions cannot be initiated due to the lack of direct private interest.[29] However, the potential danger of this regime is that there may be overlaps and conflicts between various actors who have a willingness to remedy the ecological damage through different approaches. One can suppose that when an environmental pollution incident occurs, NGOs are allowed to bring a PIL, while competent authorities (such as Environmental Protection Bureaus, EPBs) could force the polluters to take removal and remedial actions through administrative orders. This situation is not hypothetical, but exactly happens in practice, for example, in the Changzhou soil pollution case.[30] On the one hand, it results in the distortion between administrative and judicial powers, since it is argued that it is inappropriate for the courts to make judgements on the remedy of environmental damage. Judges generally lack relevant expertise on highly technical issues, compared to the officials from EPBs. Moreover, the courts are also facing strong criticism for intervening in the domain of administrative powers, as environmental protection is traditionally deemed as a state obligation, clearly written in the Constitutions of many jurisdictions.[31] This brief overview shows that in China, a pluralistic system is in place, requiring a coordination between the administrative branch, the judicial branch, the prosecutorial branch and civil society for its effectiveness. Table 9.1 shows that all of these parties can, in principle, play a role in seeking remedies for environmental damage. In the following sections there will now be a focus on the specific role of NGOs in remedying ecological damage; particular attention will also be paid to their interaction with the other actors in environmental enforcement.

9.3 The Rise of NGOs Before the Courts in China

9.3.1 Testing the Water: Local Experiments of PIL Before 2012

From a historical perspective, the idea of NGOs or social organizations[32] is not new to Chinese society, as some scholars traced the origins of NGOs in China back to the Qing Dynasty.[33] It is observed that the development of NGOs has a turbulent history

[29] Cao and Wang (2011), pp. 217, 211.

[30] Friends of Nature (FON) and China Biodiversity Conservation and Green Development Foundation (CBCGDF) v. Jiangsu Changlong Chemical Ltd. and Changzhou Changyu Chemical Ltd, Second Instance, Jiangsu High People's Court, (2017) SuMinzhong No. 232.

[31] See Wang (2016), pp. 49–68.

[32] In China's legal and policy documents, China has several types of organizations referred to as NGOs in English, such as social organizations, foundations, and non-enterprise units. To avoid confusion, the concept of NGOs is used in this chapter.

[33] Simon (2013).

in China and the roles of NGOs under China's authoritarian context have experienced dramatic changes during the past decades.[34] At the very beginning, NGOs were deemed a foe, threatening the ruling legitimacy of the state, and consequently, their activities were under very strict control and supervision.[35] The recent decades witnessed a change in attitude by the Chinese government towards NGOs from confrontation to cooperation, particularly in the field of environmental protection. Many explanations could be provided to justify this dramatic shift, particularly as Chinese policymakers have widely realized their limitations in the face of worsening environmental conditions. Chinese leadership tends to be more tolerant of the rise and role of environmental NGOs compared to those that focus on other sensitive issues, and environmental NGOs are among the most active and influential ones in China.[36] With the first environmental NGO (Friends of Nature) founded in 1993,[37] the number of NGOs has increased significantly in China.[38] It was reported that by the end of 2005 there were 2768 environmental NGOs in China[39] and the number reached 7881 in 2013.[40] Despite the rapid growth in numbers, the impact of NGOs on the government as well as the public has become increasingly visible, ranging from organizing environmental education programs to raising public awareness, participating in the drafting of environmental protection legislation, exposing illegal business activities and administrative actions or omissions, to bringing a lawsuit before the court on behalf of a public environmental interest.[41]

Until the early part of this century, China remained quite cautious and kept NGOs out of the courts on the grounds of the lack of legal standing. In fact, before PIL was officially incorporated into the Civil Procedural Law in 2012, both policymakers and scholars started to explore the feasibility of establishing effective PIL in China. The year 2005 is regarded as a landmark in the development of environmental PIL in China. On 3 December 2005, the State Council issued the Decision on the Implementation of the Outlook on Scientific Development and Strengthening Environmental Protection, which explicitly called for strengthening environmental law enforcement by "allowing social groups to play their role, encouraging reporting on and exposing illegal environmental acts, and promoting environmental public interest litigation".[42] Just a few days later, on 7 December 2005, six environmental

[34] Percival and Zhao (2014), p. 146.

[35] Simon (2013).

[36] Percival and Zhao (2014), p. 152.

[37] See Schwartz (2007), pp. 28–49. In some literature, China's first environmental NGOs is regarded to be the Chinese Society for Environmental Sciences, founded in 1978, which is a research-oriented institution rather than an activity-oriented organization.

[38] Percival and Zhao (2014), p. 146.

[39] State Administration of Environmental Protection (2007), p. 23.

[40] People's Daily (2013).

[41] Percival and Zhao (2014), p. 146.

[42] State Council, Decision on the Implementation of the Outlook on Scientific Development and Strengthening Environmental Protection, 3 December 2005, art 27.

scholars and students from Peking University filed a lawsuit against the Jilin branch of the China National Petroleum Corporation (CNPC) for the ecological damage caused by an explosion in a chemical factory.[43] This case quickly attracted social media attention due to the appearance of three non-human plaintiffs: sturgeon fish, Songhua River and Taiyang Island. Although this case was rejected by the court due to a lack of legal standing,[44] it became a stark example of the difficulties of a PIL system which was not underpinned by a legal standing proposition.

Another important impetus for China's PIL development was the pilot trial of PIL cases by local specified environmental courts.[45] Since the first environmental court was established in Qingzhen of Guizhou Province in 2007, environmental courts or tribunals have proliferated around the country in recent years,[46] which provides an important avenue for NGOs to bring PIL cases.[47] For example, in China's first environmental PIL case,[48] the Wuxi Intermediate Court accepted the case brought by an individual victim and an NGO (All-China Environment Federation, ACEF) against a container company for causing air and water pollution. This case was later cited as one of the guiding environmental cases by the SPC due to its important attempt to liberalize the standing of NGOs before the courts.[49] In the following years, several environmental PIL cases were accepted and trialed by the local environmental courts located in Guiyang, Wuxi and Changzhou. In addition to judicial adjudication practices, local authorities were also active in carrying out important attempts at establishing PIL by issuing guidelines and policy initiatives, making a sharp contrast to the precautious approach of the central government in this domain. For example, the Kunming Intermediate Court and Kunming Procuratorate jointly issued the Opinions on the Trial of Civil Public Interest Litigation Cases in 2010.[50]

It is generally observed that during this period, the local judicial systems, especially the specified environmental courts, were relatively open to PIL cases. On the one hand, all these local endeavors and experiments are beneficial for

[43] Wang et al. (2008), pp. 273–290.

[44] Wang et al. (2008). It was reported that this case was quickly refused by the High People's Court of Heilongjiang Province on the ground that the plaintiffs did not have direct interests, but this decision was delivered orally, not in a written statement.

[45] Capenter-Gold (2015), pp. 258–260.

[46] According to the report on Environmental and Resources Adjudication in China (2019), by the end of 2019, there were 1353 specified environmental adjudication organs in China, including 513 environmental tribunals, 749 collegial panels and 91 environmental courts.

[47] Wang and Gao (2010), p. 377.

[48] *Zhu and All-China Environment Federation (ACEF) v. Jiangyin Port Container Ltd.*, www.chinacourt.org, https://www.chinacourt.org/article/detail/2014/07/id/1329692.shtml.

[49] Stern (2014), p. 64.

[50] Interpretations of the SPC on Several Issues concerning the Application of Law in the Conduct of Environmental Civil Public Interest Litigations (first issued in 2015 and revised in 2020), Interpretation No. 20 of the SPC.

promoting PIL practices, and on the other hand, they also accumulate valuable experience for preparing PIL-related legislation at the central level.

9.3.2 *Locus Standi of NGOs Before the Courts*

The standing of NGOs before the court has been a key issue when establishing the legal framework for PIL. Deeply influenced by civil law traditions, there is a strict division between public and private laws. The starting point of China's environmental liability regime is established on the private law or right-based approach, which means that pollution victims who want to bring a suit for compensation need to demonstrate the infringement of individual interests caused by polluting activities. As a result, NGOs were denied access to the court due to lack of legal standing simply because they are not the direct victim of environmental pollution or ecological damage incidents. Traditionally, the protection of the public interest has been regarded as a state duty and hence should not be remedied through private laws. Therefore, the kernel of establishing PIL was to liberalize the 'direct interest' standard in determining the plaintiff's standing.

In the course of the development of PIL in China, a watershed moment was the second amendment of the Civil Protection Law in 2012, in which NGO's standing to sue was finally legalized in a national law promulgated by China's top legislative body—the National People's Congress (NPC). According to Article 55 of the Civil Procedure Law (2012), administrative organs prescribed by law and other relevant organizations were allowed to initiate PIL actions on environmental grounds. This provision has been regarded as a major breakthrough in the Chinese legal system by expanding the plaintiffs' standing, even in the absence of a direct interest. Although the decision to grant legal standing to NGOs was eagerly awaited, the expected outburst of PIL cases flooding into the courts did not occur after the adoption of the Civil Procedure Law. It was reported that in 2013, no PIL case was accepted by the court.[51] An important reason behind this interesting phenomenon was the ambiguity as to which organizations were qualified to bring a PIL case as prescribed by Article 55 of the Civil Procedure Law. Accordingly, the qualifications of NGOs became the focal point when the courts decided whether to accept or refuse PIL by NGOs. Intriguingly, the courts that, before 2012, used to be friendly to PIL by NGOs changed their attitudes towards PIL and were more conservative in interpreting the PIL provision.[52] For instance, as one of the most prominent environmental NGOs in China, ACEF brought eight environmental PIL cases in 2013, but none of these cases were accepted by the courts.[53]

[51] Li (2016), p. 258.

[52] Capenter-Gold (2015), p. 264.

[53] Ibid.

Therefore, the further clarification of NGOs' standing in PIL became a key issue in the revision of the EPL, which started in 2013. During the drafting process, there were significant divergences regarding what type of organizations would be eligible to bring a PIL, which can be proven by the unusual four review drafts of the EPL instead of the regular three times as stipulated by the Legislation Law.[54] In the second review draft of the EPL, the Chinese legislators apparently wanted to restrict the standing to sue to a single established NGO, by prescribing that the ACEF and its branches established in provinces, autonomous regions, and municipalities be permitted to bring a PIL for polluting activities that harm the public interest.[55] Not surprisingly, this provision was strongly opposed due to the simple fact that it would exclude nearly all NGOs from initiating PIL actions. In the third review draft, legislators put forwards three conditions to determine the standing of NGOs before the court: first, registration with the civil affairs department under the State Council (i.e., Ministry of Civil Affairs); second, specialization in the field of environmental protection for more than five consecutive years; third, a good reputation.[56] It was said that around the country no more than ten NGOs would be qualified to bring a PIL.[57] This also greatly reduced the possibilities of the competent NGOs in PIL. In the final review draft, which was later passed at the 8th session of the standing Committee of the 12th NPC, the legislators further loosened the grip on the restrictions of NGOs. According to Article 58 of the EPL, qualified organizations were only required to meet two requirements: registration with civil affairs departments at the municipal level or above and being a specialist in the field of environmental protection for more than five consecutive years and having no record of violations. This is now formulated in the current text that was adopted in the newly revised EPL. Consequently, the standing of NGOs is further relaxed under the new amendment to the EPL. It is estimated that nearly 700 environmental NGOs are now eligible to bring a PIL, with a more balanced geographic distribution. This is expected to provide full coverage of environmental interest protection, given China's vast territory.[58] However, it should be noted that the relatively large number of qualified NGOs does not necessarily mean that all of them would bring PIL cases in practice. In reality, NGOs are confronted with various obstacles that hamper their willingness to initiate judicial action.[59] The standing of NGOs is further clarified in the judicial interpretation issued by the SPC[60] to avoid legal uncertainties in practice. Problems arose, for example, in the well-known Tengger Desert case. This concerned a case

[54] Zhu and Wu (2017), pp. 389, 395.

[55] National People's Congress (2013).

[56] Percival and Zhao (2014), p. 174.

[57] Gong and An (2017), pp. 189–190.

[58] Xing and Jin (2015).

[59] See Sects. 9.3.2 and 9.5.

[60] Interpretations of the SPC on Several Issues concerning the Application of Law in the Conduct of Environmental Civil Public Interest Litigations (first issued in 2015 and revised in 2020), Interpretation No. 20 of the SPC, 1 January 2021, arts. 1–5.

where the China Biodiversity Conservation and Green Development Foundation (CBCGDF) brought, in August 2015, a series of PIL cases against various industrial operators for causing pollution to the Tengger Desert. These lawsuits were all rejected and dismissed by the local courts due to a lack of standing.

It is reported that between 2015 and 2018, only 32 NGOs filed PIL cases, and CBCGDF, FON and ACEF were the top 3 NGOs in the number of PIL cases filed.[61] It is also important to note that under the current EPL, local NGOs that are registered at the country level are excluded from filing PIL cases. In fact, compared to national NGOs (such as CBCGDF and ACEF) or those based in more developed areas (such as FON), local NGOs may be best suited to litigate against violators, since they have better information about local issues. Therefore, it is argued that the exclusion of local NGOs may further undermine the effectiveness of PIL by NGOs in enhancing environmental law enforcement.[62]

The basic standpoint of the Chinese legislator on the standing of NGOs can be explicitly observed from the legislative process of PIL. On the one hand, the necessity and importance of public participation in China's environmental governance are generally recognized by Chinese policymakers.[63] On the other hand, under the traditional authoritarian context in China, the government remains quite prudent of ceding state powers to civil society.[64]

9.4 The PIL Cases Brought by NGOs in Practice

As mentioned above, China has made impressive progress in boosting the involvement of NGOs in environmental governance, particularly in the field of environmental litigation. This equally resulted in an increase of the NGOs' activity before the courts in the first year after the amendments. The data in this chapter are collected from the reports published by the MEE and the SPC and also from the court decisions available online.[65]

A remarkable trend in the field of PIL is the substantial increase in the number of PIL cases by NGOs, which can be clearly vindicated by some data in Fig. 9.1. It was reported that from 1995 to 2014, only 72 PIL cases (by NGOs, procurators, administrative agencies and individuals) were accepted by the courts around the country.[66] This number increased to 38 in 2015 after the amendment of the EPL took effect. Since then, there has been a steady rise in the number of PIL cases brought by

[61] Wang (2020), pp. 137–169.

[62] Gao and Whittaker (2019), pp. 345–346.

[63] Zhu and Wu (2017), p. 389.

[64] Ma (2019), pp. 269–291.

[65] The data of court decisions is publicly available on the China Judgments Online. https://wenshu.court.gov.cn.

[66] Li (2016), p. 257.

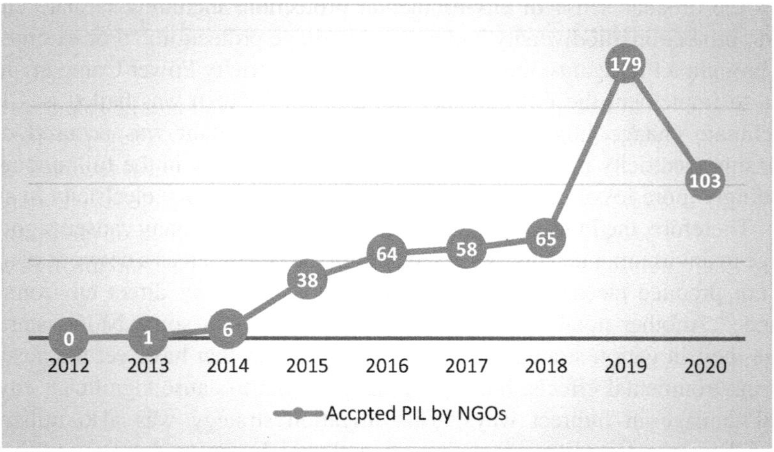

Fig. 9.1 Number of PIL cases by NGOs (by author)

NGOs. In spite of some limitations in data collection, one can still notice several major trends in the development of PIL by NGOs in China.

The first trend is the diversification of NGOs to file PIL cases.[67] As mentioned above, before 2015, the majority of PIL cases were brought by a few well-established NGOs, particularly those government-organized or approved NGOs (such as ACEF). In contrast, the grassroots NGOs were largely kept out of the court due to strict regulatory restrictions or lack of financial support. Even in the cases where the NGO could meet the strict regulatory requirements, it is striking that it was mostly the claims brought by governmental NGOs that were accepted, while those brought by grassroots NGOs were rejected.[68] The situation began to change when the requirements for standing were clarified in the EPL 2014. According to the statistics, 9 NGOs participated in PIL practice in 2015, and the number increased to 14 in 2016.[69] Despite the emergence of entrants in the domain of PIL, government-oriented NGOs remain as the major force for bringing PIL cases, such as the ACEF and CBCGDF.

Secondly, it is observed that some NGOs have made genuine efforts to expand the scope and the context of PIL by venturing into new areas of environmental protection.[70] At the primary stage of development, many NGOs have chosen to litigate about environmental pollution incidents, the damage of which is more visible and easily tracked, such as air pollution, water pollution or solid waste pollution.[71] After several years of experience, some NGOs started to explore possibilities of initiating

[67] Zhang and Mayer (2017), pp. 213–216.
[68] Gong (2019), p. 20.
[69] Li (2016), p. 335.
[70] Xie and Xu (2021), pp. 22–23.
[71] Zhang and Mayer (2017), pp. 217–219.

PIL in the broader sense of environmental protection, including climate change, wildlife protection, biodiversity and cultural heritage protection.[72] For example, the FON brought a PIL against the State Grid Gansu Electricity Power Company for the failure to implement the full purchase obligations,[73] which was hailed as China's first climate change litigation. In this case, the defendant was accused of not purchasing electricity produced from wind and solar energy in the full amount and accordingly more fossil energy would be consumed to produce electricity in substitution. Therefore, the FON claimed that the state-owned company caused significant damage to environmental interests, even though the defendant as a network company does not produce electricity and hence does not cause any direct environmental damage.[74] Another notable finding from this case is that some NGOs started to change their litigation strategy by targeting companies that have yet to cause little direct environmental effects, but have great potential to cause significant environmental damage in indirect ways. This litigation strategy was also utilized by CBCGDF in the lawsuit against Shenzhen Sumei Environmental Protection Ltd and Zhejiang Taobao Network Ltd., which was later listed as one of the most influential PIL cases of 2019.[75] The essence of this case lies in the fact that two defendant companies were not engaging in polluting activities per se. Sumei is an environmental protection company producing so-called 'defeat devices' for cheating on diesel-emission tests. Another defendant, Taobao Network, as one of China's largest e-commerce companies, provided a platform to sell those devices. The CBCGDF sued the two companies for disturbing air pollution control devices and hence impacting the environment. After two trials, CBCGDF's claims were partially supported by the Hangzhou Intermediate People's Court and were upheld by the Zhejiang High People's Court.[76]

The third remarkable trend in PIL by NGOs is the promotion of preventive PIL lawsuits. In many cases, PIL is initiated as a response to environmental harm that has already occurred, and in this sense, PIL in nature performs the ex-post function of damage compensation and recovery. However, given the irreversible nature of environmental damage, such as climate change and biodiversity loss, it is important to allow NGOs to bring preventive PIL before any harm occurs to the environment.

---

[72] According to Article 2 of the EPL 2014, natural and cultural heritage are also defined as components of the environment. Therefore, the protection of cultural heritage can be classified in the broad category of environmental interest.

[73] Zhao (2019), pp. 167, 190.

[74] This case is still under consideration and no judgement has yet been rendered by the end of May 2021.

[75] SPC, CBCGDF v. Shenzhen Sumei Environmental Protection Ltd and Zhejiang Taobao Network Ltd, 8 May 2021, https://www.chinacourt.org/article/detail/2020/05/id/5195797.shtml.

[76] Ibid. Hangzhou Intermediate People's Court of Zhejiang decided the case as follow: Sumei had to pay the amount of 3.5 million RMB for remedying air pollution, to cover the amount of 150,000 RMB for attorney fees, travel expenses and was forced to apologize to the public via national media. The court decided that Taobao had fulfilled its due diligence duties and therefore did not assume joint responsibility with Sumei.

China's first preventive PIL, also well known as the 'Green Peafowl' case, was filed by the FON against the operator of a hydropower project in 2017, which was also hailed as the first PIL concerning wildlife protection. The FON accused the operator of threatening the habitat of an endangered bird - the green peafowl - and several disputes were involved in this case concerning China's environmental impact assessment system. This preventive PIL dealt with the fundamental conflict between economic growth and environmental protection. The Kunming Intermediate People's Court finally ruled in favor of the NGOs by issuing a suspension order due to the significant risk of this project to the green peafowl.

As a result of concerted efforts during the past few years, the recent years have witnessed a steady increase in the number of NGOs involved in the field of PIL. This provides an important legal weapon for China's civil society, mainly represented by NGOs, to play a role in enhancing environmental law enforcement and safeguarding the public environmental interest, particularly where government regulation may fail.

9.5 The Decline of NGOs Before the Courts in China

As the data presented in the previous section showed, there has more recently been a decline in the number of PIL cases brought by NGOs. This section will explain the three different reasons for this decline. The first reason is the role that the State plays in restricting the activity of NGOs in PIL (Sect. 9.5.1). Second, there are serious institutional obstacles for NGOs in China to launch a PIL that have recently become even more egregious (Sect. 9.5.2). Finally, there may also be structural reasons within the organization of the NGOs themselves why they may not always be inclined to bring the PIL in the public interest (see Sect. 9.5.3).

9.5.1 The Shadow of the State Over NGOs in PIL

The potential fall-off of NGOs' access to the courts becomes clearly illustrated by the relevant figures of PIL cases during the past years. The data show that the number of procuratorial cases increased quickly since the pilot run of procuratorial PIL commenced in 2015. Taking the data from 2019 as an example, 170 PIL cases by NGOs were accepted by the court, and 58 of them were closed.[77] Meanwhile, 2309 procuratorial PIL cases were accepted, and 1895 of them were closed.[78] A closer examination of the data indicates that the majority of procuratorial cases are

[77] SPC (2020).

[78] Ibid.

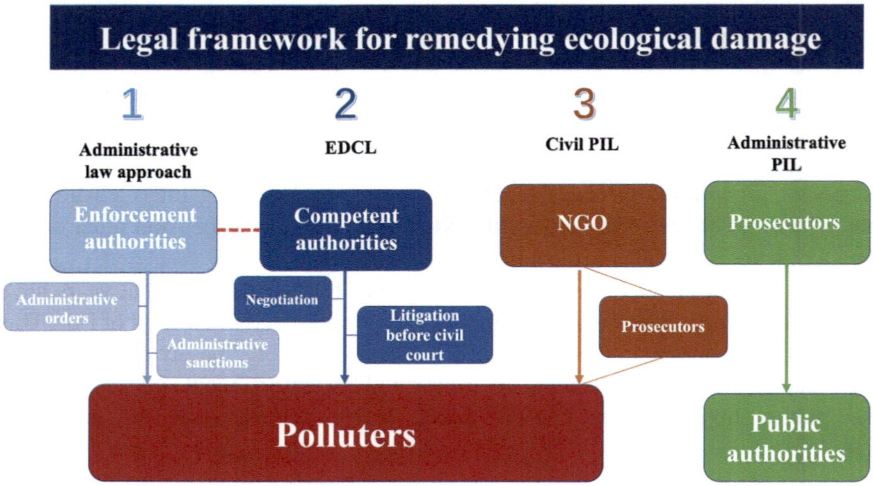

Fig. 9.2 Legal framework for remedying ecological damage (by author)

incidental (civil) PIL to a prosecution, comprising 71.1% of the total cases,[79] followed by 355 administrative PIL cases and 312 civil PIL cases.[80] In addition, the number of ecological damage compensation litigation (ECDL) cases continued to grow in 2019, as 49 cases were accepted and 36 of them were closed. From this perspective, the number of PILs by NGOs is trivial compared to the large number of procuratorial cases and ECDL cases. Some scholars have expressed concern that the role of NGOs in vindicating environmental interests has been gradually superseded by procuratorial organs and public authorities. The development of PIL indicates a remarkable trend of nationalization under China's unique context of 'state-civil society dichotomy'.[81] The root cause of this nationalization trend is the regulatory design for vindicating environmental interests in China. As mentioned above,[82] to better safeguard environmental and ecological interests, the Chinese legislators opted to establish a legal framework with the involvement of both state and non-state actors, which provides multiple possibilities for various actors to enforce environmental legislation. As a result, there are five pathways for remedying environmental damage (see Fig. 9.2) and overlaps and conflicts between different approaches could occur under several circumstances.

The first possibility for remedying ecological damage is (see Table 9.1) the administrative law approach. The problem in that case is that public authorities

[79] In 2019, 1642 incidental (civil) PIL of criminal cases were accepted by the courts and 1370 were closed.

[80] SPC (2020). It was reported that in 2019, 277 of 355 administrative cases were closed, and 248 of 312 civil PIL cases were closed.

[81] Chen and Zhou (2019), pp. 70–79.

[82] See Sect. 9.2.

have dual identities under the current legal framework for remedying environmental damage: being both regulator and potential plaintiff. As regulators, public authorities are accordingly equipped with regulatory powers and instruments to enforce environmental laws,[83] which is generally referred to as administrative enforcement of environmental law. Moreover, public authorities are also granted standing to seek remedies against polluters via the EDCL system. An issue which has arisen in practice is whether public authorities are allowed to initiate ECDL actions in place of utilizing public enforcement powers available to them. This issue is becoming increasingly important, due to the potential duplication of powers. Particularly, in most cases, public authorities are reluctant to perform their duties to correct and punish environmental violations, given their concerns with regard to the adverse effects on local economic development or the potential risks of being challenged in administrative litigation or review.[84] When facing the choice between administrative enforcement and ECDL, public authorities clearly prefer ECDL, thus bypassing the use of regulatory powers and instruments. In this way, it is argued that they provide free information to the judiciary, and a blurring of the boundaries between the administrative and the judicial branches emerges.[85] Therefore, it is critical to determine when it is appropriate for public authorities to initiate judicial actions opposed to waving public enforcement powers. An important premise to answer this question is to clarify the role of public authorities in China's environmental governance system. In modern society, the government is widely regarded as the prioritized and definitive defender of the public interest, and environmental protection is clearly defined as the state obligation in China's Constitutional Law.[86] Therefore, some scholars argue that priority should be given to administrative enforcement to induce public authorities to diligently implement their duties concerning environmental protection.[87] Moreover, administrative authorities, especially EPBs, may be far better qualified than judges to determine whether environmental harm occurs, to what extent it should be remedied and what types of remedial measures are suitable. The lower information cost and the higher expertise constitute another strong argument in favor of an administrative law enforcement approach over EDCL.[88]

A second problem that might emerge relates to potential conflicts that also arise between public authorities and NGOs. One can notice that there are significant overlaps between PIL by NGOs and ECDL by public authorities in both substantive and procedural aspects, but only a slight difference in terms of standing. One can suppose that, after incidents such as major environmental pollution or ecological damage, both NGOs and local governments could take action on behalf of environmental interests. Under the current legal framework, the NGOs' choice of pathway is

[83] Gao and Whittaker (2019), p. 335.

[84] Lu (2021), pp. 133–151.

[85] Ma (2019).

[86] Constitution Law of China, art. 26.

[87] Lu (2021), pp. 133–151.

[88] Ibid, pp. 141–142.

restricted to bringing a civil PIL. In contrast, public authorities could achieve the goal of environmental restoration via either administrative enforcement or the ECDL. Accordingly, there could be competition between NGOs and public authorities in particular situations since they both have standing before the courts. This might cause an unnecessary waste of limited resources and could subsequently undermine legal certainty. That could more particularly be the case if the claims by NGOs and public authorities were not handled simultaneously, and it would even be worse if they were handled by different courts. This could potentially lead to different verdicts regarding the same facts.

Since the entry of the ECDL into the field of environmental litigation in 2015, the relationship between PIL and ECDL has triggered heated discussion in both academic and empirical circles. As a response, the SPC issued Interpretations to clarify the coordination between these two approaches, by stipulating that the PIL trial by NGOs was to be suspended until the closure of ECDL by local governments.[89] As a result, the promotion of ECDL has pushed NGOs into a further weakened position in PIL, although NGOs are allowed to resume the PIL for the residual claims uncovered by the ECDL judgement. It is perceived as a turning point in the development of China's PIL because it sends a clear signal that Chinese policymakers have changed their attitudes towards the involvement of NGOs - from the initiators and the main protagonists to the substitutes and even bystanders in vindicating environmental interests. Some scholars argue that the role of NGOs in the current legal framework has been marginalized, and NGOs have struggled to coordinate their involvement in PIL in various ways, such as co-plaintiffs or supportive plaintiffs.[90] Another problem that might arise concerning a potential competition between NGOs and public authorities relates to the possibility that NGOs might bring a PIL before the court, while public authorities choose to rely on their regulatory powers and instruments. According to the Civil Procedure Law and the EPL, qualified NGOs could bring PIL directly to the courts without notifying the competent authorities. In practice, there are a few cases in which the relevant authorities have taken removal and remedial actions and the damaged environment has been or is under restoration, while PIL proceeding is taking place. In this case, doubts arise with regard to the necessity of initiating PIL by NGOs. As NGOs do not have to announce their intention to bring a PIL (for example, via a pre-litigation notice to the administrative authorities), there is according to some a risk that the PIL brought by an NGO might run counter to the action of the government in environmental law enforcement.[91] The Chinese legislature tried to fix this problem by requiring the court to notify the administrative authorities that a PIL had been brought after the court accepted the case. But there are

[89] SPC, Several Provisions of the SPC on Trying Ecological Compensation Cases (for trial implementation).

[90] Chen and Zhou (2019), pp. 70–79.

[91] Interpretations of the SPC on Several Issues concerning the Application of Law in the Conduct of Environmental Civil Public Interest Litigations (first issued in 2015 and revised in 2020), Interpretation No. 20 of the SPC, 1 January 2021, art. 12.

no detailed rules explaining exactly how this should be done, which may undermine the effectiveness of this arrangement.[92]

A third issue that can affect the scope for NGOs to participate in environmental litigation is related to the presence of the procuratorates. It is clear under the current legal framework that procurators play a unique role in PIL, because they are allowed to bring both civil PIL against polluters and administrative PIL against public authorities for inaction or abuse of powers. Moreover, procuratorates could bring an incidental civil PIL action while prosecuting environmental crimes. The large number of procuratorial cases mentioned above also explicitly proves that procuratorates are far more active in PIL than NGOs. With regard to civil PIL, as clearly stipulated in Article 55 of the Civil Procedure Law, the procuratorte should play a complementary role to fill the gap left by NGOs when there are no suitable NGOs or if NGOs refuse to litigate. Procutoriates are required to issue a notice to find a qualified NGO as plaintiff before coming to the court. However, in practice, this regulatory design seemingly does not serve to encourage NGOs to participate in PIL for several reasons. It is argued that the procuratorates could bypass the pre-litigation notice if the PIL claim could be attached to criminal proceedings. It should be noted that under the current legal framework, qualified environmental groups can start a civil action directly against the polluters, but they are not entitled to challenge the competent authorities when they fail to enforce environmental regulations. As mentioned before, NGOs are not required to notify the administrative authorities before bringing PIL.

These three described phenomena all contribute to the potential decline of the role of NGOs in remedying ecological damage, as they are facing various formal restrictions in legislation and practical problems as well. To fully exploit the potential of NGOs in environmental governance, the kernel of the reform of China's PIL is the configuration of different pathways, in order to avoid any overlaps and conflicts in practice.

9.5.2 Institutional Obstacles for NGOs in China

Despite recent legislative and judicial developments, many scholars still contend that NGOs have been repressed in many ways, which hinders the potential that NGOs could play in addressing China's environmental problems.[93] Although they have been empowered to bring a PIL before the courts, NGOs find themselves confronted with several formidable hurdles under China's current institutional context.

The first insuperable obstacle for Chinese NGOs is the lack of legitimacy due to the strict registration and fundraising regulations in China. This is also known as the

[92] Lu (2021), p. 147.

[93] Zhai and Chang (2018), p. 383; Simon (2013).

'identity crisis' for NGOs.[94] In China, the first step to operate an NGO in China is to acquire a legal identity. However, many environmental groups encounter problems registering as NGOs. Under China's current legal framework, a social organization shall be legally registered with the civil affairs departments, which are subordinate to local governments at or above municipal levels.[95] However before submitting the materials required for registration to the civil affairs department, social organizations need to obtain approval from the relevant department subordinate to or authorized by the State Council and local government at or above the country level in relation to industry, discipline or other professional domains.[96] In other words, NGOs in China are operated under the 'dual management regime', where the registration of NGOs is managed by the civil departments (*dengji guanli bumen*), and the professional supervisory units (*yewu zhuguan bumen*) are in charge of supervising NGOs' activities. In practice, the approval from the professional supervisory unit is very difficult to obtain, especially for grassroots NGOs, since the relevant authorities are reluctant to take such supervisory responsibilities due to a lack of clear guidance on the professional supervisory units.[97] Not surprisingly, public authorities lack incentives to take the guardianship of NGOs. The requirements became even more stringent for foreign NGOs operating in China after the adoption of the Law on the Administration of Activities of Overseas NGOs (hereinafter Overseas NGOs Law) in 2017. According to the Overseas NGOs Law, a foreign NGO is required to register with the public scrutiny department subordinate to the State Council or with the provincial government.[98] Like domestic NGOs, foreign NGOs are also required to get approval from the professional supervisory units before registration.[99] As a consequence, a large number of domestic and foreign organizations are unable to become officially registered as NGOs under China's legal system, which significantly restricts the development of NGOs and also prevents them from playing an active role in environmental governance. Some scholars condemn the lack of true freedom and independence of NGOs in China, due to these strict regulatory requirements.[100]

The second obstacle for Chinese NGOs is the lack of financial stability. Although the registration regulations do not set a high financial threshold for establishing an NGO,[101] it is widely observed that many NGOs are struggling to raise funding to

[94] Hsu and Jiang (2015), pp. 100–122.

[95] Regulation on the Administration of the Registration of Social Organizations, Order No. 666 of the State Council, 6 February 2016, art. 6, para 1.

[96] Ibid, art. 6, para 2.

[97] Gao and Whittaker (2019), p. 344.

[98] Law on the Administration of Activities of Overseas NGOs, Chairman Order No. 81, 5 November 2017, art. 6, para 1.

[99] Ibid, art. 6, para 2.

[100] Simon (2013).

[101] According to Article 10 of Regulation on the Administration of the Registration of Social Organizations, the financial requirement for establishing a national NGO is above 100,000 RMB and for a local or trans-administrative regions NGO is above 30,000 RMB.

sustain their daily operations.[102] According to a survey conducted by the ACEF and the Natural Resources Defence Council (NRDC) in 2013, only 23% of environmental NGOs have financial assets and membership dues that provide a relatively stable source of funding and nearly half have less than 500,000 Yuan as their annual budget.[103] The lack of financial stability undermines the enthusiasm of NGOs to play an active role in PIL, given that litigation costs are far beyond their budget, particularly in environmental litigation. Litigation costs have been a controversial issue for the development of PIL by NGOs. In practice, NGOs are confronted with tremendous financial burdens to bring PIL, which consist of court fees, attorney fees, judicial authentication fees and other relevant expenses. The widespread nature and complexity of environmental damage makes this situation even more problematic, by posing substantial financial risks for NGOs. In the current legal framework, PIL is to a large extent based on traditional civil procedural rules. According to the Civil Procedure Law, the plaintiff is required to pay the case acceptance fees in order to have the case accepted and tried by the court.[104] Moreover, the case acceptance fee is calculated on the amount of money claimed by the plaintiff.[105] Although the court fees will eventually be borne by the losing party, the potential risk of losing and having to pay high litigation costs still constitutes a significant obstacle for NGOs to initiate judicial actions. The allocation of litigation fees has been a focal issue in the Changzhou soil pollution case.[106] In 2016, the FON and the CBCGDF brought a PIL against three chemical factories for contaminating soil and underground water, and claimed ecological restoration or monetary compensation if the damaged environment could not be remedied, public apology and the reimbursement of all expenses incurred from this case, including the investigation fees, pollution inspection fees, damage assessment fees, attorney fees, travel expenses, expert consulting fees, and case acceptance fees. In the proceedings, both the FON and CBCGDF claimed 370 million RMB for ecological restoration. In addition, the FON claimed 413,675.6 RMB to cover their attorney fees and travel expenses and CBCGDF asked for 108,460 RMB for all relevant expenses. At first instance, the court did not decide in favor of the NGOs, and they were required to jointly bear the court fees, which amounted to nearly 1.9 million RMB, in accordance with the 'loser pays principle' in the allocation of litigation costs.[107] Such costs were astronomical figures, even for these two well-established NGOs. Not surprisingly, both the FON and CBCGDF filed an appeal and one of the controversial issues was the allocation

[102] Zhu et al. (2018), pp. 1176–1189.

[103] ACEF and NRDC (2014).

[104] Civil Procedure Law, art. 118.

[105] Measures on the Payment of Litigation Cost, Order of State Council No. 481, 19 December 2006, art. 13.

[106] FON and CBCGDF *v.* Jiangsu Changlong Chemical Ltd. and Changzhou Changyu Chemical Ltd, Second Instance, Jiangsu High People's Court, (2017) SuMinzhong No. 232.

[107] Measures on the Payment of Litigation Cost, Order of State Council No. 481, 19 December 2006, art. 29.

of court fees. The appeal court held that this case should be treated as a non-monetary case, since the plaintiffs' claims of environmental restoration and public apology are in nature non-monetary. As a result, the appeal court partially supported the claims of the FON and CBCGDF, requesting that the three chemical factories undertake the attorney fees and travel expenses incurred by the two NGOs as well as the court fees of 100 RMB for the first instance and the appeal. The final decision was accepted by the FON and CBCGDF, since they did not have to pay the 1.9 million RMB court fees.[108] However, this case triggered a heated discussion on the court fees of PIL by NGOs, given that high litigation costs significantly dampen the enthusiasm of NGOs, especially those at the grassroots level, to bring a PIL before the court. Therefore, scholars called for the reform of court fees for PIL by NGOs. However, this principle of allocating litigation costs to the losing party is also important to prevent vexatious litigation, since the risks of having to pay high litigation costs may push NGOs to exhaust other possible measures without necessarily resorting to the court.

9.5.3 Voluntary Failure of NGOs: Atruism or Self-Interest?

Although the possibilities for NGOs to intervene in PIL have increased, the way in which NGOs use PIL in practice has also raised particular questions. The first disputed issue is the lack of representative legitimacy of NGOs. Theoretically, a fundamental argument in favor of PIL by NGOs is to break the state monopoly as the incarnation of public interest defenders, by providing an alternative mechanism to supervise law enforcement and multiple possibilities to safeguard environmental interests. NGOs have been dubbed as the 'self-appointed guardian' of the public interest, because NGOs are not elected to represent the interests of the public.[109] This argument may further put NGOs at a distinct disadvantage when conflicts arise between NGOs and public authorities. With the maturing of PIL in China, there are a few examples in which the motivations and practices have been questioned, which puts NGOs under public criticism. The first issue is the selected litigation strategy adopted by NGOs. It has been observed that NGOs have chosen to litigate over major environmental incidents that have attracted broad attention from the media and the public, or to pursue the well-known or famous names. From the standpoint of NGOs, it is not surprising and even reasonable for NGOs to adopt this litigation strategy, particularly given the formidable challenges they are facing as discussed above. However, it remains questionable whether this is good for protecting environmental interests, since the selection of PIL cases by NGOs is based on several factors, including the success rate as plaintiffs and the sensational effect of bringing

[108] FON and CBCGDF v. Jiangsu Changlong Chemical Ltd. and Changzhou Changyu Chemical Ltd, Second Instance, Jiangsu High People's Court, (2017) SuMinzhong No. 232.
[109] Gao and Whittaker (2019), p. 341.

such a case, rather than the necessity and urgency of remediating the environment. In some cases, a PIL is initiated even though prompt administrative measures have already been taken to address the environmental problems. This was the exact situation that also arose in the Changzhou soil pollution case. NGOs may therefore in some cases, pursue their private interests (for example, selecting high profile cases attracting a lot of publicity), rather than always pursuing the public interest with PIL.

Moreover, NGOs' motives are further questioned when they claim a large amount of monetary compensation for environmental damage. As stipulated in the Civil Code, NGOs are allowed to bring a monetary claim in PIL, covering the costs of environmental restoration and the loss of ecological functions.[110] In practice, there are several cases in which significant monetary compensation has been ordered by the court in favor of NGOs, such as the amount of 569 million RMB environmental damage compensation in the Tengger Desert Case. Due to the lack of detailed guidance with respect to the management of environmental compensation, local governments started to investigate the management of environmental compensation and there are generally three modes: the money allocated by the court could be placed in a specific account: it could go to the treasury account of the local government or to an account aimed at environmental protection funding. However, recent studies indicate that a large amount of environmental compensation is not properly spent to address the target of environmental restoration.[111] As NGOs are explicitly forbidden to pursue self-interest by initiating PIL cases,[112] NGOs lack the legitimacy and legal basis to use the large amount of monetary compensation that may be awarded by the court. But that equally has a flipside. After the NGO has won a verdict, it consequently pays relatively little attention to what is done with the environmental compensation it has received. This obviously goes against the logic of PIL as the claim was brought in the public interest to restore ecological damage. This is becoming increasingly problematic as judges in China now demonstrate a tendency to make monetary compensation judgements instead of directly ordering environmental restoration.[113]

9.6 Conclusion

NGOs have been the major force in bringing PIL cases since the PIL provision was integrated in Civil Procedural Law 2012 and in the Environmental Protection Law 2015. The report commissioned by the Supreme People's Court of China on environmental justice showed that there has been a sharp increase in the number of PIL cases since 2012, and NGOs have brought by far the most cases. However, a

[110] Civil Code of China, Art. 1235.

[111] Hu and Cui (2019), p. 14.

[112] EPL 2015, art. 58.

[113] Hu and Cui (2019), p. 16.

turning point occurred in 2017 when procuratorial organs and administrative authorities were granted the right of access to courts for safeguarding public environmental interests, through the amendments to the Administrative Procedural Law and the Civil Procedural Law, and the establishment of the ecological compensation regime. As a result, different enforcement mechanisms initiated by different actors co-exist under the current policy framework. This regulatory design could provide a wide variety of enforcement choices, thus better protecting ecological interests. However, the danger is that there may be overlaps and even conflicts concerning the coordination between different actors and approaches. As a result, the decline of NGOs in PIL is widely observed. This phenomenon can be attributed to many reasons, including the lack of legitimacy of NGOs in China, the lack of financial independence, the restrictions imposed on NGOs in challenging administrative authorities, and the ascendancy of administrative authorities in bringing ecological compensation suits. Some scholars have described PIL in China as 'castles made of sand'.[114] Indeed, the development of PIL in China is in its infancy. As this chapter has described, there are still fundamental obstacles and challenges that NGOs face. Only when these obstacles have been adequately addressed can NGOs provide an important contribution to environmental governance in China.

Acknowledgements This contribution builds further on Mengxing Lu, "The Dilemma of NGOs in Vindicating Environmental Interests before the Courts: Normative and Empirical Observations from China", The Chinese Journal of Comparative Law (2022) Vol.2 No.2, pp.222-245. The author would like to acknowledge the funding support provided by the Ministry of Justice of China (21SFB3028).

References

ACEF, NRDC (2014) The role and function of social organization in environmental public-interest litigation. China Development Brief, 11 November

Cao M, Wang F (2011) Environmental public interest litigation in China. Asia Pac Law Rev 19: 211–217

Capenter-Gold D (2015) Castles made of sand: public interest litigation and China's new environmental protection law. Harv Environ Law Rev 39:258–260

Chen H, Zhou H (2019) Reflection on the nationalization of public interest litigation. North Leg Sci 6:70–79

Economy E (2014) Environmental governance in China: state control to crisis management. Dædalus J Am Acad Arts Sci 143(2):184–197

Faure M, Partain R (2019) Environmental law and economics: theory and practice. Cambridge University Press

Gao Q, Whittaker S (2019) Standing to sue beyond individual rights: who should be eligible to bring environmental interest litigation in China? Transit Environ Law 8(2):335–346

General Office of the CPC Central Committee and General Office of the State Council (2020) Guiding Opinions on Building a Modern Environmental Governance System. 3 March

[114] Capenter-Gold (2015), p. 241.

Goldman P (2007) Public interest litigation in China: lessons learned from the U.S. experience. Vermont J Environ Law 8:255

Gong G (2019) Empirical analysis on China's environmental public interest litigation in 2015. Law Sci 6:20

Gong G, An R (2017) Progress and obstacles in environmental public-interest litigation under China's new environmental law: an analysis of cases accepted and heard in 2015. Climate Law 7:189–190

Guiding Opinions on Establishing Modern Environmental Governance System. 3 March 2020., http://www.gov.cn/zhengce/2020-03/03/content_5486380.htm

Guttman et al (2018) Environmental governance in China: interactions between the state and "nonstate" actors. J Environ Manag 220:126–135

Hsu CL, Jiang Y (2015) An institutional approach to Chinese NGOs: state alliance versus state avoidance resources strategies. China Q 22:100–122

Hu J (2016) The functions of public interest litigations by NGOs—reflections on the judicial interpretations on environmental public interest litigation. Law Rev 4:173

Hu J, Cui M (2019) Feasibility study on responsibility performance of ecological environment restoration under binary litigation mode. J China Univ Geosci (Soc Sci Ed) 6:14

Khan MI, Chang C (2018) Environmental challenges and current practices in China—a thorough analysis. Sustainability 10:2547

Li S (2016) Review of public interest litigation in environmental protection. Law Press

Lu M (2021) On the coordination of different relief mechanisms for the ecological environmental damage. Comp Law 1:133–151

Lu M, Faure M (2020) Shifts in compensation for environmental damage: reflections on China's new soil pollution law. Asia Pac J Environ Law 23(2):136–153

Ma Y (2019) Vindicating environmental public interests in China: a balanced approach to institutional interaction in public interest litigation system. 21(4):269–291

National People's Congress (2013) Second Reading Draft for the Amendment of Environmental Protection Law. 17 July 2013. http://www.npc.gov.cn/npc/c1481/201307/c9beeb740a014530 8fba722306b72207.shtml

People's Daily (2013) The number of China's environmental NGOs has grown by nearly 40% over the past five years, reaching 8000 5 December. http://www.chinanews.com/gn/2013/12-05/ 5584508.shtml. Accessed 26 May 2022

Percival R, Zhao H (2014) The role of civil society in environmental governance in the United States and China. Duke Environ Law Policy Forum 24:146

Reform Plan on an Ecological Environmental Damage Compensation System, 17 December 2017., http://www.gov.cn/zhengce/2017-12/17/content_5247952.htm

Report on National Environmental Condition (2016) p 14

Report on National Environmental Condition (2017) p 14

Report on National Environmental Condition (2020) pp 18–19

Schwartz J (2007) Environmental NGOs in China: roles and limits. Pac Aff 77(1):28–49

Simon KW (2013) Civil society in China: the legal framework from ancient times to the "new reform era". Oxford University Press

State Administration of Environmental Protection (2007) Report on China's Environmental NGOs. Xuehui 3:23

State Council (2005) Decision on the Implementation of the Outlook on Scientific Development and Strengthening Environmental Protection. 3 December, Art 27

Stern R (2014) The political logic of China's new environmental courts. China J 72:64

Supreme People's Court of China (2020) White Paper of Environmental and Resources Adjudication in China 2019. Court Press

Wang A (2007a) The role of law in environmental protection in China: recent developments. Vermont J Environ Law 8:196–223

Wang C (2007b) Chinese environmental law enforcement: current deficiencies and suggested reform. Vermont J Environ Law 8:159–193

Wang M (2016) On the development direction of EPIL in China: analysis based on the relationship between administrative and judicial power. China Leg Sci 5:49–68
Wang Y (2020) Reflections and improvement on the adjudication of environmental public interest litigation: empirical analysis based on data between 2015 and 2018. Wuhan Univ Environ Law Rev 4:137–169
Wang A, Gao J (2010) Environmental courts and development of environmental public interest litigation in China. J Court Innov 3:377
Wang J et al (2008) Reflections from the transboundary pollution of Songhua River. In: Michael Faure M, Ying S (eds) China and international environmental liability: legal remedies for transboundary pollution. Edward Elgar Publishing, pp 273–290
Xi J (2017) Secure a decisive victory in building a moderately prosperous society in all aspects and strive for the great success of socialism with Chinese characteristics for a new era, Report Delivered at the 19th National Congress of the Communist Party of China, 18 October 2017
Xie L, Xu L (2021) Environmental public interest litigation in China: a critical examination. Transit Environ Law:22–23
Xing S, Jin Y (2015) SPC: about 700 NGOs are eligible to bring environmental PIL. BJNews, 7 January
Zhai T, Chang Y-C (2018) Standing of environmental public interest litigant in China: evolution, obstacles and solution. J Environ Law 30:383
Zhang R, Mayer B (2017) Public interest litigation in China. Chin J Environ Law 1:207–219
Zhao H (2019) Prioritizing access of renewable energy to the grid in China: regulatory mechanisms and challenges for implementation. Chin J Environ Law 3:167–190
Zhu X, Wu K (2017) Public participation in China's environmental lawmaking: in pursuit of better environmental democracy. J Environ Law 29:389–395
Zhu J, Ye S, Liu Y (2018) Legitimacy, board involvement, and resource competitiveness: drivers of NGO revenue diversification. 29:1176–1189

Chapter 10
Regulating Through Litigation: Regulatory Functions of NGO-Led Civil Environmental Public Interest Litigation in China

Juan Chu

Abstract As a tort-based framework, civil environmental public interest litigation (EPIL) established in China over the past decade authorizes environmental non-governmental organizations (NGOs) to challenge any "acts of polluting or damaging the environment that have harmed the public interest." Treating civil EPIL as a legal pathway to enter the regulatory process, this chapter examines how the broad scope and flexible remedies of civil EPIL have allowed environmental NGOs to use it as a powerful regulatory tool. It demonstrates that NGO-led civil EPILs perform valuable regulatory functions by providing an alternative enforcement mechanism when government enforcement fails, filling various regulatory gaps when a statutory remedy is unavailable or weak, challenging the legitimacy of government decisions, and setting the agenda in the regulatory process. Meanwhile, this chapter identifies several pitfalls of utilizing civil EPIL as a regulatory tool: threatening the primacy of government enforcement, displacing statutory remedies, unreasonable calculation of environmental damages, and the misuse of civil EPIL. This suggests that environmental NGOs should avoid the inappropriate use of civil EPIL so as to increase its legitimacy and regulatory significance in China's environmental governance system.

Keywords China · Civil EPIL · Environmental NGOs · Tort-based framework · Regulatory functions

10.1 Introduction

As environmental degradation threatens sustained economic growth and social stability, Chinese leadership views environmental protection fundamentally as a matter of regime legitimacy.[1] Over the last decade, the Chinese government has

[1] See Wang (2013).

J. Chu (✉)
School of Economic Law, East China University of Political Science and Law, Shanghai, China
e-mail: chujuan@ecupl.edu.cn

X. Wang et al. (eds.), *Environmental Public Interest Litigation in China*,
https://doi.org/10.1007/978-3-031-26526-6_10

launched an extraordinary range of legal and institutional reforms in the course of promoting ecological civilization.[2] Among these reform efforts, the establishment of civil environmental public interest litigation (EPIL) in 2012 is seen as an important initiative to improve China's environmental governance system. The civil EPIL framework that authorizes environmental non-governmental organizations (NGOs) and other qualified parties to bring lawsuits on behalf of the public interest has attracted increasing academic interest. Existing literature on civil EPIL tends to be descriptive or doctrinal, and some pay special attention to the standing rule and its limitations.[3] Other empirical studies examine the practice of this new type of litigation and analyze the legal and practical obstacles confronting environmental NGOs in initiating civil EPIL suits.[4] A different strand of scholarship explores how environmental NGOs have used various frames to influence the "support structure" in the process of mobilizing civil EPIL from the lens of social movements.[5]

While existing studies advance our understanding of the rules and implementation of civil EPIL, few have attempted to examine the function of civil EPIL from a regulatory perspective. As a groundbreaking development of China's environmental governance system, the civil EPIL provides a legal pathway for environmental NGOs to enter into the regulatory landscape traditionally occupied by bureaucrats.[6] This chapter explores how the broad scope and flexible remedies of civil EPIL have allowed environmental NGOs to utilize it as a powerful regulatory tool. Section 10.2 of this chapter first provides an overview of the legal framework for EPIL in China. It then points out that civil EPIL is a tort-based framework that can provide flexible remedies for a variety of environmental harms irrespective of existing statutory requirements. By conducting an empirical analysis of judicial practice, Sect. 10.3 investigates four important regulatory functions of NGO-led civil EPILs: providing an alternative enforcement mechanism; filling regulatory gaps; challenging the legitimacy of government decisions; and setting the agenda in the regulatory process. Section 10.4 identifies potential pitfalls of invoking civil EPIL as a regulatory tool and offers preliminary suggestions for avoiding these risks. Section 10.5 provides a conclusion.

[2] See Kitagawa (2017) and Wang (2018).

[3] See e.g., Carpenter-Gold (2015), McCallum (2017), Zhai and Chang (2018), Gao and Whittaker (2019) and Jiang et al. (2020).

[4] See e.g., Zhang and Mayer (2017), Sun and Tuholske (2017), Gao (2018), Liu (2019) and Xie and Xu (2021a, b).

[5] See Ren and Liu (2020).

[6] See Van Rooij et al. (2016).

10.2 Theoretical Foundation of Civil Environmental Public Interest Litigation in China

10.2.1 The Legal Framework for Environmental Public Interest Litigation

Policy entrepreneurs from various fields began to call for establishing EPIL in China in the 2000s.[7] Chinese scholars generally understand EPIL as a framework that allows certain parties without a "direct interest" to bring lawsuits to defend the environmental public interest, with the liberal standing rule as its defining character.[8] EPIL is further classified into two categories: civil EPIL against polluters and administrative EPIL against administrative agencies.[9] After years of experimentation by local environmental courts,[10] the Standing Committee of the National People's Congress (NPC) formally established the civil EPIL by passing an amendment to the Civil Procedure Law (CPL) in 2012. Article 55 of the 2012 CPL authorizes "relevant organs and organizations prescribed by law" to initiate lawsuits against "acts of polluting the environment that have harmed the public interest." However, because of the ambiguous language of the law, the 2012 CPL did not bring about an increase in EPIL suits as anticipated.[11]

On 24 April 2014, the Standing Committee of the NPC revised the Environmental Protection Law (2014 EPL). Article 58 of the 2014 EPL extends the scope of civil EPIL to "acts of damaging the environment that have harmed the public interest" and clarifies the standing requirements for NGOs. While earlier drafts of the 2014 EPL adopted a restrictive approach to NGOs' standing, the final version relatively relaxed the standing rule as a result of persistent lobbying and advocacy by environmental NGOs and liberal academics.[12] According to Article 58, NGOs may acquire standing if they (1) are registered with civil affairs departments at or above the level of a city with districts and (2) have specialized in environmental protection activities for at least five consecutive years without any record of legal violations.

After a two-year pilot program that empowers procuratorates to initiate EPIL suits in 13 provincial regions,[13] the national legislature officially granted standing to procuratorates to bring both civil and administrative EPIL by amending the CPL

[7] See Wang (2010), pp. 6–8.

[8] See Cao and Wang (2011), p. 221.

[9] See Zhai and Chang (2018), p. 372.

[10] See Wang and Gao (2010), Stern (2013), pp. 115–122 and Sun and Tuholske (2017), pp. 10498–10499.

[11] See Zhang and Mayer (2017), p. 205 and Gao and Whittaker (2019), pp. 342–343.

[12] See Liu (2015) and McCallum (2017), pp. 69–72.

[13] See Decision of the Standing Committee of the National People's Congress on Authorizing the Supreme People's Procuratorate to Launch the Pilot Program of Initiating Public Interest Actions in Certain Areas (2015). For implementation of the pilot program, see Ding and Xiao (2021) and Zhai and Chang (2018), pp. 379–383.

and the Administrative Litigation Law (ALL) on 27 June 2017. The CPL as amended in 2017 added a subsection to Article 55, which authorized procuratorates to file civil EPIL suits provided that no eligible public authorities or environmental NGOs file a suit. Under the administrative EPIL system created by Article 25(4) of the 2017 ALL procuratorates may sue agencies in charge of environmental protection and natural resource preservation if their illegal exercise of authority or inaction causes damage to the national or public interest. Unfortunately, the 2017 ALL does not allow environmental NGOs to file administrative EPIL lawsuits. This chapter addresses only the regulatory function of civil EPIL.

10.2.2 The Nature and Scope of Civil Environmental Public Interest Litigation

To provide guidance for courts and other practitioners to implement civil EPIL, the Supreme People's Court (SPC) issued a judicial interpretation on 6 January 2015 (2015 EPIL Interpretation).[14] The 2015 EPIL Interpretation further lowered the standing bar by liberally interpreting each element of the standing requirements set forth by the 2014 EPL.[15] Concerning remedies, the 2015 EPIL Interpretation adapted remedies provided by tort law to civil EPIL cases. Specifically, plaintiffs may request "cessation of infringement, removal of obstacle, elimination of danger" to prevent threatened ecological damage or avoid further damage,[16] "restoration to the original status" to restore the injured environment (or pay the costs of restoration alternatively),[17] and "damages" to compensate for the interim losses of ecological services and functions during recovery.[18]

The extension of traditional tort remedies to civil EPIL has spawned intense debate among scholars. Many scholars insist that ecological damage is different from traditional types of damage (i.e., personal injury, property damage, and economic loss) and does not fit properly in the tort law system that has conventionally focused on private interest.[19] Given the conceptual distinctions between "restoration of the property" and "restoration of the environment," the Civil Code passed on 28 May 2020 authorizes plaintiffs of civil EPIL to request "restoration of the environment" and damages where ecological damage is caused in violation of the law.[20] Following the promulgation of the Civil Code, the SPC amended the 2015 EPIL Interpretation

[14] SPC Interpretation on Several Issues Concerning the Application of Law in Environmental Civil Public Interest Litigation (2015) (2015 EPIL Interpretation).

[15] See Zhai and Chang (2018), pp. 377–378.

[16] See Art.19 2015 EPIL Interpretation.

[17] See Art.19 2015 EPIL Interpretation.

[18] See Art.21 2015 EPIL Interpretation.

[19] See, e.g., Li (2014), Xu (2010) and Zhang and Wang (2012).

[20] Arts.1234, 1235 Civil Code (2021).

on 23 December 2020, replacing the controversial "restoration to the original status" with "restoration of the environment."[21] In other words, the civil EPIL system in China has been crafted as an expanded tort liability regime that protects the public interest as opposed to the private interest.

It is widely believed that the Chinese civil EPIL was inspired by the U.S. citizen suit, which are known as "the most pervasive, prominent, and continuing innovation in the modern environmental era."[22] In fact, the U.S. citizen suit and the Chinese civil EPIL have different theoretical justifications and functions.[23] Citizen suit provisions in the U.S. environmental statutes typically authorize "any person" to commence a suit against any person (including state and federal agencies) alleged to violate statutory requirements.[24] The purpose of the U.S. citizen suit is to provide a goad to government enforcement and an alternative means of enforcement when agencies fail to enforce due to lack of sufficient resources or political will.[25] Citizen suits derive their theoretical validity from the violation of precautionary regulations that are suited to address systemic environmental harms and impose strict liability for noncompliance.[26] Therefore, the scope of citizen suit is limited to specific violations defined by the underlying statutes, such as emission standards, permits, licenseses, orders or procedural requirements.[27] To ensure that citizen suits do not supplant government enforcement actions and flood courts, plaintiffs are required to notify relevant enforcing agencies and the alleged violator at least 60 days prior to filing suit.[28] In citizen enforcement actions, the operative issue is whether an environmental standard has been violated irrespective of whether the conduct has caused an actual injury. The result of citizen enforcement is identical to government enforcement: injunctions that compel compliance and/or fines paid to the government.[29]

Unlike the U.S. citizen suit, the Chinese civil EPIL derives its theoretical validity from tort law and the showing of environmental harm regardless of whether the

---

[21] Arts. 18, 20 2015 EPIL Interpretation (amended 2020).

[22] Thompson (2000), p. 185.

[23] Discussion on the differences between the U.S. citizen suit and Chinese civil EPIL is drawn from Chu (2019).

[24] See, e.g., Clean Water Act, 33 U.S.C. § 1365(a)(1) (2012); Resource Conservation and Recovery Act, 42 U.S.C. § 6972(a)(1)(A) (2012). Citizen suit provisions also allow plaintiffs to challenge agencies who fail to perform a nondiscretionary duty under the statutes. *See, e.g.*, Clean Air Act, 42 U.S.C. § 7604(a)(2) (2012); Clean Water Act, 33 U.S.C. § 1365(a)(2) (2012). Since this chapter only focuses on civil EPIL, this type of agency-forcing actions will not be discussed.

[25] See Miller (1983), p. 10310.

[26] See Percival and Goger (2001), p. 129.

[27] See, e.g., Clean Air Act 42 U.S.C. § 7604(a)(1) (2012); Emergency Planning and Community Right-to- Know Act, 42 U.S.C. § 11046(a)(1)(A) (2012).

[28] See, e.g., Clean Water Act, 33 U.S.C. § 1365(b)(1)(A) (2012); Clean Air Act, 42 U.S.C. § 7604 (b)(1)(A) (2012).

[29] See, e.g., Clean Water Act, 33 U.S.C. § 1365(a)(2) (2012); Clean Air Act, 42 U.S.C. § 7604(a) (3) (2012); Resource Conservation and Recovery Act, 42 U.S.C. § 6972(a)(2) (2012).

conduct has been formally prescribed by statute. By allowing environmental NGOs to challenge any "acts of polluting or damaging the environment that have harmed the public interest," Chinese civil EPIL cannot only cover harms caused by regulatory violations but also remediate harms when an effective regulation does not exist. While remedies in citizen suits are restricted to abatement and/or civil penalties, remedies in civil EPIL suits can include injunctive relief and damages for ecological injuries. Therefore, compared with the U.S. citizen suit, the Chinese civil EPIL is a broader framework that can provide expansive and flexible remedies for a broad spectrum of environmental problems.

10.3 Civil Environmental Public Interest Litigation as a Multifunctional Regulatory Tool

The civil EPIL system established in 2012 essentially liberalized the traditional "direct interest" standing rule. As an expanded tort-based liability regime, civil EPIL can be applied to a variety of activities that allegedly caused harm to the public interest. Because of its broad scope and availability of a wide range of remedies, civil EPIL has been utilized by environmental NGOs to perform different regulatory functions.

10.3.1 Providing an Alternative Enforcement Mechanism

The enforcement of environmental laws in China is notoriously weak.[30] As many scholars have observed, insufficient capacity and resources of environmental agencies, the divergence between central policies and local interests, and capture-like risks have contributed to China's poor environmental enforcement.[31] In addition, government enforcement failures may also result from the lenient sanctions (e.g., low penalties) for non-compliance and the inherent inefficiency of government bureaucracy.[32] While the U.S. citizen suit is intended only to overcome government enforcement failures due to lack of will or resources, civil EPIL can address both types of government enforcement failures. When government enforcement is hampered by resource shortages or political pressure, or when government enforcement fails to abate the regulatory violations due to limited deterrent effects, environmental NGOs can use the civil EPIL as an alternative mechanism to enforce environmental laws.

[30] See Wang (2013) and Ryan (2013).

[31] See e.g., Economy (2004), Ma and Ortolano (2000), Van Rooij (2012), Wang (2007), Van Rooij and Lo (2010), McElwee (2011) and Xu and Faure (2016).

[32] See, e.g., Sitaraman (2007), Van Rooij (2012), Carpenter-Gold (2015) and Chu (2019).

In practice, environmental NGOs have brought civil EPIL to mainly address violations of statutory discharge standards.[33] For example, in the first air pollution case filed after the 2014 EPL took effect, the All-China Environment Federation (ACEF), a leading environmental NGO, sued Zhenhua Company for causing serious air pollution by repeatedly emitting numerous air pollutants in excess of statutory emission standards (the *Zhenhua* case).[34] Before the suit was filed, the local Ecology and Environment Bureau (EEB) had taken multiple enforcement actions against the defendant and fined the company 500,000 yuan. However, the penalty was not a sufficient deterrent to the defendant, who refused to come into compliance. Similarly, in another air pollution case ranking among ten Model EPIL cases released by the SPC in 2017, the defendant Jinling Chemical Company in Shandong Province had been found to continuously violate emission standards since 2014 (the *Jinling* case).[35] Moreover, the defendant violated relevant regulatory requirements by tampering with monitoring equipment and performing the operation before the pollution treatment facilities were checked and approved by the EEB. A series of administrative sanctions imposed by the EEBs at the county, municipal, and provincial levels of Shandong Province obviously failed to correct violations by the defendant, leading to the suit filed by Friends of Nature, another reputable environmental NGO in China. In these two cases, the plaintiff NGOs sought rectification of illegal emissions and other violations, as well as compensation for damage caused to the air. Since the defendant in the *Zhenhua* case had closed the plant before the court ruled, the court awarded substantial damages (21,983,600 yuan, approximately 3,136,000 US dollars) to fund improvement programs for local air quality. In the *Jinling* case, the parties reached a settlement in which the defendant agreed to pay 3 million yuan for the ecological damage caused by air pollution.

Civil EPIL is also used by environmental NGOs to enforce effluent standards. In a water pollution case brought by the ACEF, the defendant Jiangshan Company discharged wastewater through a rainwater pipe and polluted water bodies in the vicinity of the outfalls (the *Jiangshan* case).[36] Local EEB tests showed that illegally discharged sewage failed to meet the discharge standards for ammonia nitrogen, total nitrogen, and total phosphorus. During the trial, the defendant installed a rain and sewage flow diversion system, and these installations were approved by the local EEB. Consequently, the court ordered the defendant to develop a feasible environmental management system to ensure compliance and submit regular environmental compliance reports to the court.

---

[33] See Sun and Tuholske (2017), pp. 10503–10504 and Xie and Xu (2021b), pp. 9–10.

[34] See Guiding Case No. 131, *All China Envtl. Fed'n v Jinghua Group Zhenhua Co., Ltd.*, https://www.court.gov.cn/fabu-xiangqing-216931.html.

[35] See Model EPIL Case No. 3, *Friends of Nature v Jinling Chem. Co., Ltd.*, http://www.xinhuanet.com//legal/2017-03/07/c_129503217.htm.

[36] See *All China Envtl. Fed'n v Jiangshan Biological Preparation Co., Ltd.*, Wuxi Intermediate People's Court, First Instance No. 2 (2014).

Similar to the U.S. citizen suit, the Chinese civil EPIL permits environmental NGOs to initiate independent enforcement suits if agencies are unable or unwilling to enforce the law, as illustrated by the *Jiangshan* case. In the *Zhenhua* and *Jinling* cases, however, EEBs have taken enforcement measures before environmental NGOs filed suit. Due to the limited deterrent effect of statutory penalties or other deficiencies embedded in government enforcement, enforcing agencies failed to stop the violation. Since courts have broad authority to enjoin statutory violations, order cleanup actions, and award monetary compensation for ecological damage, civil EPIL can address repeated or continuing violations more effectively than government enforcement. In sum, environmental NGOs may resort to civil EPIL as an alternative enforcement mechanism to cure government enforcement failures caused either by political pressure or resource shortages or by the limited deterrence offered through government enforcement.

10.3.2 Filling Regulatory Gaps

Environmental laws in any country will never form a seamless web covering all types of environmental problems. While China has achieved remarkable progress in enacting a wide range of environmental laws since the late 1970s, there are inevitable limitations and gaps in existing regulatory laws. For example, some issue areas or pollutants (e.g., management of toxic chemicals, light pollution, persistent organic pollutants) have not been effectively regulated by legislation. Even if an environmental concern is regulated by law, compliance with weak environmental standards is usually insufficient to protect the environment.[37] A major gap in Chinese environmental laws is that both traditional tort law and environmental regulatory laws fail to provide an effective remedy for ecological damage sustained from pollution.[38] Since the 2015 EPIL Interpretation adapted traditional tort law remedies to address ecological damage, environmental NGOs have brought civil EPIL to seek relief for past ecological injuries where a statutory remedy is currently unavailable or incomplete.

As the first case heard after the 2014 EPIL took effect, *Friends of Nature et al. v Xie Zhijin et al.* filed at the Nanping Intermediate People's Court in Fujian Province was notable for the broad scope of ecological damages awarded (the *Nanping* case).[39] In this case, two environmental groups, Friends of Nature and Green Home Environment-Friendly Center, alleged that the four individual defendants had engaged in mining activities without obtaining relevant permits, causing extensive damage to the area's vegetation and ecological system. The plaintiffs filed the

---

[37] See Chu (2019), pp. 507–509.

[38] See Chu (2019), pp. 493–495 and Faure and Liu (2014).

[39] See Model Environmental Tort Case No. 1, *Friends of Nature et al. v Xie Zhijin et al.*, https://www.court.gov.cn/zixun-xiangqing-16396.html.

suit to seek restoration of the destroyed land and compensation for the loss of interim ecological services. The first-instance court held that the defendants' illegal mining operations had caused ecological damage and constituted harm to the public interest.[40] Ruling in favor of the plaintiffs, the court ordered the defendants to restore the site by replanting trees, or pay 1.1 million yuan (approximately 0.16 million US dollars) for site remediation if they failed to perform the restoration. The court also ordered the defendants to pay 1.27 million yuan for ecological interim losses during recovery. On appeal, the ruling of the trial court was upheld by both the Higher People's Court of Fujian Province and the SPC.[41] Because the *Nanping* case is the first case where plaintiffs requested damages for ecological injuries, its ruling is seen as an important victory for environmental NGOs.

Another case that resulted in significant damages for ecological injury is the landmark *Taizhou Environmental Protection Association of Jiangsu Province v. Taixing Jinhui Chemical Company, et al.* (the *Taizhou* case).[42] In this case, the plaintiff group alleged that six chemical companies illegally dumped thousands of tons of chemical waste into two rivers in Taizhou, causing severe ecological damage to the Yangtze River Basin. The first-instance court found six defendants liable for the ecological damage resulting from their illegal disposal of acid waste and assessed about 160 million yuan (approximately 23 million US dollars) in restoration costs, making this case famous for the "sky-high" award of damages.[43] This case was also appealed to the SPC, which upheld the trial court's ruling.[44]

In addition to overcoming the weaknesses of statutory remedies for ecological damage, civil EPIL has also been invoked to fill other gaps in existing laws. For example, due to a lack of a meaningful legal framework for addressing transboundary pollution disputes,[45] environmental NGOs have fielded a civil EPIL suit to seek redress for transboundary pollution. In *Jianshi Huangchangping Mining Co., Ltd. v. Green Volunteer League of Chongqing* (the *Jianshi* case),[46] the defendant was a pyrite processing plant located in Jianshi County of Hubei Province. On 10 August 2014, the plant discharged untreated wastewater and tailings into a nearby karst depression that seeped through underground caves and eventually reached and contaminated the Qianzhangyan reservoir. The Qianzhangyan reservoir that is located in Wushan County, Chongqing, borders Jianshi County, and is the source of drinking water for Wushan County. The relevant authorities in Chongqing and

---

[40] See Nanping Intermediate People's Court, First Instance No. 38 (2015).

[41] See Higher People's Court of Fujian Province, Second Instance No. 2060 (2015); Supreme People's Court, Retrial No.1919 (2016).

[42] See Model EPIL Case No. 1, *Taizhou Envtl. Prot. Ass'n of Jiangsu Prov. v Taixing Jinhui Chem. Co., Ltd. et al.*, http://www.xinhuanet.com//legal/2017-03/07/c_129503217.htm.

[43] See Bie (2015).

[44] See Supreme People's Court, Retrial No. 1366 (2015).

[45] See Tyler (2006).

[46] See Guiding Case No. 134, *Jianshi Huang Changping Mining Co., Ltd. v Green Volunteer League of Chongqing*, https://www.court.gov.cn/fabu-xiangqing-216961.html.

Hubei Province immediately took emergency actions to respond to the pollution incident. The EEB of Hubei Province imposed a penalty of 1 million yuan and ordered the plant to cease the discharge of pollutants and undertake cleanup actions according to the Water Pollution Prevention and Control Law (WPPCL). While the plant paid the penalty, it failed to address the cleanup of the pollution as required by the administrative order. Additionally, the source county's EEB also did not take any rigorous action to enforce its order.

Because of territorial jurisdiction, the EEB of the affected county had no power to take enforcement actions against a pollution source located outside its jurisdiction. Green Volunteer League of Chongqing, a local environmental group in Chongqing, filed a civil EPIL suit against the plant located in Hubei Province. Rejecting the notion that the payment of an administrative penalty by the defendant barred the plaintiff's action, the court ruled that civil EPIL liability can operate in parallel with administrative and criminal liability. While the environmental impact assessment (EIA) report for the plant had been approved by the competent EEB of Hubei Province, the court found that the original EIA report failed to examine the anticipated environmental effects of the plant on the Qianzhangyan reservoir, which was designated as the source water protection area by the Chongqing government. Finding that the construction of the plant in the karst area posed a significant risk of contamination and the plant's discharges had led to transboundary pollution, the court granted the plaintiff's request for "cessation of infringement" by issuing an injunction prohibiting the defendant from operating until it had completed a new EIA. Moreover, the court required the defendant to remediate existing contamination or pay 991,000 yuan for restoration if the defendant failed to perform the remediation.[47]

The *Nanping, Taizhou* and *Jianshi* cases demonstrate how environmental NGOs have utilized civil EPIL to seek relief for ecological damage caused by pollution or other events (e.g., illegal mining operations in the *Nanping* case) and address transboundary pollution. Statutory remedies seem absent or inadequate in these cases, and civil EPIL brought by environmental NGOs provides an independent or complementary remedy for the environmental harm. Since there are always various gaps in existing environmental laws, NGO-led civil EPIL will continue to play a useful gap-filling role whenever statutory remedies for ecological damage or other environmental harm are not available or are incomplete.

10.3.3 Challenging the Legitimacy of Government Decisions

As mentioned above, the administrative EPIL framework established by the 2017 ALL does not authorize environmental NGOs to challenge government actions that

---

[47] See Wanshan People's Court of Chongqing, First Instance No. 00001; Second Intermediate People's Court of Chongqing, Second Instance No. 772 (2016).

violated the law. Nevertheless, environmental NGOs have found ways to circumvent legal barriers by filing civil EPIL suits. This innovative strategy is illustrated by a high-profile case seeking the protection of endangered species. In March 2017, environmental groups found that an ongoing hydropower project on the Jiasa River of Yunnan Province would destroy the core habitat of the green peafowl, an endangered species classified as National First-class Protected Wild Animal in China and listed as endangered on the International Union for Conservation of Nature's Red List.[48] The EIA report for the 270 MW Jiasa River hydropower project was approved by the then Ministry of Environmental Protection (MEP) in August 2014. To save the green peafowl, environmental groups tried to convince the MEP to revoke EIA approval by submitting petition letters and organizing academic seminars.[49] After these efforts failed, Friends of Nature sued the project proponent Xinping Company and the EIA firm to block construction of the dam (the *Green Peafowl* case).[50]

Between July 2017 and April 2018, Friends of Nature organized a group of scientists specializing in flora and fauna to conduct several biodiversity baseline studies, collecting first-hand data on the distribution and number of rare species along the Jiasa River.[51] In the trial, Friends of Nature argued that the hydropower project would submerge the largest and most intact natural habitat of the green peafowl in China and would likely result in the local extinction of the green peafowl. In addition to impacts on peafowl habitat, the power plant would also cause serious damage to several rare and endangered species (e.g., Cycas trees) and the monsoon forests in the reservoir area. Meanwhile, Friends of Nature argued that the project's EIA report was not supported by scientific evidence and field study because it concluded that the construction would not affect the living and breeding of the green peafowl and failed to evaluate its effect on the Cycas trees. In March 2020, the court of first instance suspended the Jiasa River hydropower project. According to the ruling, competent authorities would decide if this project would proceed after the Xinping company conducted an ex-post EIA. Both the plaintiff and defendant appealed, and the court of second instance dismissed the appeals and upheld the original decision. Although a temporary injunction disappointed the plaintiff, the defendant abandoned the Jiasa River project permanently because construction of the dam would violate the management requirements of Yunnan's ecological red lines delineated in 2018.[52]

---

[48] See Wang (2020).

[49] See Friends of Nature (2017) and interview with Friends of Nature.

[50] See Guiding Case No.173, *Friends of Nature v China Hydropower Eng'g Consulting Group Xinping Dev. Co., Ltd. & Power China Kunming Survey and Design Inst. Co., Ltd.*, https://www.court.gov.cn/fabu-xiangqing-334691.html; Higher People's Court of Yunnan Province, Second Instance No. 824 (2020).

[51] See Wang (2020).

[52] Ecological red lines refer to "the strictly controlled boundaries legally drawn for key areas of ecological functions and ecologically sensitive or fragile areas". See UNEP (2016).

Friends of Nature used the same strategy to file another suit to protect wild bird habitat in Lianyungang (the *Lianyungang* case).[53] Coastal wetlands in Lianyungang, including the Linhong Estuary, are natural habitats of numerous endangered species of shorebirds listed as National Key Protected Wild Animals (e.g., spoon-billed sandpiper, semipalmated sandpiper). The Linhong Estuary is also known as an important site for shorebirds migrating along the East Asian-Australasian Flyway. The "Blue Bay" coastal restoration project, which is located in the Linhong Estuary, is intended to restore damaged wetlands and shoreline ecosystems. However, Friends of Nature alleged that the project would have detrimental and irreversible impacts on coastal wetlands and destroy the habitat of protected wild bird species. Additionally, Friends of Nature claimed that relevant EIA reports of the project were seriously flawed and inadequate in the analysis of its impact on wild birds and their habitat. On 24 May 2021, Friends of Nature filed a civil EPIL suit against the developer of the "Blue Bay" project and the EIA firm, requesting that the defendants stop construction activities and restore the damaged areas.

"A chronic lack of capacity" has long hindered the implementation of EIAs in China.[54] In particular, owing to a lack of awareness of biodiversity and its value and crucial baseline data among EIA practitioners, EIAs in China generally perform poorly in assessing likely ecological damage caused by a proposed development project, let alone provide useful alternatives.[55] In both the *green peafowl* and *Lianyungang* cases, the plaintiff NGOs argued that the EIA reports of the projects were flawed because they failed to provide accurate foundational data and adequately analyze the effect of the proposed projects on protected species and their habitats. According to EIA-related laws and regulations in China, EEBs could not approve the EIA report if the baseline data were evidently inaccurate, the content was significantly flawed or omitted, or the EIA conclusion was incorrect or unreasonable.[56] Since environmental NGOs are denied access to courts to challenge agency actions, it is impossible for them to seek judicial review of EEBs' approval of EIA reports in the service of the public interest. By bringing civil EPIL directly against project developers and EIA firms, the *green peafowl* and *Lianyungang* cases essentially challenged the merits of the EEBs' approval of the EIA reports. The rationale was that a deficient EIA report should not have been approved by the EEB in the first place; if the court is persuaded that a project based on problematic EIA analysis constitutes a harm to the public interest and orders a halt to the project, the civil EPIL suit has the effect of overturning the EEB's substantive decision on the EIA report. In the *green peafowl* case, relying on crucial baseline information and persuasive assessment of direct and indirect impacts on habitats and biodiversity in

[53] For the detailed information about this case, see Zhang (2022), Yang (2021) and Ma (2021).

[54] See Kim and Jones (2006), p. 101.

[55] See Beach et al. (2006).

[56] See Art.11 Regulation on the Management of Environmental Protection of Construction Projects (1998, revised 2017); Arts. 28, 32, 34 Environmental Impact Assessment Law (2002, amended 2016 and 2018).

the project areas, Friends of Nature successfully convinced the court that the dam would pose a significant risk of habitat and biodiversity loss. While the court could not invalidate the government approval of the EIA report in a civil EPIL suit, it did acknowledge that the EIA report of the project was deficient.

It should be noted that while civil EPIL provides an indirect way to challenge the legitimacy of government decisions, the suit can only be filed after construction has begun. As shown by the *green peafowl* and *Lianyungang* cases, by the time the suit was filed, the construction activities had caused some damage to the environment. In fact, much of the irreparable ecological damage occurs in the early phase of the construction process. Moreover, it is practically impossible for the court to disregard the sunk costs already invested in the project when deciding on the case and fashioning the remedy. Therefore, the better way is to grant standing to environmental NGOs to bring administrative EPIL so that the court can review the substantive and procedural legality of government decisions before damages occur.

10.3.4 Setting the Agenda in the Regulatory Process

While procuratorates have brought more civil EPIL cases than environmental NGOs in recent years,[57] as commented on by the vice-president of the Environmental and Resources Tribunal of the SPC, NGO-led civil EPIL has involved diverse types of cases, attracted wide public attention, and had a far-reaching impact.[58] Procuratorates prefer to pick low-hanging fruits (e.g., small and weak defendants, minor violations, and low-value damages),[59] whereas environmental NGOs are more willing to pursue cases involving powerful enterprises, complicated facts, and high-value damages.[60] A batch of NGO-led cases were selected as various types of model cases and guiding cases released by the SPC for their innovation. Intensive media coverage of many high-profile NOG-led lawsuits has not only raised public awareness of civil EPIL but has also placed issues such as biodiversity protection and climate change on the regulatory agenda.

Climate change litigation provides an important illustration of civil EPIL's role in setting the agenda in the regulatory process. While litigation has become an increasingly important tool to deal with climate change issues in many jurisdictions around the world, climate change litigation is relatively new in China. The breakthrough occurred in September 2016, when Friends of Nature filed two civil EPIL suits

[57] See Xie and Xu (2021a), pp. 10–11, Xie and Xu (2021b), pp. 17–19 and Gao and Whittaker (2019), pp. 337–341.

[58] See Xi and Hou (2018).

[59] See Ding and Xiao (2021), Xie and Xu (2021b), pp. 11–13 and Shi and Van Rooij (2016), pp. 49–51.

[60] See Xie and Xu (2021b).

against grid enterprises to tackle the issue of wind and solar curtailment.[61] The rapid development of renewable power in China has given rise to high curtailment rates since 2000. Curtailment was particularly severe for 2016, with national average curtailment rates of wind and solar power reaching 17% and 10%, respectively.[62] Wind curtailment rates were extraordinarily high for certain wind-rich provinces such as Gansu (43%), Xinjiang (38%) and Jilin (30%).[63] In order to mitigate curtailment, Article 14 of the Renewable Energy Law (REL) required prioritized grid access for renewable energy generators and guaranteed the full purchase price of renewable energy inputs by grid enterprises. However, these requirements have not been effectively enforced in practice for a variety of reasons.[64] Citing Article 14 of the REL as a legal basis, Friends of Nature sued branches of the State Grid Corporation of China in Ningxia and Gansu for failing to purchase in full for the electricity generated by wind and solar power within their geographic boundaries (the *Gansu* case and the *Ningxia* case, respectively).[65] According to Friends of Nature, the curtailed wind and solar power was replaced by coal-fired power, which in turn increased the emission of air pollutants and greenhouse gases. Asserting that wind and solar curtailment caused harm to the public interest, Friends of Nature requested the defendants to fulfil their full purchase of renewable energy obligations and pay for ecological damages caused by electricity generation from coal-fired plants. While acknowledging that grid enterprises were not fully responsible for the poor enforcement of Article 14 of the REL, Friends of Nature insisted that the purpose of initiating civil EPIL suits was to stimulate the public debate over the causes and solutions of curtailment.[66]

On 14 August 2018, the *Gansu* case was dismissed by the Lanzhou Intermediate People's Court because the grid enterprise itself did not commit the act of polluting or damaging the environment.[67] On 24 January 2019, the appellate court reversed the trial court's decision, ruling that the requirements for filing a civil EPIL suit were satisfied and ordering Gansu Kuangqu People's Court to hear the case.[68] The *Ningxia* case was accepted by the Yinchuan Intermediate People's Court on

---

[61] Curtailment refers to the deliberate abandonment of electricity generation of effective power capacity usually due to system-wide oversupply or transmission constraints.

[62] See National People's Congress Standing Committee Law Enforcement Inspection Team, Report on the Inspection of the Implementation of the Renewable Energy Law, 24 December 2019, http://www.npc.gov.cn/npc/c30834/201912/2b7568de01944c33b9326c325dcd498f.shtml.

[63] See National Energy Administration, Accessing to the Grid Network by Wind Power in 2016, 26 January 2017, http://www.nea.gov.cn/2017-01/26/c_136014615.htm.

[64] See Zhang (2019b) and Liu and Xu (2018).

[65] See Diao (2018a).

[66] See Friends of Nature, Our Aim Is to Draw Attention to the Issue of Wind and Solar Curtailment and Deal with This Problem, 12 April 2018, https://mp.weixin.qq.com/s/MqfZRVZGYx2ASCuDDFhxzw.

[67] See Diao (2018b).

[68] See *Friends of Nature v State Grid Gansu Electric Power Co.*, Higher People's Court of Gansu Province, Second Instance No. 679 (2018).

11 January 2018. The latest development in the *Ningxia* case is that the plaintiff and the defendant had an exchange of evidence on 19 July 2020.[69]

As the first two cases that explicitly raised climate-specific arguments, the *Gansu* and *Ningxia* cases gained widespread public attention. The *Gansu* case was selected as one of the Top Ten Public Interest Lawsuits in China in 2018.[70] Mainstream media reports on the *Gansu* and *Ningxia* cases also generated publicity for climate change litigation and stimulated serious academic discussion.[71] More importantly, climate change litigation has risen on the judicial agenda of the SPC. In the White Paper on China's Environmental and Resources Trial (2019), the SPC now requires courts at all levels to hear climate change cases in accordance with law and specifically mentions the *Gansu* and *Ningxia* cases when discussing mitigation litigation.[72] In February 2021, the SPC issued an opinion on implementing the Yangtze River Protection Law, which again called for the acceptance of trials of climate change cases.[73] These moves suggest that climate change litigation has emerged as a field of litigation that is of increasing significance in China's judicial practice.

The years since the *Gansu* and *Ningxia* cases were filed have also seen increased regulatory efforts to address wind and solar power curtailment. For instance, the National Development and Reform Commission (NDRC) and the National Energy Administration issued "Implementation Measures to Resolve the Curtailment of Hydro, Wind and Solar Energy" in 2017 and "Clean Energy Consumption Action Plan (2018–2020)" in 2018; both policy initiatives developed a range of mechanisms and strategies to reduce curtailment in China.[74] The average curtailment rates of wind and solar power began to reduce gradually since 2018 and decreased to 3.1% and 2%, respectively, in 2021.[75] It is not suggested that there is a clear causal link between regulatory responses and climate change litigation. However, it is fair to argue that the *Gansu* and *Ningxia* cases, in combination with other factors, played a role in altering the regulatory environment to address climate change.

---

[69] See Friends of Nature, Ningxia Wind Curtailment case: Creating Bright Futures for Children, 5 August 2020, http://www.fon.org.cn/action/domain/content/160.

[70] See Zhang (2019a).

[71] For scholarly work on the *Gansu* and *Ningxia* cases, see Chen (2020), Zhang (2020), He (2021) and Zhu (2022).

[72] See SPC, White Paper on China's Environmental and Resources Trial (2019), https://www.court.gov.cn/zixun-xiangqing-228341.html.

[73] See SPC Opinion on Implementing the Yangtze River Protection Law (2021), https://www.court.gov.cn/fabu-xiangqing-287881.html.

[74] See NDRC and NEA, Implementation Measures to Resolve the Curtailment of Hydro, Wind and Solar Energy (2017), http://zfxxgk.nea.gov.cn/auto87/201711/t20171113_3056.htm; NRDC and NEA, Clean Energy Consumption Action Plan (2018–2020) (2018), https://www.ndrc.gov.cn/xxgk/zcfb/ghxwj/201812/t20181204_960958.html?code=&state=123.

[75] See NEA, Press Conference of NEA on Accessing to the Grid Network by Renewable Energy in 2021, 28 January 2018, http://www.gov.cn/xinwen/2022-01/29/content_5671076.htm.

10.4 Pitfalls of Using Civil Environmental Public Interest Litigation as a Regulatory Tool

While NGO-led civil EPIL performs valuable regulatory functions, its role is limited by various factors. For example, the conservative standing rule effectively excludes a large number of environmental NGOs from filing suits and imposes unnecessary restrictions on their ability to sue.[76] Despite the SPC's relatively liberal approach to NGOs' standing in the Tengger Desert pollution case,[77] some lower courts have begun to demand additional documents from environmental NGOs to prove standing. In addition, high litigation costs also make civil EPIL an expensive regulatory tool for underfunded environmental NGOs.[78] In addition to institutional constraints, internal factors such as a lack of technical knowledge and legal expertise have limited environmental NGOs' capacity to deploy litigation to bring regulatory change.[79] Since these obstacles have been well analyzed in existing literature, this section turns its attention to the potential pitfalls of utilizing NOG-led civil EPIL as a regulatory tool.

10.4.1 Threatening the Primacy of Government Enforcement

Because civil EPIL covers "acts of polluting or damaging the environment that have harmed the public interest," environmental NGOs may file suit to enforce environmental violations when agencies fail to enforce due to lack of will or resources or the limited deterrence provided by government enforcement. When civil EPIL serves as an alternative mechanism for enforcing environmental laws, it should not replace government enforcement actions and burden the courts. To preserve the government's primary role in enforcing environmental laws, the U.S. citizen suit requires a 60-day notice before filing a suit, which gives the violator an opportunity to correct the violation and avoid the need for litigation.[80] More importantly, the notice will ensure that the enforcing agency is aware of the alleged violation and is provided an opportunity to take enforcement actions.[81] A citizen suit is barred if the agency has

[76] See, e.g., McCallum (2017), pp. 79–80, Zhai and Chang (2018), pp. 383–385, Gao (2018), pp. 64–67 and Gao and Whittaker (2019), pp. 341–346.

[77] See McCallum (2017), pp. 73–74, Sun and Tuholske (2017), p. 10506 and Liu (2019), pp. 230–231.

[78] See, e.g., Zhang and Mayer (2017), pp. 224–225, Zhai and Chang (2018), pp. 385–386, Gao (2018), pp. 70–72 and Xie and Xu (2021a), pp. 16–20.

[79] See, e.g., Zhai and Chang (2018), p. 386 and Xie and Xu (2021b), pp. 18–19.

[80] See Reisinger et al. (2010), p. 11.

[81] See *id.*

commenced or is diligently prosecuting a civil action to require compliance.[82] The civil EPIL framework does not contain a similar 60-day notice mechanism. Therefore, many scholars are concerned that plaintiffs may inappropriately invoke the judiciary to infringe upon an agency's primary responsibility for environmental law enforcement.[83]

In fact, the 2015 EPIL Interpretation does contain a notice requirement. The court is required to notify the agency that holds the regulatory power over the defendant's conduct within 10 days after accepting a case.[84] If the competent agency has performed its regulatory duty so that all the claims of the plaintiff become moot, the court should permit the withdrawal of the suit by the plaintiff.[85] However, this notice does not function the same way as the 60-day notice in citizen suits. First, it is the court, not the plaintiff, that is responsible for notifying the enforcing agency. The alleged violator is not given an opportunity to correct the violation. Second, compared to the 60 days in U.S. citizen suit, the 10 days in civil EPIL cannot provide sufficient time for agencies to take effective enforcement actions. Third, while an agency's enforcement action might render the plaintiff's claims moot, it does not automatically block the suit unless the plaintiff voluntarily withdraws the suit. In sum, the 10-day notice requirement is insufficient to allow government agencies to take the lead role in enforcing environmental laws.

To ensure that civil EPIL supplements rather than supplants government enforcement, the 2015 EPIL Interpretation should explicitly incorporate a 60-day notice requirement. When agencies fail to enforce due to lack of will or resources, the function of the 60-day notice in civil EPIL would be similar to that in US citizen suits. However, when diligent government enforcement turns out to be unsuccessful in bringing the recalcitrant polluters into compliance during the 60-day notice period, environmental NGOs are not prevented from utilizing civil EPIL as an alternative venue to address the violation. In either case, the 60-day notice would preserve the government's role as the primary enforcer of environmental laws.

10.4.2 Displacement of Statutory Remedies

As an adjunct to government regulation, civil EPIL can provide flexible remedies for environmental harms that occurred or are left uncured due to limitations and gaps in existing environmental laws. Since civil EPIL is supposed to play a gap-filling role, it is justified when a statutory remedy is unavailable or incomplete. However, in some civil EPIL lawsuits seeking remedies for ecological damage, whether there is a

---

[82] See, e.g., Clean Water Act, 33 U.S.C. § 1365(b)(1)(B) (2012); Clean Air Act, 42 U.S.C. § 7604(b)(1)(B) (2012).

[83] See, e.g., Wang (2016), Gong (2017), pp. 109, 123 and Luo (2017), pp. 253–254.

[84] Art. 12 2015 EPIL Interpretation.

[85] Art. 26 2015 EPIL Interpretation.

lack of a statutory remedy deserves careful analysis. For example, in the *Nanping* case discussed above, the defendants violated the Forestry Law by conducting a mining project without obtaining approval for the use of forest lands.[86] According to the regulation on the implementation of the Forestry Law, competent forestry authorities can impose administrative fines on parties who changed the use of forest land without a permit and order the violator to restore the forest land to its original state.[87] It is not clear what the objective of restoration is (e.g., how to define "original state") and how this enforcement tool has been applied in practice. Therefore, it is difficult to assess whether the Forestry Law and its implementing regulations provided a sufficient remedy for the damage to forest land resulting from defendants' illegal mining activities.

Similarly, in the *Taizhou* case, Article 85 of the WPPCL authorizes EEBs to impose a penalty for discharging any oil, acid, or alkaline solution and highly toxic waste liquid into a water body and/or issue an order requiring correction and cleanup within a specified time period. Where a violator fails to clean up pollution, EEBs can perform the cleanup at the expense of the violator. Again, the content and the level of required cleanup are ambiguous and there is little guidance on when a cleanup order can be issued. Therefore, it is not clear whether the remedies provided by the WPPCL can fully redress ecological damage caused by the defendants' illegal dumping as in the *Taizhou* case.

When a statutory remedy is absent or weak, NGO-led civil EPIL provides a viable tool to secure cleanup and restoration of an injured environment. However, if applicable environmental statutes can provide the very relief requested by environmental NGOs, the competent agency should be given an opportunity to exhaust the available statutory remedy before a civil EPIL suit is initiated. Otherwise, there is a risk that NGO-led civil EPIL would overstep its role of filling gaps in the existing statutory framework. In order to maintain the primacy of statutory remedies and avoid agencies from abdicating their responsibilities, courts need to engage in an analysis of whether statutory laws can adequately remedy the specific type of environmental harm alleged by environmental plaintiffs. If the environmental problem in question is covered by a developed regulatory scheme, the plaintiff should follow the 60-day notice procedure. If the court determines that there is no adequate statutory remedy for the alleged environmental harm, the court should allow the civil EPIL suit to proceed without going through the 60-day notice procedure. By requiring the court to review whether there is a real statutory gap, it can ensure that NGO-led civil EPIL does not displace statutorily-created remedies that are intended to be primary in addressing environmental harm.

[86] The Forestry Law requires approval by relevant forestry authorities if prospecting, mining, and various construction projects need to occupy woodlands. See Art. 37(1) Forestry Law (1984, amended 1998 and 2009, revised 2019).

[87] See Art. 43(1) Regulation on the Implementation of the Forestry Law (2000, revised 2011, 2016 and 2018).

10.4.3 Unreasonable Calculation of Environmental Damages

By extending tort law remedies to civil EPIL, the 2015 EPIL Interpretation authorizes courts to provide not only injunctive relief (i.e., cessation of infringement, removal of obstacles, elimination of danger, restoration of the environment) but also damages for ecological injuries. Regarding the calculation of damages, Article 23 of the 2015 EPIL Interpretation stipulates that when it is difficult or costly to assess the amount of damages for ecological injuries, courts may determine the restoration cost by taking into consideration opinions of relevant administrative agencies and experts as well as factors such as the scope and extent of pollution and ecological damage; scarcity of the ecological resources; difficulty of restoration; costs of operating pollution control facilities; economic benefits gained by the defendant; and fault on the defendant. In practice, the so-called "virtual cost of pollution treatment"—the amount of money needed to remove pollutants released into the environment with current available treatment technologies[88]—is widely used to calculate damages for ecological injuries.

Relying on this technical expediency, environmental NGOs have successfully secured substantial amounts of damage from ecological injuries caused by defendants' excessive discharges. For example, damages in the *Zhenhua* and *Taizhou* cases were all calculated by this convenient virtual cost of pollution treatment method. In these cases, environmental NGOs did not need to prove the scope and extent of the injury, the causal link between the defendants' conduct and the injury, and the necessity and feasibility of restoration. The underlying assumption is that illegal discharges necessarily cause an injury to the environment itself that requires compensation, regardless of whether it is possible to prove, quantify, and remediate the ecological injury.

"Restoration of the environment" and "damages" in the civil EPIL are specifically formulated to remediate ecological injuries. This liability regime is quite similar to the liability system established under certain U.S. environmental statutes. For example, the Comprehensive Environmental Response, Compensation and Liability Act (CERCLA) establishes a comprehensive scheme for providing broad federal authority to respond promptly to releases or threatened releases of hazardous substances, and imposing liability upon responsible parties for costs of such response and natural resource damage resulting from releases.[89] CERCLA expands the scope of recovery by authorizing the federal government to recover response costs incurred.[90] Moreover, the natural resource damages provisions of CERCLA incorporated the goal of the common tort law "to make whole the injured plaintiff" but replaced "the common-law damage theories of market valuation and least cost solutions" with a new statutory approach to damages is "restoration of an injured

[88]See Ministry of Environmental Protection, Recommended Methods for Ecological Damage Appraisal and Assessment (II)] (2014), Appendix A, A.2.3.

[89]42 U.S.C. §§ 9601–9675 (2012).

[90]See Latham et al. (2011), p. 756.

natural resource".[91] To recover damages for injuries to natural resources, the designated trustees must be able to demonstrate an injury to the affected natural resources,[92] determine that there is a pathway linking the release to the injury,[93] quantify the effect of the release in terms of a reduction in natural resource services from the baseline condition,[94] and once restoration is warranted, convert the proven injury into an appropriate amount of damages. Under CERCLA and its regulations, damages include two components: (1) restoration costs—the costs to restore, rehabilitate the injured resources, or the costs to replace and/or acquire the equivalent of resources; and (2) compensable value—the value of services lost to the public for the time period from the release until the return of baseline level of resources/services.[95]

Because the procedures and capabilities for assessing ecological damage in China are still underdeveloped, the 2015 EPIL Interpretation allows for the application of the virtual pollution treatment cost method. While environmental NGOs tend to favor such an eco-centric theory of damages, it may be resisted by defendants. For instance, in the *Zhenhua* case, one defense raised by the defendant was that it was difficult to attribute air pollution to the defendant's emissions given diverse sources of air pollution. Similarly, the defendant in an air pollution case brought by Friends of Nature contended that local air quality data showed no decline in air quality despite the defendant's excessive emissions, so the plaintiff's claim for environmental damages is not supported by facts and law.[96] Given these controversies, policy makers and legal practitioners should re-examine the advantages and disadvantages of measuring damages by the virtual cost of pollution treatment. While an injunction is an appropriate remedy for continuing or repeated harmful activities, damages calculated by the virtual cost of pollution treatments are incompatible with traditional tort damages theory where it is difficult to define the affected area, quantify the injury directly caused by the defendant's discharge, and actually restore the injured environment (e.g., air pollution cases).

10.4.4 The Misuse of Civil EPIL

With its sweeping scope and flexibility, civil EPIL has been invoked by environmental NGOs to challenge a variety of acts that allegedly constitute a harm to the public interest. For example, environmental NGOs have brought suits against

---

[91] See Lee and Bridgen (2002), p. 9.

[92] Injury is broadly defined as a measurable adverse change in the chemical or physical quality or the viability of a natural resource resulting directly or indirectly from exposure to a release of a hazardous substance. See 43 C.F.R. § 11.14 (v) (2017).

[93] See *id.* § 11.63.

[94] See *id.* § 11.70 (a).

[95] See *id.* § 11.80 (b).

[96] See Diao (2016).

Volkswagen for using software to rig emission tests and against other automobile manufacturers for selling certain types of vehicles that failed to meet emission standards.[97] Sellers on China's leading eCommerce website, Taobao, were also sued for causing air pollution by selling devices that could assist vehicle owners to pass the annual exhaust inspection,[98] and for causing damage to soil ecosystems by selling earthworm electronic shock devices.[99] Similar to the *Ningxia* and *Gansu* cases, defendants in these cases did not directly discharge pollutants or damage the environment. Nevertheless, the environmental plaintiffs maintained that defendants' conduct harmed the public interest and sought environmental damages measured by the virtual cost of pollution treatment. This raised concerns about the possible misuse of civil EPIL.

A recent case on the protection of endangered plants epitomizes the possible misuse of civil EPIL. Relying on a study report on the Yalong River hydropower development project, the China Biodiversity Conservation and Green Development Foundation (CBCGDF) alleged that the planned Yagen cascade hydropower plant on the Yalong River in Sichuan Province posed a serious threat to the existence of an endangered species, the acer pentaphyllum (the *Yalong* case).[100] The plaintiff requested that the court to order the hydropower company to (1) take appropriate measures to ensure that the proposed hydropower projects would not jeopardize the survival of the acer pentaphyllum; (2) suspend the construction of the Yagen hydropower plant and its auxiliary facilities (e.g., roads) until the measures taken are sufficient to eliminate the threat to the acer pentaphyllum; and (3) bear the plaintiff's travel expenses, investigation expenses, and attorney's fees incurred.

During the trial, the court found that the Yagen hydropower plant was still in the pre-feasibility study stage and had not yet been approved by the NDRC. Contrary to the plaintiff's allegations, construction of the Yagen hydropower station and its auxiliary works have not begun. Due to the lack of reliable sources, the court did not accept the data provided by the CBCGDF on the number and distribution of the acer pentaphyllum in the Yalong River basin. Despite these factual findings, the court issued a verdict in favor of the plaintiff. Since the construction of the Yagen hydropower station had not yet started, the court denied the CBCGDF's request to suspend the construction. Nevertheless, the court ruled that the defendant should focus on analyzing the impact of the hydropower plant on the existence of the acer pentaphyllum when conducting the EIA, and that only after the EIA report is approved by the environmental authority could the defendant proceed to the next

---

[97] See Spring (2015); 2019 Model Environmental and Resources Case No. 33, *Friends of Nature v Hyundai Motor*, https://www.court.gov.cn/zixun-xiangqing-228361.html.

[98] See 2019 Model Environmental and Resources Case No. 31, *China Biodiversity Conservation and Green Development Foundation v Shenzhen Sumei Co. & Taobao*, https://www.court.gov.cn/zixun-xiangqing-228361.html.

[99] See Yuan (2020).

[100] See *China Biodiversity Conservation and Green Development Foundation v Yalong River Hydropower Development Co., Ltd.*, Sichuan Ganzi Tibetan Autonomous Prefecture Intermediate People's Court, First Instance No. 45 (2015).

step. Given that the CBCGDF filed the lawsuit for the purpose of protecting the public interest, the court partially supported the plaintiff's request for reimbursement of litigation costs.

The ruling in this case is somewhat surprising especially when comparing it with the *green peafowl* case. In the *green peafowl* case, the ongoing construction of the hydropower project based on a flawed EIA report posed a significant threat to the green peafowl's habitat. Yet, the defendant in the *Yalong* case has not engaged in any conduct detrimental to the public interest, as acknowledged by the court. With respect to the remedy, it seems meaningless to order the defendant to conduct the EIA and analyze the impact of hydropower project on the acer pentaphyllum, since the defendant would have been required by the law to complete the EIA before starting construction in any event.

Because of the amorphous and mutable quality of civil EPIL, plaintiffs can challenge almost anything that constitutes a harm to the public interest. Compared with procuratorates, environmental NGOs are more willing to test the boundary of civil EPIL. While many cases represent applaudable innovative use of civil EPIL, environmental groups should be cautious about invoking civil EPIL when the defendant did not engage in obvious harmful conduct and when there is an attenuated line of causation to the eventual injury alleged. At the same time, the courts should also exercise judicial restraint to avoid the misuse of civil EPIL and a waste of judicial resources.

10.5 Conclusion

While environmental NGO activism in China had featured environmental education, awareness-raising, research and advocacy in the early days due to a lack of formalized channels,[101] civil EPIL provides a legal pathway for environmental NGOs to become involved in the environmental regulatory process. By examining the theoretical foundation and judicial practice of civil EPIL, this chapter suggests that NGO-led civil EPIL can perform an important regulatory function in China's environmental governance. With its broad scope and availability of flexible remedies, civil EPIL has been invoked by environmental NGOs to provide an alternative enforcement mechanism when government enforcement fails for a variety of reasons; it also fills various gaps in regulatory schemes when a statutory remedy is absent or incomplete and enables an NGO to challenge indirectly the legitimacy of government actions.

As an independent operating system, civil EPIL is supposed to provide an alternative or complementary remedy when the government fails to enforce the law or the existing environmental legislation fails to provide an adequate remedy.

---

[101] See Ho (2008), Thompson and Lu (2006), Yang (2005), Schwartz (2004) and Lo and Leung (2000).

Therefore, NGO-led civil EPIL suits should respect the primary enforcement responsibility of the government and avoid displacing statutory remedies. Moreover, while damages calculated by the virtual cost of pollution treatment may serve as a stronger deterrence, this approach should be applied with particular caution in air pollution cases since it deviates from conventional damages theory and raises the issue of equity. Finally, despite the great potential offered by the wide scope of civil EPIL, it is critical that environmental plaintiffs do not misuse civil EPIL to target those whose conduct has a less than direct effect on the alleged environmental harm.

By functioning as an independent mechanism to vindicate public environmental interest, NGO-led civil EPIL has become an integral part of the environmental governance system in China. With China's environmental laws still plagued by weak government enforcement, various regulatory gaps, and the prevalent illegal exercise of government power, NGO-led civil EPIL will continue to play a valuable regulatory role. In expanding the scope of civil EPIL to address different environmental problems, environmental NGOs should minimize its inappropriate use so as to increase the legitimacy and regulatory significance of civil EPIL in China's environmental governance system.

References

Beach M, Bleish B, Yang S (2006) The role of public participation in ecological impact assessment (ECIA) and environmental impact assessment (EIA) in China. China Environ Ser:117–121

Bie T (2015) Successful experience of environmental public interest litigation—analysis of the Taizhou "sky high award" case. China Environment News

Cao M, Wang F (2011) Environmental public interest litigation in China. Asia Pac Law Rev 19: 217–235

Carpenter-Gold D (2015) Castles made of sand: public-interest litigation and China's new Environmental Protection Law. Harv Environ Law Rev 39:241–274

Chen W (2020) Comparative study of climate change litigation: analysis based on two environmental public interest lawsuits on wind and solar curtailment. J Law Appl:80–87

Chu J (2019) Vindicating public environmental interest: defining the role of environmental public interest litigation in China. Ecol Law Q 45:485–532

Diao F (2016) Petrochina Jilin Petrochemical Company was sued for pollution, defending that "closing the plant would cause more harm to the environment." In: The Paper. https://www.thepaper.cn/newsDetail_forward_1563840

Diao F (2018a) Environmental NGO claims 0.3 billion for wind and solar curtailment; Ningxia Grid: we are not able to purchase the electricity in full amount. In: The Paper. https://www.thepaper.cn/newsDetail_forward_2069105. Accessed 28 June 2022

Diao F (2018b) Lanzhou Intermediate People's Court dismissed Friends of Nature's suit against State Grid Gansu for wind and solar curtailment: case acceptance criteria were not met. In: The Paper. https://www.thepaper.cn/newsDetail_forward_2364374. Accessed 28 June 2022

Ding C, Xiao H (2021) A paper tiger? Prosecutorial regulators in China's civil environmental public interest litigations. Fordham Environ Law Rev 32:323–380

Economy EC (2004) The river runs black: the environmental challenge to China's future. Cornell University Press, Ithaca

Faure MG, Liu J (2014) Compensation for environmental damage in China: theory and practice. Pace Environ Law Rev 31:226–309

Friends of Nature (2017) To save the endangered species of green peafowl, environmental organizations submitted a letter to the Ministry of Environmental Protection. https://mp.weixin.qq.com/s?__biz=MjM5OTY1MTc4MA==&mid=2652700674&idx=2&sn=84d830dbb9aaf411dadc34dc47e6ce29&chksm=bcd190a98ba619bf153f15b6fa92eabcc310c3e1eee09b9089b0e6dbde3d155afd726540b0bf&scene=21#wechat_redirect. Accessed 28 June 2022

Gao Q (2018) "Public interest litigation" in China: panacea or placebo for environmental protection? China: Int J 16:47–75

Gao Q, Whittaker S (2019) Standing to sue beyond individual rights: who should be eligible to bring environmental public interest litigation in China? Transnatl Environ Law 8:327–347

Gong G (2017) Just look like twins: comparative study of environmental public interest litigation in China and the United States. J Comp Law:105–125

He X (2021) Mitigation and adaptation through environmental impact assessment litigation: rethinking the prospect of climate change litigation in China. Transnatl Environ Law 10:413–439

Ho P (2008) Self-imposed censorship and de-politicized politics in China: green activism or a color revolution? In: Ho P, Edmonds RL (eds) China's embedded activism: opportunities and constraints of a social movement. Routledge, London; New York, pp 20–43

Jiang H, Blazey P, Wang Y (2020) China's new approach to environmental governance and environmental public interest litigation. Asia Pac J Environ Law 23:39–73

Kim MJ, Jones RE (2006) Public participation with Chinese characteristics. China Environ Ser:98–102

Kitagawa H (2017) Environmental policy under President Xi Jinping leadership: the changing environmental norms. In: Kitagawa H (ed) Environmental policy and governance in China. Springer Japan, Tokyo, pp 1–15

Latham M, Schwartz VE, Appel CE (2011) The intersection of tort and environmental law: where the twains should meet and depart. Fordham Law Rev 80:737–773

Lee VA, Bridgen PJ (2002) The natural resource damage assessment deskbook: a legal and technical analysis. Environmental Law Institute

Li Z (2014) Judicial discretion on the environmental restoration. J China Univ Geosci (Soc Sci Ed) 14:20–27

Liu T (2015) China's revision to the Environmental Protection Law: challenges to public interest litigation and solutions for increasing public participation and transparency. Geo Wash J Energy Environ Law 6:60–70

Liu Y (2019) Friends of nature and public interest environmental litigation. Chin J Environ Law 3: 225–232

Liu D, Xu H (2018) The politics of curtailment: multi-level governance and solar photovoltaic power generation in China. Environ Polit 27:852–871

Lo CWH, Leung SW (2000) Environmental agency and public opinion in Guangzhou: the limits of a popular approach to environmental governance. China Q 163:677–704

Luo L (2017) The problems and solutions of China's environmental public interest litigation system. China Leg Sci:244–266

Ma X (2021) Wetland restoration project costing hundreds of millions: how social organizations play a monitoring function. In: The Paper. https://www.thepaper.cn/newsDetail_forward_13391734. Accessed 28 June 2022

Ma X, Ortolano L (2000) Environmental regulation in China: institutions, enforcement, and compliance. Rowman & Littlefield, Lanham

McCallum K (2017) Changing landscapes: enforcing environmental laws in China through public interest litigation. Asia Pac J Environ Law 20:57–93

McElwee CR (2011) Environmental law in China: mitigating risk and ensuring compliance. Oxford University Press, New York

Miller JG (1983) Private enforcement of federal pollution control laws part I. Environ Law Rep 13: 10309–10323

Percival RV, Goger JB (2001) Escaping the common law's shadow: standing in the light of Laidlaw. Duke Environ Law Policy Forum 12:119–154
Reisinger W, Dougherty T, Moser N (2010) Environmental enforcement and the limits of cooperative federalism: will courts allow citizen suits to pick up the slack. Duke Environ Law Policy Forum 20:1–62
Ren X, Liu L (2020) Building consensus: support structure and the frames of environmental legal mobilization in China. J Contemp China 29:109–124
Ryan E (2013) The elaborate paper tiger: environmental enforcement and the rule of law in China. Duke Environ Law Policy Forum 24:183–240
Schwartz J (2004) Environmental NGOs in China: roles and limits. Pac Aff 77:28–49
Shi Y, Van Rooij B (2016) Prosecutorial regulation in the global south: environmental civil litigation by prosecutors in China compared to Brazil. Regul Gov 10:44–57
Sitaraman S (2007) Regulating the belching dragon: rule of law, politics of enforcement, and pollution prevention in post-Mao industrial China. Colo J Int Environ Law Policy 18:267–336
Spring J (2015) Chinese environmental group sues Volkswagen over emissions scandal. In: Reuters. https://www.reuters.com/article/us-volkswagen-emissions-china-idUSKBN0TY0CS20151216. Accessed 28 June 2022
Stern RE (2013) Environmental litigation in China: a study in political ambivalence. Cambridge University Press, New York
Sun Q, Tuholske J (2017) An exploration of and reflection on China's system of environmental public interest litigation. Environ Law Rep 47:10497–10510
Thompson BH Jr (2000) The continuing innovations of citizen enforcement symposium: innovations in environmental policy. Univ Ill Law Rev 2000:185–236
Thompson D, Lu X (2006) China's evolving civil society: from environment to health. China Environ Ser:27–39
Tyler Z (2006) Transboundary water pollution in China: an analysis of the failure of the legal framework to protect downstream jurisdictions. Colum J Asian Law 19:572–614
UNEP (2016) Green is gold: the strategy and actions of China's ecological civilization. Reliefweb. https://reliefweb.int/report/china/green-gold-strategy-and-actions-chinas-ecological-civilization. Accessed 28 June 2022
Van Rooij B (2012) The people's regulation: citizens and implementation of law in China. Colum J Asian Law 25:116–179
Van Rooij B, Lo CW-H (2010) Fragile convergence: understanding variation in the enforcement of China's industrial pollution law. Law Policy 32:14–37
Van Rooij B, Stern RE, Fürst K (2016) The authoritarian logic of regulatory pluralism: understanding China's new environmental actors. Regul Gov 10:3–13
Wang A (2007) The role of law in environmental protection in China: recent developments. Vt J Environ Law 8:195–223
Wang AL (2010) Environmental courts and public interest litigation in China. Chin Law Gov 43:4–17
Wang AL (2013) The search for sustainable legitimacy: environmental law and bureaucracy in China. Harv Environ Law Rev 37:365–440
Wang M (2016) On the future development of environmental public interest litigation in China: analysis based on the relationship between administrative power and judicial power. China Leg Sci:49–68
Wang AL (2018) Symbolic legitimacy and Chinese environmental reform. Environ Law 48:699–760
Wang L (2020) The battle between 200 green peacocks and 3 billion hydropower plant. In: Caijing. https://news.caijingmobile.com/article/detail/415666. Accessed 28 June 2022
Wang AL, Gao J (2010) Environmental courts and the development of environmental public interest litigation in China survey of environmental tribunals and regulatory schemes. J Court Innov 3:37–50

Xi D, Hou X (2018) 54.82 million in damages: the first case of government suing enterprise for environmental damages was decided. In: CCTV.com. http://news.cctv.com/2018/08/27/ARTIo7Y9i41XmKZ6tTtbJiZG180827.shtml. Accessed 28 June 2022

Xie L, Xu L (2021a) Environmental public interest litigation in China: a critical examination. Transnatl Environ Law 10:441–465

Xie L, Xu L (2021b) Environmental public interest litigation in China: findings from 570 court cases brought by NGOs, public prosecutors and local government. J Environ Law:1–29

Xu X (2010) Analysis of the liability for environmental pollution—relationship between tort law and environmental laws. Leg Forum:17–20

Xu G, Faure M (2016) Explaining the failure of environmental law in China. Colum J Asian Law 29:1–99

Yang G (2005) Environmental NGOs and institutional dynamics in China. China Q 181:46–66

Yang K (2021) When Lianyungang's coastal dream meets migratory birds: who gives way when the snipe and the harbor are in competition. In: Southern Weekend. http://www.infzm.com/contents/210964. Accessed 28 June 2022

Yuan Y (2020) Environmentalists call for ban on 'electric earthworm machines.' In: Sixth Tone. https://www.sixthtone.com/news/1006023/environmentalists-call-for-ban-on-electric-earthworm-machines. Accessed 28 June 2022

Zhai T, Chang Y-C (2018) Standing of environmental public-interest litigants in China: evolution, obstacles and solutions. J Environ Law 30:369–397

Zhang G (2019a) Results of the eighth "top ten public interest lawsuits in China" were announced. In: Legal Weekly. https://www.163.com/dy/article/ECVDV14B0512CN4M.html. Accessed 28 June 2022

Zhang H (2019b) Prioritizing access of renewable energy to the grid in China: regulatory mechanisms and challenges for implementation. Chin J Environ Law 3:167–202

Zhang R (2020) Towards a breakthrough in China's climate change litigation: environmental public-interest litigation filed by NGOs. In: He X, Zhang H, Zahar A (eds) Climate change law in China in global context. Routledge, London, pp 162–187

Zhang X (2022) Controversy over wetland restoration: protection or encroachment? China Newsweek

Zhang R, Mayer B (2017) Public interest environmental litigation in China. Chin J Environ Law 1: 202–228

Zhang Z, Wang L (2012) Limitations of private law relief for injuries to natural resources in China and solutions. Law Sci Mag:56–62

Zhu M (2022) The rule of climate policy: how do Chinese judges contribute to climate governance without climate law? Transnatl Environ Law 11:119–139

Chapter 11
From Advocators to Rule Makers: Exploring the Role of Chinese Lawyers in Environmental Law Making and Public Interest Litigation

Jun Zhao and Jinye Han

Abstract As the course of China's environmental rule of law has evolved, the role that Chinese environmental lawyers can play has changed dramatically. The environmental legislation process has gradually improved, and the voice of environmental lawyers in the legislative process has been more fully guaranteed. Through seminars and acting as legal advisors to the government, they have been able to provide feedback on environmental issues encountered in practice. The construction of the environmental litigation system has gradually become clearer, and environmental lawyers are now able to participate as non-government litigators. In specific terms in Environment Public Interest Litigation (EPIL), lawyers are now able to represent clients for compensation where ecological damage has occurred and to assist NGOs and government agencies in their environmental litigation activities. With the introduction of new environmental policies and the gradual benchmarking of China's environmental standards with international standards, the need for environmental lawyers to interpret policies and promote environmental protection concepts is increasing.

Keywords Chinese environmental lawyers · Environmental protection · Environmental legislation · Public interest litigation

11.1 Introduction

The history of the rule of law in environmental protection in China has undergone a process of change and development, from a marginalized and external administrative regulatory model to a broader conception that now includes the role of Chinese private lawyers. The legislative process that, underpinned this change had its birth in the 1970s and the initial reason for the drafting of environmental legislation was

J. Zhao (✉) · J. Han
Shanghai University of Political Science and Law, Shanghai, China
e-mail: zhaojun@shupl.edu.cn

X. Wang et al. (eds.), *Environmental Public Interest Litigation in China*,
https://doi.org/10.1007/978-3-031-26526-6_11

extensive pollution across the country which had begun to cause unrest in local communities. Nevertheless, the role of Chinese lawyers in environmental law making during this period was minimal.[1] Because there was little room for a realistic application of the legislation, there was naturally no endogenous initiative to study the environmental legal provisions and related systems, let alone to promote and publicize the relevant laws. This situation gradually changed after 2012, as China's environmental legislation system and associated management systems gradually improved.

11.2 Environmental Lawyers' Suggestions on Legislation

Environmental lawyers play a comprehensive role in the whole legislative process. At the beginning of the legislative process, lawyers can make suggestions[2] and provide preliminary ideas due to their professional knowledge and practical experience. Environmental lawyers can directly participate in the drafting and consideration of draft environmental laws, regulations and rules. In addition, some lawyers can promote the development of local environmental laws and regulations. After the initial drafting of a law is completed, environmental lawyers can be commissioned to evaluate it, or they can convene a seminar for professionals in the environmental field so that they can make further legislative proposals and promote legislative improvements. Active participation in the development of new laws is an important way for lawyers to participate in the environmental arena. Lawyers who have been engaged in environmental protection cases for a long time and have the corresponding qualifications can obtain the status of legal experts,[3] so that they can participate in the seminars for formulating new environmental protection laws.

11.2.1 Participation in the Formulation of National Laws

Environmental lawyers have established the Environmental and Resources Law Specialized Committee of the All China Lawyers Association (ERL), whose members can participate in the formulation of new laws. The ERL can convene seminars on the formulation of new environmental laws as legal experts and issue expert opinions on new laws. Since its establishment, the ERL has participated in seminars and issued expert opinions on a number of newly enacted or amended laws,

[1] Zhanglu (2021), pp. 27–40.

[2] Many environmental lawyers were invited to advise on the development of *Environmental Protection Law, Environmental Impact Assessment Law, Soil Pollution Prevention and Control Law, Water Pollution Prevention and Control Law*, etc.

[3] Energy Resources and Environment Business Research Committee (2022).

including the *Law on Energy Conservation*[4] and the *Tort Liability Law*.[5] The *Tort Liability Law* (which was converted into a part of China's *Civil Code* in 2020) established an environmental liability as a civil tort. It provides for environmental tort liability in four articles in Chapter 8, Environmental Pollution Liability. During the development of the draft legislation of the *Tort Liability Law*, the legislature specifically invited the ERL to participate in the discussion and to issue expert opinions. For specific provisions, Article 65 of the draft *Tort Law* provided that if the discharge meets the prescribed standards and causes harm to others, the discharger shall bear the corresponding liability for compensation. The experts of the ERL believed that the tort liability flows from an environmental infringement and had nothing to do with whether the emissions meet the standards or not. They suggested deleting the phrase "discharge in accordance with the prescribed standards". The legislature adopted the opinion of the ERL, and Article 65 of the *Tort Liability Law*, which was promulgated and implemented, reads, where damage is caused by pollution of the environment, the polluter shall bear tort liability.[6]

In the gradual improvement of environmental protection laws, new environmental laws and regulations are emerging, and environmental lawyers are involved in their formulation. For example, during its drafting, the *Yangtze River Protection Law*[7] legislators convened a seminar in a Shanghai law firm at the beginning of its formulation, with the participation of leaders from the National People's Congress, the National Development and Reform Commission, the Ministry of Ecology and Environment, the Ministry of Natural Resources, the Yangtze River Water Conservancy Commission of the Ministry of Water Resources, the Zhejiang Provincial People's Congress, the Shanghai Municipal Procuratorate and experts from the Environmental Resources and Energy Law Research Center of major universities. All the parties involved in the meeting made suggestions and recommendations for the *Yangtze River Protection Law* legislation based on different perspectives, which laid a good foundation for the eventual introduction of the *Yangtze River Protection Law*.

In addition to participating in the formulation of new environmental laws and regulations, environmental lawyers can also participate in the revision of environmental laws and regulations, such as the *Water Pollution Prevention and Control Law* and the *Law on the Prevention and Control of Pollution by Solid Waste*. In each instance the environmental lawyers provided relevant legal opinions through seminars and other discussions. As an example, civil environmental public interest litigation (EPIL) has had a far-reaching impact on the environmental field and the ERL participated in the seminar on the revision of the *Civil Procedure Law* held by the Supreme People's Court (SPC), the Supreme People's Procuratorate (SPP), the Legal Work Committee of the National People's Congress, and the former Ministry

[4]The National People's Congress of the People's Republic of China (2018).

[5]The National People's Congress of the People's Republic of China (2009).

[6]Wang and Wang (2010), pp. 103–105.

[7]The National People's Congress of the People's Republic of China (2020).

of Environmental Protection (MEP). Among them, environmental civil public interest litigation experts' opinions were issued on the relevant clauses. In the negotiation on the subject matter, a proposal was put forwards to widen the qualifications of litigation subjects and expand the eligible litigation groups.

11.2.2 Participation in the Formulation of Local Laws and Regulations

When local legislation related to the environmental field is considered the local legislature may now consult with environmental protection lawyers in the bar association. Environmental lawyers are involved in the legislative process in a more complete and detailed manner in their role as legal advisors to the government and relevant environmental protection departments. Government legal advisors take on the responsibility of participating in the review of the legality of draft government regulations and rules, fully implementing the lawyer's legal rights to information, participation and supervision. In Shanghai, for example, the professional lawyers of the Environment and Resources Law Committee of the Shanghai Lawyer Association[8] (ERL Association hereinafter) has played an active role in the formulation of local regulations such as *Shanghai Environmental Protection Regulations* and *Shanghai Standardization Regulations* by participating in expert symposiums, conveying public opinion and specific case experiences. The group put forwards many practical legislative and regulatory proposals. They have used their professional knowledge and industry practice to give feedback on specific regulations such as the main body of the ecological red lines and management requirements, the emission permit system, the environmental impact assessment system, and the standards for resourceful reuse of hazardous waste.

The ERL Association holds an annual meeting to exchange ideas on legislation by inviting experts from all walks of life who practice in the environmental field. Experts, scholars, lawyers, lawmakers and government officials in the environmental field speak freely and exchange opinions at the annual meeting. This enables lawmakers and government officials to hear environmental protection suggestions from practitioners and provides important points of reference for scientific legislation. In Guangxi, for example, the ERL recently held its annual meeting in Guangxi and discussed legislation on the protection of the Li River, issued the *Li River Declaration*,[9] and discussed and adopted the *Report on the Current Situation of the Li River* and *Legislative Proposal for the Li River Environmental Protection Regulations*. Guangxi TV, South China Morning Post, Guangxi Daily, Guangxi Politics and Law News, China Lawyer Network, Guangxi News Network and other news media reported it prominently and in prime time, which has caused a strong

---

[8] Shanghai Bar Association (2016).

[9] Sina News (2007).

reaction in Guilin and even in many other cities. After the meeting, the committee and Guangxi law association discussed the annual meeting through *Report on the Current Situation of the Li River* and *Legislative Proposal for the Li River Environmental Protection Regulations* submitted to the Guilin government, the Guangxi district government and relevant departments. After that, the *Li River Environmental Protection Regulations*[10] were listed in the 2008 legislative plan of the Guangxi regional government and the Local People's Congress, and on 5 July 2008, the legal office of the Guangxi legal affairs office of government officially commissioned the Guangxi Lawyers Association to draft expert commentaries about the *Li River Environmental Protection Regulations*. In July 2010, the Guangxi law association officially submitted the draft to the legal affairs office of the government.[11]

In summary, environmental lawyers play an indispensable role in China's environmental legislation by acting as government legal advisors and organizing and participating in seminars and annual meetings. Whether it is national legislation or local legislation, legal submissions from practitioners are always considered as part of the review process. As China's legislative process accelerates, more attention has been given to the environmental field, and the environmental code is increasingly on the agenda. It is foreseeable that Chinese environmental lawyers will play an even greater role in the field of environmental legislation in the future.

11.3 The Role of Chinese Lawyers as a 'Think Tank' in Litigation

11.3.1 Supporting Role in Environmental Public Interest Litigation

The All-China Environment Federation (ACEF) has established a strategic partnership to better promote environmental litigation, the training of environmental lawyers and the creation of career paths in environmental management and litigation. The members of the Committee are of course volunteer lawyers of the ACEF and can be chosen as litigation agents when the ACEF initiates litigation. Members of the Legal Expert Committee of the ACEF[12] are actively engaged in the practice of environmental public interest litigation, filing civil public interest litigation and administrative public interest litigation as a plaintiff organization. For example, the first "haze" environmental public interest case[13] was undertaken jointly by private law firms in partnership with specialized government departments. In this case, the

[10]Standing Committee of Guangxi People's Congress (2011).

[11]Wang and Wang (2010), pp. 103–105.

[12]All-China Environment Federation (2017).

[13]Chinese Society of Environmental and Resources Law (2020a, b).

plaintiffs referred to *the Ambient Air Quality Standards (gb3095-2012)*,[14] *the Recommended Method for Environmental Damage Identification*[15] *and Technical Specifications for Emergency Monitoring in Environmental Accidents (HJ 589—2021)*[16] *Phase of Environmental Incidents*, and other corresponding norms. This led to a fine for allowing pollutants to be discharged in excess of the limit, together with a public apology to society published in the media.

Lawyers also play a significant role in confirming the status of NGOs as eligible joint partners in the conduct of litigation. In the case of *China Biodiversity Conservation and Green Development Foundation v Ningxia Ruitai Technology Co.*,[17] one of the controversies focused on whether the Green Development Foundation could initiate this lawsuit as a social organization "specializing in environmental protection public welfare activities". The environmental lawyers distinguished between the concepts of "environment" and "environmental protection" from a legal professional perspective and corrected the original judgment that the purpose and scope of business of the Green Development Foundation did not include engaging in environmental protection public welfare activities. The environmental lawyer pointed out that the Green Development Foundation has been actually engaged in environmental protection public welfare activities, which confirms the purpose and scope of the business of environmental protection public welfare activities as defined in its constitution. In the end, the SPC ruled that the decision in the first and second instance viz. that the Green Development Foundation was not qualified to initiate civil environmental protection lawsuits was the result of an improper understanding of the Environmental Protection Law, the Judicial Interpretation of Environmental Public Interest Litigation and the meaning of environmental protection. and that the application of the law was wrong which should be corrected. With the joint efforts of environmental lawyers and the Green Development Foundation, the Court ruled that the Green Development Foundation was indeed qualified to initiate environmental public interest litigation, which led to the subsequent development of its role in the ecological and environmental field to a greater extent and depth.

On 17 April 2016, CCTV reported that hundreds of students at Changzhou Foreign Language School suffered health damage due to environmental pollution,[18] so the Green Development Foundation and other environmental organizations filed an environmental public interest lawsuit with three chemical companies as defendants. This is the famous *Changzhou Toxic Land Case.*[19] However, in the first instance judgment, the environmental groups' entire claim was rejected, and the environmental groups were ordered to bear the huge litigation costs of more than $1.89US million. After a three-year litigation process, the case was finally partially

[14] MEP (2016).
[15] MEP (2014).
[16] Ministry of Ecology and Environment of the People's Republic of China (2022).
[17] SPC (2017).
[18] Legal Daily (2017).
[19] Chinese Society of Environmental and Resources Law (2020a, b).

won. The court reversed the first instance judgment and ruled that the three polluters should bear the responsibility for environmental infringement, apologize to the public for their polluting behavior in national media, and bear the related litigation costs. The case came to an end in 2019 when the three polluting companies publicly apologized to the public in the *Legal Daily*.

In the course of the case issues such as which party would receive a costs order, which appellant should be held responsible for an environmental pollution tort, the allocation of the burden of proof, the costs of environmental civil public interest litigation and other crucial issues, the group of lawyers comprehensively collected evidence, carefully wrote the agent's statement, overcame multiple difficulties such as geographical area for many collective discussions, and finally made a significant contribution to environmental protection. This case is of great importance to China's soil pollution management and environmental public interest litigation, because unlike air and water pollution, soil pollution is largely concealed and has a long latent period, with years or even decades between the occurrence and outbreak of pollution problems, and the difficulty and cost of subsequent environmental management is high. Many of the problems encountered in this case are relevant to all soil pollution cases.[20]

11.3.2 Auxiliary Role in Ecological Damage Litigation

Although environmental damage lawsuits can only be filed by the authorities prescribed by law, lawyers can participate in the litigation process as litigation agents. The government can rely on lawyers to carry out litigation activities, and the cooperation between the two will make the progress of ecological environmental damage litigation smoother. Government agencies usually assume the responsibility for discovering pollution and acting as a competent party to commence the litigation. The gathering of the basic facts and establishing what charges should be leveled is a joint undertaking by the lawyers and relevant experts in the field. The same group of lawyers will further analyze the issues in dispute, calculate the probability of winning the lawsuit, and consider the issue of liability from a more professional perspective.

Environmental lawyers not only help NGOs in public interest litigation with expertise to deal with relatively strong enterprises, but also help government departments in ecological and environmental damage compensation litigation. This is reflected in the *Jiujiang Municipal People's Government v Jiangxi Zhengpeng Environmental Protection Technology Co., Ltd.*[21] and other ecological and environmental damage liability disputes. The case involved several problematic issues such as the division of responsibility among multiple defendants to transfer and dump

[20] Cao, Chinese Lawyer (2019), pp. 36–38.

[21] China Court News (2019).

sludge, the sharing of responsibility between the operator and the illegal disposer, and the transfer of sludge waste across provinces. The case took place near the Yangtze River Economic Zone, which caused great damage to the coastal ecological environment. Environmental lawyers put together a complete chain of evidence in ecological and environmental damage compensation litigation, and dissected the relationship between the company, and the illegal disposer. They also played their role in the pre-litigation negotiation stage, together with the Jiujiang Municipal People's Government, to make some of the compensation obligors reach an agreement and actively fulfil their restoration obligations and the lawyers solved the problem of sharing the restoration cost of environmental damage by multiple people. In the case of cross-provincial multi-pollution, the problem of scattered pollution locations was solved by separately investigating and taking samples and requesting judicial appraisal institutions to help issue corresponding environmental assessment reports.

In the ecological damage compensation cases proposed by the government, environmental lawyers also participate as litigation agents to promote the smooth resolution of the cases and to promote the establishment of an ecological civilization in a cooperative manner with the local government.

11.4 The Timely Follow-Up of Environmental Policies and the Promotion of Environmental Protection Consciousness

11.4.1 Research and Dissemination of the Carbon Peaking and Carbon Neutrality Goals

China recently unveiled a plan to promote the concept of a circular economy for the next five years, aiming to basically establish a resource recycling system in the country by 2025 as part of its efforts to achieve carbon peak and neutrality goals. However, when President Xi Jinping first raised the issue of "carbon peaking" and "carbon neutrality" in the general debate of the 75th session of the United Nations General Assembly, it was not well known to everyone.[22] At that time, even environmental lawyers, as a group that needed to keep up with current environmental issues, heard of this conception for the first time however they quickly gathered together the various elements of the program with an international perspective. On this basis, the attorneys began to popularize this policy to businesses and communities. In Shanghai, for example, a team of environmental protection lawyers from Shanghai would go into the community to analyze the relevant data in the form of law enforcement activities and hold small classes on environmental protection. This

[22] Qiushi (2021).

was done to inform the general public about the goal of the legislation. They called on the community to start by using one less plastic bag and one less car trip, so that each resident would be conscious of the need to achieve their personal carbon reduction goals. In addition, training sessions and lectures would be held at the corporate level to educate the relevant department members about the necessity and urgency of reducing carbon emissions, and to popularize the relevant policies from the corporate side, which is the largest emitter of carbon dioxide.

In order to implement the carbon peaking and carbon neutrality goals, relevant departments have issued a series of laws and regulations and implemented new policies. On 16 July 2021, the national carbon market was launched in Beijing, Shanghai and Wuhan, and the national carbon trading market was officially opened.[23] For enterprises, carbon emissions trading is a new thing, so there is no specific in-depth understanding. Environmental lawyers learned it long before the opening of the carbon emission market, and have developed a thorough understanding of which entities can trade carbon emissions, how to trade in the carbon market, and what legal issues need to be paid attention to in the process of carbon trading. On this basis, they have conducted relevant training and lectures with enterprises to popularize the relevant knowledge and operational processes. For example, the lawyers will point out the key points and emphasize the increased penalties to prevent enterprises from violating carbon emissions illegally. Lawyers emphasize the more detailed provisions in the penalty section.

11.4.2 The New Environmental Laws and Regulations Publicity

On 29 April 2020 the new *Law on the Prevention and Control of Pollution by Solid Waste*[24] (hereinafter referred to as the *New Solid Waste Law*) was adopted by the Standing Committee of the National People's Congress, effective from 1 September 2020. For the focus of the new revision of the law usually by environmental lawyers to promote the popularization, by combing the emphasis on the the New Solid Waste Law once again strengthen the pollution prevention and control responsibilities of waste-producing units, requiring waste-producing units to establish and improve the whole process of solid waste pollution and an environmental prevention responsibility system. The law established the following principles:

(1) that waste-producing enterprises should not only review the main qualifications of the third party entrusted with the disposal of solid waste, but also verify the technical capacity of the third party.

[23] CNR News (2021).

[24] Qiushi (2021).

(2) They emphasized that if the enterprise violates the relevant legal requirements, it may lead to legal risks of civil compensation, administrative penalties and even criminal penalties.
(3) Through the comparison of old and new key laws and regulations, combined with the current status of solid waste generation and disposal in China and the criminal cases related to solid waste pollution in the past five years.
(4) The legal obligations that enterprises as waste producers need to pay attention to fulfil.
(5) In addition, on the enterprise side, environmental lawyers are to provide project services for relevant government departments and compile law enforcement manuals and administrative guidelines based on the new *Solid Waste Law*. The law also implemented review steps and clarified review priorities, so that eventually new law enforcement officers could understand the guidelines for solid waste disposal. In essence, environmental lawyers provide a detailed structural plan for both supervisors and supervisees for the implementation of the new *Solid Waste Law*.

The Emission Permit Management Regulations came into effect on 1 March 2021, and the *Catalogue of Stationary Sources Subject to Pollution Discharge Permit Control 2019*[25] also came into effect on 11 July 2019. Therefore, according to the latest requirements of the Ministry of Ecology and Environment (MEE) and local ecological and environmental departments, enterprises should gradually complete the application and registration of emission permits, and there are more new requirements in the registration of emission permits. The regulations give a step-by-step explanation of the application for an emission permit, including the application timeline, who to apply to, and how to apply and to the issuance of emission permits.

11.5 Conclusion

In recent years, Chinese environmental lawyers have played a large role in promoting environmental legislation and environmental protection. They provide comments at the early stage of the draft legislation,[26] provide practical feedback during the implementation process, and participate in seminars during the revision process. However, there are still problems such as the lack of influence of environmental lawyers, their inability to act as qualified environmental litigants, and the high cost of the environmental rights protecting.

[25] MEE (2019).

[26] Legal Daily (2022).

References

All-China Environment Federation (2017) Admin, http://www.acef.org.cn/a/fzjg/2019/0327/19738.html. Accessed 29 Sep 2022

Cao J, Chinese Lawyer (2019) Be a good "green guardian" in environmental public interest litigation -- on the role of environmental lawyers from the perspective of two cases, 2019(08): 36–38

China Court News (2019) Shen Wushuang, https://www.chinacourt.org/article/detail/2019/09/id/4491318.shtml. Accessed 29 Sep 2022

Chinese Society of Environmental and Resources Law (2020a) The secretariat, http://cserl.chinalaw.org.cn/portal/article/index/id/375/cid/10.html. Accessed 29 Sep 2022

Chinese Society of Environmental and Resources Law (2020b) Zhang Xiaoxue, https://cserl.chinalaw.org.cn/portal/article/index/id/273/cid/25.html. Accessed 29 Sep 2022

CNR News (2021) Lin Fuyu, https://www.cnr.cn/shanghai/tt/20210717/t20210717_525537213.shtml. Accessed 29 Sep 2022

Energy Resources and Environment Business Research Committee (2022). http://www.acla.org.cn/home/toMenu?menuIdStr=100. Accessed 29 Sep 2022

Legal Daily (2017) Ye Quan, https://news.cctv.com/2017/02/08/ARTISbDotTtoLiA2l8t41wHd170208.shtml. Accessed 29 Sep 2022

Legal Daily (2022) Du Yang, http://epaper.legaldaily.com.cn/fzrb/content/20220908/Articel07003GN.htm. Accessed 29 Sep 2022

MEE (2019). https://www.mee.gov.cn/gzk/gz/202112/t20211211_963806.shtml. Accessed 29 Sep 2022

MEE (2022). https://www.mee.gov.cn/ywgz/fgbz/bz/bzwb/other/qt/202202/t20220228_970076.shtml. Accessed 29 Sep 2022

MEP (2014). https://www.mee.gov.cn/gkml/hbb/bgt/201411/t20141105_291159.htm. Accessed 29 Sep 2022

MEP (2016). https://www.mee.gov.cn/ywgz/fgbz/bz/bzwb/dqhjbh/dqhjzlbz/201203/t20120302_224165.shtml. Accessed 29 Sep 2022

Qiushi (2021) xueershixi, http://www.qstheory.cn/zhuanqu/2021-09/17/c_1127873054.htm. Accessed 29 Sep 2022

Shanghai Bar Association (2016) Energy Resources and Environment Business Research Committee, https://www.lawyers.org.cn/info/5bb80305ae394981b3d1ddba0b6c0355. Accessed 29 Sep 2022

Sina News (2007) Guangxi News, https://news.sina.com.cn/o/2007-11-18/193812925055s.shtml. Accessed 29 Sep 2022

Standing Committee of Guangxi People's Congress (2011) Guangxi News, http://www.gxnews.com.cn/staticpages/20111225/newgx4ef66964-4497368.shtml. Accessed 29 Sep 2022

The National People's Congress of the People's Republic of China (2009). http://www.npc.gov.cn/npc/c12488/200912/2a632e056df44478a4637e161af76ec9.shtml. Accessed 29 Sep 2022

The National People's Congress of the People's Republic of China (2018). http://www.npc.gov.cn/npc/c12435/201811/045c859c5a31443e855f6105fe22852b.shtml. Accessed 29 Sep 2022

The National People's Congress of the People's Republic of China (2020). http://www.npc.gov.cn/npc/c30834/202012/1626d0bc5284485588222995e712c434.shtml. Accessed 29 Sep 2022

The Supreme People's Court of The People's Republic of China (2017) Fang Fang, https://www.court.gov.cn/shenpan-xiangqing-34322.html. Accessed 29 Sep 2022

Wang J, Wang J (2010) The Experience of the Environmental Resources Committee of the National Law Association in Promoting Environmental Legislation in China. Proceedings of the Conference on Economic Development and Environmental, Resource and Energy Law Practice 2010:103–105

Zhanglu (2021) The orientation transformation and administrative supervision transformation of environmental law in China. J China Univ Geosci (Soc Sci) 21(02):27–40. https://doi.org/10.16493/j.cnki.42-1627/c.2021.02.004

Chapter 12
Three-Dimensional Model of Expertise-Input for EPIL in China

Jianwei Zhang and Yanni Luo

Abstract The input of expertise in Chinese judicial proceedings relies on three types of subjects with specialized knowledge, which form the "Three-dimensional Model" for the specialized fact-finding issue. The process of collecting, identifying, and testing evidence for EPIL is very complex and highly professional; the identification of pollutants, as well as the causation and assessment of pollution remediation, are highly specialized. In EPIL, parties and judges often extensively rely on the work of these three types of experts in terms of collecting and identifying fact-related evidence. This chapter presents a typology of how the "three-dimensional model" of expertise-input is running in China and analyzes in detail the role and problems of the three types of professionals with specialized knowledge in EPIL.

Keywords Three-dimensional model · Specialized issue · Accreditation body · Auxiliary expert · Expert Jurors

12.1 Introduction

With the development of science and technology and the increasing impacts of technology on people's lives and work, all kinds of litigation inevitably involves various specialized fact-finding issues. Specialized fact-finding issues is a collective term.[1] It refers to facts within the scope of proof of the case that cannot be answered directly as affirmative or negative by judicial officers within the realms of common knowledge.[2] As a result, the resolution of specialized issues in litigation has become an extremely common and realistic requirement in the process of judicial governance. To resolve specialized issues, it is necessary to rely on the expertise of specialists and not just on that of the judges or litigants. The input of expertise in

[1] Li et al. (2017), pp. 39–43.
[2] Chen (2014), p. 277.

J. Zhang (✉) · Y. Luo
Law School of Beijing University of Chemical Technology, Beijing, China
e-mail: 2022500009@buct.edu.cn

X. Wang et al. (eds.), *Environmental Public Interest Litigation in China*,
https://doi.org/10.1007/978-3-031-26526-6_12

Chinese judicial proceedings relies on three types of subjects with special knowledge: accreditation bodies, auxiliary experts, and jurors with specialized knowledge. This forms the "Three-dimensional Model" for the specialized fact-finding in China. The "three-dimensional model" of expertise-input refers to three types of persons with expertise who assist the court in solving special fact-finding problems encountered in the trial of cases.

Compared with other forms of civil or criminal litigation, environmental public interest litigation (EPIL) is a legal mechanism with "environmental characteristics", which puts new requirements on the judicial philosophy of environmental judges and the application of procedural and substantive laws. In EPIL, the process of collecting, identifying, and testing evidence is highly complex and professional. The identification of pollutants, causation and issues concerning the assessment of pollution remediation are highly specialized. Difficulties in obtaining evidence, identification, and cross-examination in environmental litigation are some of the key factors that restrain environmental litigation from functioning effectively. In practice, judges generally lack knowledge of the relevant environmental-related professions. Most judges do not have the expertise to address complex environmental disputes, which may affect the decisions that are made.

Therefore, in EPIL, parties and judges often rely extensively on the work of the three types of experts to collect and identify fact-related evidence. However, this three-dimensional model still faces many obstacles and creates quite some chaos.[3] Indeed, environmental trials, especially EIPL in China, are still in a continuous process of improvement. This chapter will present a typology of how the "three-dimensional model" of expertise input in China has addressed specialized issues and review whether the existing institutions adequately consider and accommodate the particularity of fact-finding issues in EPIL, with the aim of improving the quality of environmental case trials and effectively enhancing judicial credibility. This chapter will first briefly introduce the general operation of the three-dimensional model of expertise-input as well as the special characteristics of the fact-finding issue in EPIL in the Chinese context.[4] Then, it will analyze in detail the functional role and problems of the three types of professionals with specialized knowledge in EPIL.

[3]For example, lack of uniform legislation concerning accreditation has occasionally led to far too expensive fees, conflicting and confusing conclusions, bias opinion, etc. Jiang and Zhao (2017).

[4]In the Chinese judicial system, the role of the expert lies solely in the area of fact finding, the application of the law being the exclusive competence of the judge. Therefore, the study of the role of professional environmental public interest litigation is in its role in fact finding.

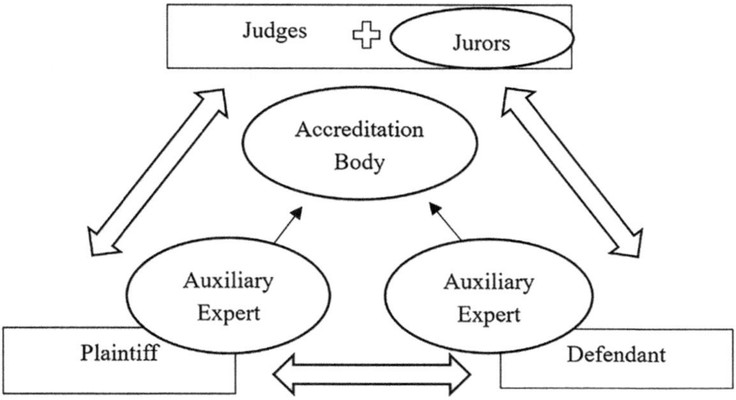

Fig. 12.1 Three-dimensional model of expertise-input (by author)

12.2 "Three-Dimensional Model" of Expertise-Input for Specialized Issue

To resolve the specialized issues and compensate for the judges' deficit of environmental professional knowledge in fact-finding in the litigation process, the conventional "three-dimensional model" of expertise-input can be applied. The relationship between the accreditation body, auxiliary expert, and jurors with specialized knowledge can be described as follows in Fig. 12.1.

In this model, the accreditation body acts as a neutral assistant of the court.[5] It should be an appraisal institute with relevant qualifications, independent, objective, and impartial to assist the court in effectively resolving specialized issues. In the vast majority of cases, the accreditation body is appointed by the judge. The appointment may be initiated by the judge himself or herself or it may be required by the parties. Its key functional role is to explain specialized issues to the prosecution (plaintiff), defendants and trial judges. Meanwhile, its opinion must be subject to effective cross-examination by both parties to the lawsuit or by the auxiliary expert appointed by one party. 'Auxiliary Expert' refers to an expert who is entrusted by one of the parties to a lawsuit to provide professional opinions on reports of the accreditation body and other specialized issues. Literally, auxiliary experts are responsible for the entrusted party, and the main role is to help the party to the litigation cross-examine the experts as well as to confront the auxiliary expert entrusted by the other party.[6] Jurors with specialties are specially appointed jurors for specific cases. They have the expertise to provide a neutral perspective on factual findings. Jurors with specialties are actually the adjudicators as well. In brief, the main role for these experts is to resolve specialized issues in litigation, including (1) to assist the court in

[5]Zheng (2019), pp. 67–77.
[6]Ibid.

examining the scene and taking evidence, (2) to analyze and test the specialized issues and give an expert opinion, (3) to appear in court to give an opinion on an existing expert opinion or specialized issue, and (4) to demonstrate and explain the technical issues in court.[7]

This arrangement under the "Three-dimensional Model" has two main functions. First, it assists judges and parties in understanding professional issues. Due to the limitations of the judge's knowledge, a judge does not always fully understand professional issues when judicial appraisal is made. Professional jurors can translate the professional issues to help the judge understand while overcoming the bias or partisanship that auxiliary experts may have. Secondly, it bridges the gaps in the accreditation system and promotes cross-examination. It can result in the factual disputes of the case being presented in a more comprehensive way by introducing auxiliary experts to cross-examine the appraisal opinions of the appraisal institutions or explaining the truth of the case to the maximum extent to help the judge make accurate fact-finding. In addition, the model can address the problem of vitiated cross-examination in trials. It can also enrich the litigation rights of the parties and balance the litigation power of both sides.[8] Overall, through the interplay of these three expert roles in the trial process, the court can jointly facilitate the presentation, examination, and certification of specialized questions and thereby maximize the search for the truth of the facts of the case.[9]

12.3 The Particularity of Specialized Fact-Finding Issues in EPIL

In recent years, the EPIL in China has been developing as an effective means of resolving environmental disputes. These cases are characterized by complexity, latency, cumulative, slow-onset, long-term and are highly specialized in fact-finding. Based on the constitutive elements incompletely addressed by scholars on EPIL,[10] the specialized fact-finding issue in EPIL can be categorized as violation/infringement, causality, a consequence of damage, restoration identification, and the like (Fig. 12.2).

To facilitate specialized fact-finding in EPIL, a special fact-finding system is required. One of the most critical issues is to combine the professional characteristics of environmental scientific knowledge to build a multi-faceted expert participation system. A special "Three-dimensional Model" for EPIL will help resolve disputes and maintain environmental justice. Moreover, to develop the expert participation system, some features of specialized fact-finding issues in EPIL should be clarified.

[7]Li (2020), pp. 37–54.

[8]Wang (2011), pp. 95–115.

[9]Zheng (2019), pp. 67–77.

[10]Li (2017), pp. 39–43.

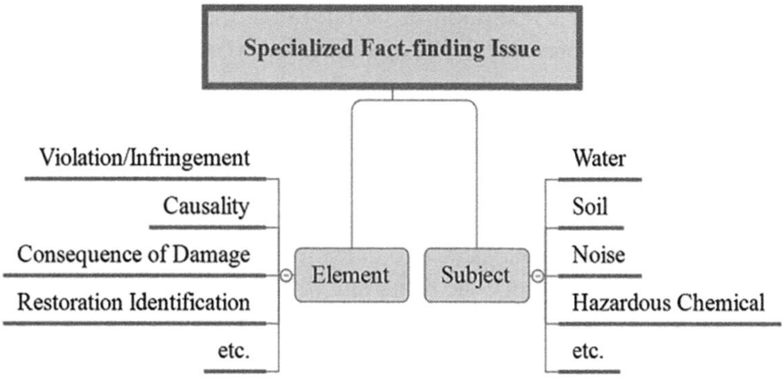

Fig. 12.2 Specialized Fact-Finding Issues in EPIL (by author)

These specialized issues can be characterized as having a wide-ranging influence, high complexity, and irreversibility of damage.

First, EPIL aims to protect the public interest, so the facts involved are wide-ranging. Generally, once the natural environment is damaged, not only the interests of individuals but also the common interests of the public at large will be affected. Under the current legal system in China, environmental pollution or damage in EPIL does not necessarily directly affect the personal interests of the litigants, i.e., the prosecutors or NGOs. The plaintiff sues to secure the health of the public environment in the sense that the public interest has been violated or may be damaged. In other words, in the context of EPIL, there is no direct legal relationship between the plaintiff and defendant. This can lead to a problem, namely, the plaintiff cannot fully understand the factual environmental damage and does not control the costs of environmental restoration. Due to this unfamiliarity as well as the non-directive nature, judicial proceedings in practice may not protect the public interest very well. This also highlights the importance of having multiple experts involved in professional issues in EPIL.

Second, factual findings in environmental cases are highly complex, and this feature is even more distinct in EPIL. It has been observed that environmental problems are characterized by destruction, extensive pollution, long response times, and irreversibility. Specifically, environmental pollution can be very complicated. For example, for most environmental cases, the causes are multiple, and more than one pollutant is generally involved. In addition, once environmental damage occurs, the remedy will not only consume huge human and financial resources but it also can be difficult to restore the environment to its original state. In other words, unlike other forms of damage to rights and interests, damage to the environmental public interest has a certain degree of irreversibility. Finally, in contrast to other types of civil cases, the damage in EPIL cases may be hidden, the contamination cannot be revealed immediately, and long and constant accumulation is normally needed for the contamination to be detected.

To confirm the pollution, it is necessary to collect a vast amount of information, such as specific pollution indicators, types and nature of pollutants, pollution pathways, construction and operation of polluting facilities, production records of the polluting enterprise, accidental damage reports, environmental impact reports, environmental impact assessment (EIA) reports, environmental incident investigation reports by local administrative authorities or environmental prosecution agencies, environmental monitoring reports of pollution sources, accident treatment reports, and audio and video recordings of relevant damage. To investigate the damaging impacts of EPIL cases, accreditation bodies or experts are generally needed to prepare environmental damage detection, testing, monitoring reports, or test reports of animals and plants, the death and disability of which are the results of polluting behavior. For the extent of damage, professional institutions are required to grant price certificates, assess various types of emergency management costs, pollution remediation costs, etc. Similarly, the causal relationship between the defendant's acts and the damaging results may require professionals involvement to provide investigation and treatment reports and identification reports. The requirement to involve professionals means that the parties will have to invest enormous time and energy in evidence collection and resort to a variety of professional data and accurate testing, which requires significant expenditure.

Statistics in 2018 show that, in judicial practice, among 116 selected cases, 100% of these cases have involved professionals or professional bodies to assist in compiling the evidence. The rate of adopting professional evidence in court was 100%.[11] In addition, researchers have compiled statistics on the guiding cases issued by the Supreme People's Court (SPC) of China between 2015 and 2021,[12] which are not binding but have a crucial impact on local courts. Among these guiding cases, only one out of 17 EPIL cases neither invited expert bodies nor sought professional opinions. This meant that in EPIL, judges and litigants are highly dependent on external expertise. The following two cases exemplify how experts played their roles in EPIL.

Case I *"People's Procuratorate of Taizhou City, Jiangsu Province v Wang Xiaopeng and 59 Others in Civil Public Interest Litigation for Ecological Damage"*[13]

In the first half of 2018, 38 people, including Dong Ruishan, individually or jointly, illegally caught and sold Yangtze eel fry in the Yangtze River's dry waters using prohibited fishing gear for profit. Wang Xiaopeng and 13 others, knowing that the Yangtze River eel fry was illegally caught, acquired them individually or established a partnership to acquire eel fry by signing partnership agreements and jointly funding the acquisition and selling of them to the public. They acquired 116,999 eel fry from seven people, including Gao Jinchu, and 38 others, including

[11] Zhu (2019) .

[12] Data is available on the website of SPC, under the column of 'Guiding Cases'.

[13] The court decision for the case *People's Procuratorate of Taizhou City, Jiangsu Province v Wang Xiaopeng and 59 Others in Civil Public Interest Litigation for Ecological Damage* is available on the legal database https://wenshu.court.gov.cn.

Dong Ruishan, who illegally trafficked or fished for them. In this case, the Court commissioned a fisheries expert to give a professional opinion. Ultimately, the court awarded the defendants 2.5 times the total price of the fish caught or sold to compensate for the ecological damage.

In this case, there was a central question as to whether the amount of fish caught by the defendants "seriously infringed upon the eel resources and other ecological resources of the Yangtze River, causing serious damage to the fishery resources and other ecological resources of the Yangtze River and resulting in the reduction of biodiversity". Experts believe that the eels belong to the key protected fish in Jiangsu Province, which thus far cannot be artificially reproduced. To catch the eel fry, open nets were set in fish and shrimp breeding sites or fish migration channels. The mesh of the nets was small and the nets were set for long periods of time. This resulted in catching an extensive range of species, rather than just the target species. It was most harmful to the marine environment. Finally, the court relied on the expert's opinion and comprehensively considered multiple factors, such as the scope and extent of the ecological damage, the scarcity of the resources, the destructive nature of the illegal act, the sensitivity of the time, and the specificity of the location, when making the decision.

Case II "*Yiyang City People's Procuratorate of Hunan Province v Xia Shun'an and 15 Others Civil Public Interest Litigation for Ecological Damage*"[14]

Xia Shun'an and others drove sand mining vessels to an undeveloped area of the Lower Lake of Dongting Lake to illegally extract sand, making illegal profits of RMB 22,433,300 yuan (approximately 3.2 million USDs). The Yiyang City People's Procuratorate of Hunan Province instituted civil public interest litigation, requesting that Xia Shun'an and others be ordered to bear joint and several liability for the ecological damage caused by their illegal sand mining activities. The People's Court commissioned a qualified institution, the Hunan Environmental Protection Scientific Research Institute Ecological and Environmental Damage Judicial Appraisal Centre, to conduct an accreditation. After the accreditation, the impact of Xia Shun'an and 15 other people's illegal sand mining on the ecological environment of the illegal sand mining area was divided into: (1) damage to the quality of the water environment, (2) damage to the structure of the riverbed, (3) damage to water connotation and (4) damage to aquatic biological resources. The total spatial extent of the ecological impact covered approximately 99,000 square meters. The loss of aquatic biological resources caused RMB 26,530 yuan worth of damage. The costs required to repair the damage to the aquatic biological resources and the damage to water source connotation was RMB 79,690 yuan and RMB 8,656,100 yuan, respectively, totaling RMB 8,735,790 yuan (approximately 1.25 million USDs). The judge adopted the conclusions of the judicial appraisal and directly ordered the defendant to pay compensation of RMB 8,735,790 yuan.

[14] SPC (2021b), Guiding Case No 176, https://www.court.gov.cn/fabu-xiangqing-334721.html. Accessed 20 August 2022.

In this case, the People's Court invited the accreditation experts to conduct an all-round accreditation of the damage to the riverbed, water connotation, biological habitat, biological resources such as fish and shrimp, and quality of the water environment in the area of sand extraction affected by illegal sand extraction. The accreditation experts, following the principle of comprehensive compensation, quantified the total amount of sand extracted, the swelling coefficient, the density of loose sand in the water, the proportion of water content, and the average fish resource yield in Dongting Lake when calculating respective damages. It is clear that accreditation plays an important, if not decisive, role in determining the facts of an environmental dispute. The introduction of scientific evidence on a technical level by a person with specialist knowledge of the relevant professional issues in an environmental case has become a central element of securing appropriate decisions.[15] Using the expertise of accreditation experts can facilitate the resolution of complex and difficult issues in environmental litigation and has become a rational judicial choice to "shine a light on environmental protection".[16]

Although the three types of experts in the "three-dimensional model" are all professionals with specialized knowledge, their roles in litigation are distinct. This chapter will examine their respective roles and problems with the accreditation scheme in EPIL.

12.4 Accreditation Body: Professional Expert as Assistant of the Court

The accreditation body plays an important role in the entire fact-finding process and is an assistant to the court.

12.4.1 Functional Role of Accreditation Bodies

Generally, the term 'accreditation body' refers to an institution that accepts appraisal[17] commissions from courts or parties involved in litigation. An accreditation body, by following the ways, methods, procedures, and relevant rules and standards stipulated by law, applies scientific technology or specialized knowledge to identify and assess the specialized issues involved in litigation and provides appraisal opinions during litigation.[18] According to Article 3 of the "Measures for

[15] Zhu (2019).

[16] Wang and Zhang (2021), pp. 657–671.

[17] It should be noted here that this chapter does not make a strict distinction between appraisal and accreditation.

[18] Ministry of Environmental Protection (2016).

the Administration of Registration of Judicial Accreditation Body", a judicial accreditation body is an institution for judicial appraisers that meets the conditions set out in the Measures. It should be registered with the provincial judicial administrative authorities holding a judicial appraisal licence and carrying out judicial appraisal activities within the scope of the registered judicial appraisal business. This means that the establishment of an accreditation body must meet specific requirements; it must have specific qualifications and be recognized by the administrative authorities. According to Article 66 of the *Civil Procedure Law* of the People's Republic of China, unlike the testimony of an auxiliary expert, which can only be used as a type of professional testimony, the findings of an accreditation body are an independent type of evidence that has a high probative value in court. Its probative power is similar to that of public documents issued ex officio by state authorities and public bodies and is pivotal to the determination of facts and legal issues in a case. Therefore, conclusions of appraisal opinions have a high degree of credibility and are usually accepted by the judge.

Judicial accreditation of environmental damage refers to the activities of an appraiser who uses the technology or expertise of environmental science, such as monitoring, testing, site investigation, experimental simulation, or comprehensive analysis, to identify and assess the specialized issues involved in environmental pollution or ecological damage litigation and s appraisal opinions.[19] Among others, specialized issues include determining the nature, extent, and degree of environmental damage and the nature of the pollutants; assessing the causal relationship between the responsible parties and the damaging results to the environment, the pollution control program and its operating costs; preventing the extension of the damage; and developing a program to restore the ecological environment.[20]

When accreditation on environmental damage is considered, several features can be identified: (1) Professional appraisal. A forensic appraiser is a specialist who has environmental-related expertise, has the ability to solve professional problems in environmental pollution cases, and has been accredited with appropriate qualification by the administrative department of forensic appraisal. (2) Specified object. Professional advice is sought regarding environmental pollution or other environmental issues that have arisen in the ecological litigation. (3) Scientific assessment technology. Current environmental science technology or related knowledge should be utilized. (4) Formal procedure. When carrying out identification and issuing the corresponding appraisal reports, the accreditation body must follow the legal procedures.[21]

[19] Zhang et al. (2021), pp. 1–9.

[20] SPC (2021a).

[21] Zhu (2019).

12.4.2 Regulations Concerning Qualification of Accreditation

There are few rules regulating the subject of environmental damage appraisal or accreditation qualifications, and these rules are mainly scattered among the normative documents issued by various administrative departments. Therefore, accreditation bodies have close links with administrative authorities. Different administrative bodies may recognize or set up different accreditation bodies. For example, the agricultural administrative authorities are responsible for the management of agriculture-related appraisal bodies. The Ministry of Agriculture divided appraisal bodies into three tiers: municipal, provincial, and national appraisal bodies. In addition, the Ministry of Agriculture stipulates that the relevant appraisal body must pass an examination by the Resource and Environment Research Centre of the China Academy of Fisheries Science (a scientific institution) in order to become a qualified appraisal body. It is also clear from this regulation that appraisal bodies in China are, on the one hand, managed by administrative authorities and, on the other hand, work closely with scientific research institutions. However, there has been no specified and state-recognized accreditation on environmental damage in China. Only three categories of judicial appraisal, i.e., forensic evidence, physical evidence, and audio and visual data, are professional state-recognized appraisals. However, this does not mean that there was no specific identification of environmental pollution in China.

However, the wide range of environmental damage accreditation, technical difficulties, and complex identification process make litigation activities regarding environmental damage forensic identification an urgent need in China. In 2011, the former Ministry of Environmental Protection issued a policy called "*Several Opinions on the Development of Environmental Pollution Damage Appraisal and Assessment Work*", marking the launch of China's formal attempt to develop environmental damage assessment. This opinion proposes the construction of a professional appraisal team on environmental damage. In December 2015, the SPC, the Supreme People's Procuratorate (SPP) and the Ministry of Justice jointly issued a "*Notice on The Environmental Damage Accreditation into The Scope of Unified Registration and Management*" (the "Notice"), which specified environmental damage accreditation as a new appraisal matter and brought it into the scope of unified registration and management. That means that accreditation on environmental damage has become an independent, state-recognized category of appraisal, which is paralleled with the other three categories (forensic; physical evidence; audio and visual data). State-recognized environmental accreditation can, to a certain extent, make appraisal conclusions more credible in the sense that constraints were imposed by state administrative or judicial authorities on accreditation bodies and the broad extent and high standards of accreditation were adopted. In 2018, the Ministry of Ecology and Environment (MEE) and the Ministry of Justice (MOJ) jointly issued the *Rules for the Registration and Evaluation of Judicial Appraisal Institutions for Environmental Damage*. In Article 3 of the Rules, legal persons or other organizations that are willing to engage in the judicial appraisal of environmental damage shall submit their application material to the provincial judicial administrative authorities for inspection and review. Administrative authorities shall also comprehensively take into consideration the appraisers, the technical level of appraisal, and the conditions of appraisal laboratory instruments and equipment when scoring the accreditation of appraisal institutions. The appraisers shall be evaluated on the basis of their situation, working ability, the results obtained, and professional ethics. It is also provided that the appraiser cannot conduct the act of appraisal alone; it must work through an appraisal institution. The rules also require accreditation bodies that existed before the rules were issued to be assessment in accordance with these rules.[22]

Several subsequent policies and regulations for the standardized accreditation of environmental damage are provided. These policies and regulations aim to facilitate the construction and management of environmental damage identification, promote the standardization, legalization, and scientific development of environmental damage identification, and effectively meet the needs of environmental litigation and environmental administrative law enforcement.[23] By providing guidance for judicial practice, they significantly improve the effectiveness of environmental dispute resolution.

12.4.3 Regulations in Relation to Accreditation

The regulation of standards and procedures for accreditation has been a key concern of the legislature. In 2011, the former Ministry of Environmental Protection (MEP, Ministry of Ecology and Environment (MEE) after 2018)[24] issued the *Recommended Methods for Calculating the Amount of Environmental Pollution Damage* (Version 1). This document, for the first time, recommended the application of the virtual treatment cost and remediation cost method; it emphasized the use of the method where the actual cost of remediation works could not be ascertained and provided for a floating period for the application of the methods. After three years of trial implementation, the Ministry of Environmental Protection issued the Recommended Methodology for Environmental Damage Assessment in the Emergency Response Phase of an Environmental Incident to refine the methodology. Article 8 and Article 9 of the Recommended Methodology for Environmental Damage Assessment in Emergency stipulate that the amount of ecological damage needs to be determined by the virtual cost of the management method.

[22] Zhang (2021).

[23] SPC (2021a).

[24] In 2018, Ministry of Environmental Protection of China (MEP) was renamed the Ministry of Ecology and Environment (MEE).

At the beginning of 2021, the Ministry of Ecology and Environment and the State Administration of Market Supervision and Management jointly published a document including a general outline, a damage survey, soil and groundwater, surface water sediment, air pollution virtual treatment cost method, and water pollution. The document proposes differentiated assessment criteria, mainly for soil and groundwater, specifying the scope and extent of environmental damage under less than differential conditions. The quantitative approach to quantifying the value of surface water resources is applied to assess the damage to surface water and sediments. Where the environmental quality baseline for surface water has been exceeded, the quantification of surface water resource values should be applied. The virtual cost of the treatment method is used to determine the extent and degree of air and water pollution. The scope and extent of air and water pollution should be determined by using the virtual cost of the treatment method. However, changes to the criteria for the identification and assessment of environmental damage to grasslands and forests are not addressed in this document.

Regarding the accreditation procedure, different requirements are specified for different types of identification. In 2014, the former Ministry of Environmental Protection issued the Recommended Methodology for Environmental Damage Assessment in the Emergency Response Phase of Sudden Environmental Incidents in the form of a normative document. Article 4 of the document defines the procedures for the assessment of environmental damage in environmental incidents in China. The main components of the assessment mechanism include the following: the preliminary preparation for carrying out the appraisal and assessment work, the initiation of the appraisal and assessment work, the acquisition of information related to ecological damage and environmental pollution, the confirmation of the extent and scope of the damage, the quantification of the environmental damage, the ecological damage itself suffered and its additional losses, the judgment on the need to initiate a medium- and long-term damage assessment and the compilation and preparation of the assessment report.[25]

12.4.4 Problems of the Environmental Damage Assessment Mechanism

In practice, the environmental damage assessment mechanism has some problems. These include, for example, the lack of a sufficient number of appraisal bodies to undertake a judicial appraisal of environmental damage, the lack of unified standards for appraisal and assessment of public interest, the high cost of an appraisal, the overlong appraisal period, and the lack of neutrality of appraisal bodies and appraisers. These, to some degree, are detrimental to the credibility of the appraisal.

[25]Zhang (2021).

12.4.5 Auxiliary Expert: Supplement of Essential Expertise as Assistant of Both Parties

In recent years, environmental disputes in China have become more frequent, and an increasing number of environmental judicial cases involve professional issues such as environmental damage assessment. The facts found in environmental cases are complex and professional, and relying solely on the appraisal system can no longer meet the demand for expertise in judicial practice. It is unable to resolve the problems of the current system, such as the difficulties in appraisal and assessment and the lack of professionalism of judges in environmental justice. In other words, the judicial appraisal of environmental disputes has many problems, as mentioned above, such as the lack of unified standards for appraisal and the high cost of an appraisal. Therefore, it is insufficient to rely solely on the appraisal system when resolving professional fact-finding problems in environmental disputes, and the auxiliary expert person system is particularly important in environmental justice.

12.4.6 Functional Role of Auxiliary Experts

The introduction of auxiliary experts in litigation process is an important development in civil environmental litigation. An auxiliary expert is a litigation participant entrusted by the parties to participate in the litigation, to cross-examine the parties about the issues in the litigation and put forward their own expert opinion. Their opinion or recommendation is equivalent to the party's statement.[26] It supplements the judicial appraisal system, which not only fills the gap in the professional knowledge of the parties and their representatives in the environmental field but also facilitates the parties to refute the appraisal opinion and obtain the result in their favor. It also helps the judge to objectively determine the facts of the case, prevent bias, and avoid over-reliance on the appraisal opinion. However, the term "auxiliary expert" is not a legal term because it does not appear directly in any of the statutory provisions. Instead, it mostly appears in research papers and court judgments. The legal basis of the "auxiliary expert" mechanism is the *Civil Procedure Law of the People's Republic of China*, Article 82, which provides that "A party may apply to the people's court for notifying a person with the expertise to appear in court to offer an opinion regarding an appraisal opinion or regarding a specialized issue."

Auxiliary experts have played an important role in the judicial practice of China in the following areas.[27] First, it makes up for the inadequacy of the accreditation system to safeguard the full and effective exercise of litigation rights of the parties. China's appraisal institutions and appraisers are subject to a registration system.

[26] Zhang and Peng (2016), pp. 729–734.

[27] Liu (2018).

Appraisers are only allowed to participate in appraisal work if they have undergone a strict qualification review. A result of such a rigorous review system is that quite a few environmental damage assessment agencies and experts have obtained the appropriate qualifications. In 2014, 2016 and 2020, the former Ministry of Environmental Protection released lists of “Recommended Environmental Damage Appraisal and Assessment Institutions”, recommending a total of 42 environmental damage appraisal and assessment institutions for citizens, legal persons, and other organizations. Considering that there are thousands of environmental lawsuits in China, a few dozen environmental assessment agencies can hardly meet the needs of society in dealing with complex environmental disputes. The long duration and high cost of environmental appraisal are also important shortcomings of the environmental appraisal system.[28] Therefore, the disproportion between the number of environmental cases and that of the appraisal and assessment institutions makes it appropriate and necessary to introduce the expert assistants’ system. The use of auxiliary experts is another form of participation in litigation. It provides an alternative channel for litigants to prove specialized issues and for judges to fully decide specialized facts. The auxiliary expert can then participate in the litigation as a technical expert, questioning the appraisal opinion or providing his or her own independent opinion on the specificity of the case for the court to accept.

Second, auxiliary experts enhance the comprehensive examination of specialized facts during the trial. The determination of such issues often involves expert opinions, test reports, other professional materials and data, scientific principles, technical analysis, and other issues that are difficult. It consists of great obstacles for cross-examination by the defendants and litigation agents. The auxiliary expert’s involvement in the proceedings, can effectively help to decipher this complex evidence for the benefit of the parties and the judge, and their supervision of the matter can also ensure that the appraisal experts are presenting neutral and objective opinions. Mutual corroboration, complementarity, discrepancies and even confrontation between appraisal experts and auxiliary experts can, to a certain extent, reduce the judges’ over-reliance on expert opinions in certain cases and achieve the effect of “balanced proof and clear hearing”. Once a professional opinion is accepted by the court and cited in a judgment, it has a substantial impact on the outcome of the proceedings and the rights and interests of the parties.[29]

In fact, expert opinions, unlike judicial opinions, are seldom adopted by judges. Moreover, there is inconsistent adoption of an auxiliary expert in civil trials. In the following two cases, the auxiliary expert’s opinion had different impacts on the court trial.

Case III From January 2012 to February 2013, six companies, including Changlong and Jinhui, paid the Jiangzhong Company and other companies to dispose of hazardous wastes (such as acid) which they had generated during their

[28] Zhang (2021), p. 8.

[29] Zhang (2016).

production process. Jiangzhong Company did not have a hazardous waste disposal licence. Jiangzhong Company and the other enterprises did not treat these wastes, but rather dumped them directly into the Rutai Canal and Gumagan River, causing serious environmental pollution. In August 2014, the Taizhou Environmental Protection Federation (NGO) filed a lawsuit against the six enterprises mentioned above. The Intermediate People's Court of Taizhou City, Jiangsu Province, found that the six defendants constituted an infringement and the polluters were ordered to pay a total of approximately RMB160,666,745 (approximately 338 million US dollars) environmental remediation costs.

The case was broadly reported in China, as it created a record for environmental damages awarded in public interest litigation in China. In the appeal, the central issue was how the consequences of the damage were determined. Other questions included, among others, whether repair was needed. Whether the method of calculating the cost of the repair was appropriate? What manner of performance would be appropriate?[30] The defendant in the first instance insisted that the river in question had a self-purifying function and that there was no need to restore or compensate. In addition, it was argued that the determination and calculation of restoration costs were incorrect. Whether the defendant should bear the responsibility for ecological restoration and the amount of damage were difficult questions to determine. The plaintiff in the first instance applied for expert involvement in the proceedings. Professor Lu Xiwu from Southeast University was invited to provide expert evidence. Professor Lu repudiated the defendant's claim that the water quality had been restored. He noted that dumping hazardous waste, in this case, could directly damage the ecological environment of the region. Although river water is mobile, damaging results have occurred and have been difficult to reverse. Even if the water quality near the dumping point was restored, damage to the water ecological environment remains.[31] Accordingly, the Jiangsu High People's Court ruled in favor of the plaintiff and held that RMB 160 million for ecological restoration should be awarded. In this case, the expert's professional opinion provided a strong basis for the court in deciding the defendant's tortious liability and the exact amount of the environmental damage costs.

Case IV On 9 February 2016, the five lobular maples were included in the List of Key Plants for Protection in Sichuan Province. On 10 August 2018, the International Union for Conservation of Nature assessed the five lobular maples as "Critically Endangered" and added this species to its Red List. However, the five lobular maples are not listed in China's National List of Key Wild Plants. According to the China Biodiversity Conservation and Green Development Foundation (the Green Development Council), the population of five lobular maples near Woloshi Village, Malangtso Township, Yajiang County, is the largest remaining population of maple in the world and is the only one with the ability to reproduce naturally. The

[30] Lv (2016), pp. 244–264.

[31] Wang (2016), pp. 1–8.

Yagen step hydroelectric power station was planned to be built in that region. There was a concern that the construction of the Yagen step hydroelectric power station and the accompanying road would directly threaten the survival of the five lobular maples and pose a direct threat to the public interest.

The Green Development Council filed a public interest lawsuit, requesting the court to rule that the hydropower station development be stopped. Finally, the court ruled that the defendant Yalong River Basin Hydropower Development Co., Ltd. should take the survival of the five lobular maples as an important element of the environmental assessment work in the feasibility study stage of the Toogan hydropower station project and that the environmental impact report should be approved by the environmental protection administrative department before proceeding to the next step.

The major issue in this case was the question of whether the development of the hydroelectric power stations would be devastating to the survival of the five lobular maples. During the trial, the defendant invited an expert, a university professor of forestry, to provide evidence. Expert evidence showed that the five lobular maples were present in many other areas and that artificial propagation was now possible. The expert argued that the development of the hydroelectric power station would not be so damaging to the five lobular maples. However, the court did not fully adopt the expert's opinion. Given the ranking of the five lobular maples in the Red List of Biodiversity and the potential risk of damage to the original living environment of the five lobular maples at the site in question, the construction of the power station and its broad impact would affect the survival of the maples, which would in turn harm the public interest. As a result, the court requested the defendant to undertake further work on the environmental impact assessment of the five small-leaved maples and fulfil the statutory requirements under the approval procedures before the project proceeded to the next step.

12.4.7 Operation Rules for Expert Support System in Environmental Litigation

There are no rules governing who can become an expert and what rules an expert should follow. However, for the smooth implementation of environmental judicial practice, courts in all provinces in China are constantly exploring specific ways of implementing the expert support system. For example, in 2014, the High Court of Fujian Province issued the *Measures on the Management of the Technical Advisory Expert Pool for Ecological and Environmental Trials* (for Trial), which explored the establishment of a technical advisory expert pool for courts in Fujian Province. In 2017, Xi'an City in Shanxi Province established a pool of technical experts for environmental trials and openly selected technical experts for environmental trials across the province. Selected experts can appear in court to express their opinions on the identification of professional issues concerning environmental and natural resource protection, to preserve evidence and to conduct on-site investigations in environmental cases.[32]

In the meantime, the court in Guiyang, the capital city of Guizhou Province, introduced the mechanism of expert participation at an early stage of environmental litigation. Courts in Guiyang now have a wide range of expert support available to assist in environmental trials. The scope of work of the supporting experts in environmental trials in Guiyang is relatively broad. The work mainly includes the following: (1) to provide guidance and advice on the preservation of evidence, such as the number of samples that should be taken, the location, scope, time, and the method of sampling; (2) to provide advice on the admissibility of the evidence; (3) to provide opinions on the scope of loss and the amount of loss; (4) to provide opinions on the determination of causality; and (5) to provide recommendations about the remedy that should be awarded.[33]

Jiangsu Province has also established a pool of environmental emergency experts. In the area of public interest litigation, procuratorial authorities can select an auxiliary expert from the pool of environmental experts for emergencies. Fujian Province has also established a pool of technical advisory experts to facilitate the selection and appointment of experts by the parties in the environmental field. It also provides selection criteria for experts, which reduces courts' burden in selecting qualified expert supporters in litigation. These selected technical experts are mainly from universities and scientific research institutions, and they have engaged and been experienced in teaching and scientific research in the fields of environmental protection, forestry, state map resources, marine, fisheries, and water conservancy. In addition, they must have professional expertise in the field of ecology and the environment in China and be of high prestige in their respective field. In addition, Fujian Province also specifies the main duties of the supporting expert, which include, but are not limited to, receiving and answering inquiries and appearing in court as an auxiliary expert to assist the court in ascertaining professional ecological and environmental facts. The expert is entitled to access the relevant case file, to observe the trial, to express an independent opinion on the professional issues in relation to the case and to participate in the mediation of the advisory case. The expert is required to give objective and clear answers to professional questions, not to evaluate legal issues and to maintain confidentiality. These specific provisions provide clear and concrete guidelines for the participation of expert supporters in litigation.

[32] Liu (2018).

[33] Li (2010), p. 26.

12.4.8 Problems of the Expert Support System in Environmental Litigation

There are some problems with the system of an auxiliary expert in practice, for example, the courts' lack of confidence in the experts' opinions. Some courts do not include the participation of expert assistants in litigation. The parties often choose to hire expert supporters only when they are not convinced by the accreditation opinion in the first instance or when there is a risk of losing the case. In some local courts, expert supporters sit with their client, whereas the practice in other courts requires that the expert supporters sit in the witness seat, directly opposite the judge. These differences in judicial practice reflect the fact that the current system still needs to be improved, and the role of auxiliary experts and the rules of operation of the system still need to be clarified.

12.5 Juror with Specialty: Fact-Finding as Adjudicator

In order to lift the veil of accreditation and auxiliary experts' expertise, to fill the "information gap" of judges and to hear cases on specialist issues, as a part of China's judicial reform, expert jurors have also been introduced to participate in the proceedings of environmental lawsuits.

12.5.1 Functional Role of Expert Jurors

Expert jurors can assist the judge during the trial by effectively controlling and directing the examination of expert opinions of the parties and their expert supporters, ensuring that the examination is focused on particular issues. They can also share with the judge their understanding of particular issues involved in the examination by translating them into plain language during the post-trial deliberations, which enables the judge to conduct free-minded testimony and accurate fact-finding.[34] The role of expert jurors are as follows.

(1) The panel of the expert jurors can play a role in regulating the presentation of evidence and cross-examination of specialized issues by both parties to the proceedings, either ex officio or based on the other party's submissions so that the cross-examination of expert opinions truly revolves around the specialized issues at issue in the matter. The sharing of power between the expert jurors and the trial judge can also alleviate the problem of duplication and multiple appraisals. That is, when there are different appraisals, the expert jurors and

[34]Zheng (2019), pp. 67–77.

the judge can jointly decide whether the case requires a new appraisal, through which they can improve the efficiency of the proceedings.

(2) Expert jurors can also guide the accreditation expert or auxiliary expert during the trial to convert the specialist language into everyday language that the judge can understand, which can therefore improve the judge's ability to understand the specialist issues, increase the efficiency of the trial, and ensure that the judge can understand the various scientific issues that might be raised in the trial.

(3) The participation of expert jurors can assist the credibility of the trial process in practice. Due to their lack of expertise, parties are prone to doubt the impartiality of court trials. This gives rise to concerns about the credibility of the judiciary. For example, in 2007, a plaintiff sued a mining and chemical company in Fujian Province for the emission of waste gas. One of the main reasons for the prolonged and repetitive trial process was the inconsistent opinions of different expert bodies. Lacking relevant environmental expertise impeded the judges in this trial from making timely and accurate judgments on expert opinions.[35] When the appraisal body provides appraisal opinions that are nonscientific, illegal, inconsistent, or even contradictory, courts usually have difficulties in making a decision on the case based on appraisal opinion. Furthermore, environmental experts involved in environmental protection litigation can only play a judicial auxiliary role but cannot participate in the hearing of environmental cases.[36] Professional jurors do not have such drawbacks and therefore are essential to environmental litigation.

12.5.2 Operation Rules of Expert Jurors

On 27 April 2018, the Standing Committee of China's National People's Congress adopted the *Law of the People's Republic of China on People's Jurors* (People's Jurors Law), which stipulates the basics of the jury system for EPIL. It lays a theoretical foundation for the development of EPIL in China, which represents a significant breakthrough in China's environmental litigation system. However, the People's Jurors Law only contains three general articles in relation to the jury system of EPIL matters. Article 16 of the People's Jurors Law provides for the formation of a seven-member panel, four of whom are jurors, in the case of major environmental pollution cases. Article 16 stipulates that the jury system must be utilized in EPIL matters, which is appropriate given the importance of the jury system in such trials. Article 14 provides that "When people's jurors and judges form a collegiate court to try cases, the judge shall serve as presiding judge. A three-member collegiate court may be formed, or three judges and four people's jurors may form a seven-member collegial panel." As seen, a panel of three professional judges and four people's

[35] Yu (2015), pp. 125–129.

[36] Sun and Yang (2022), pp. 1–8.

jurors must form the panel for EPIL. This arrangement ensures that the people's jurors have a numerical advantage over the professional judges in making findings of fact, thus making it possible for the complex facts in EPIL to be understood in the most realistic way. It also ensures that the professional judges can listen to as many voices representing different interest groups in the factual trial, so the final verdict can effectively harmonize the interests of all parties and increase the acceptability of the verdict. This guarantees that the final judgment will effectively reconcile the interests of all parties, thus increasing the acceptability of the decision and bringing about positive legal and social effects.[37] Article 22 of the People's Jurors Law stipulates that "People's jurors shall sit in a seven-member collegial court and independently express their opinions on fact-finding and vote jointly with the judge; they may express their opinions on the application of law but do not take part in the voting." This law distinguishes "legal trial" and "factual trial" and establishes that people's jurors have the same "factual trial" function and voting rights as judges in EPIL.

It should be noted that the use of expert/professional jurors already existed in judicial practice prior to the adoption of the People's Jurors Law. For example, in the case "*China Environmental Protection Federation, Guiyang Public Environmental Education Centre v Dingpao Paper Mill in Wudang District, Guiyang City*" in 2010, the Qingzhen Environmental Court invited two environmental protection experts to be expert jurors. One of the jurors was a senior environmental protection engineer and director of the Guiyang City Two Lakes and One Reservoir Administration, and the other juror, Dr Liu, holds a PhD degree in environmental science. The expert jurors' opinions provided substantial support for the identification and testing of the discharges.[38]

12.5.3 Problems of the Expert Jurors System

For years, the juror system has solved the problem of insufficient trial personnel in China. However, in practice, the phenomenon of "inviting but not attending, accompanying but not hearing, and joining but not deliberating" often occurs, and jurors regard this democratic right to participate in justice as an "act of help" to the court. However, the provisions of the Jury Law are too abstract and vague, and there are even certain omissions, which makes it an inadequate response to the special characteristics of the jury system for EPIL. To form procedural rules that meet the needs of EPIL, a series of issues, such as juror selection, trial participation, and the system of expert jurors in EPIL, needs to be resolved through judicial interpretation and subsequent legislation.[39]

---

[37] Yu (2018), pp. 75–81.

[38] Sun and Yang (2022), pp. 1–8.

[39] Yu (2018), pp. 75–81.

References

Chen G (2014) Expert draft of the criminal evidence law of the People's Republic of China. China Legal Publishing House

Jiang H, Zhao S (2017) The Judicial Dilemma of the environmental damage appraisal system and the path to break it - 10 cases of environmental civil public interest litigation as an example. J Guizhou Univ (Soc Sci Edn) 4:11–18

Li X (2020) The solution to the problem of specialization in litigation - also on the improvement of China's identification system and statutory forms of evidence. Polit Legal Forum 6:37–54

Li Y (2010) The practice and improvement of expert participation mechanism in environmental litigation, Dissertation. Suzhou University

Li Y, Huo Y, Liu Z (2017) An empirical analysis of the mechanism for resolving specialized issues in environmental public interest litigation. Environ Protect 24:39–43

Liu H (2018) Research on the system of expert supporters in environmental justice, Dissertation. Chongqing University

Lv Z (2016) Environmental judicial rationality cannot stop at "Sky-High" compensation: a review of the Taizhou environmental public interest litigation case. China Law 3:244–264

Ministry of Environmental Protection (2016). The General Outline of The Technical Guidelines for The Identification and Assessment of Damage to The Ecological Environment, (Huanban Zhengfa) [2016] No 67

SPC (2021a) White Paper on Forensic Identification of Environmental Damage, http://www.moj.gov.cn/pub/sfbgw/fzgz/fzgzggflfwx/fzgzggflfw/202106/t20210604_426692.html. Accessed 20 August 2022

SPC (2021b) Guilding Case No 176, https://www.court.gov.cn/fabu-xiangqing-334721.html. Accessed 20 August 2022.

Sun Y, Yang F (2022) A study of the expert Juror system in the context of the specialization of environmental justice. China Forensic Sci 1:1–8

Wang C, Zhang T (2021) On the probative value of professional opinions in environmental litigation. Sci Evid 06:657–671

Wang J (2011) The integration of the forensic model and the expert witness model - the direction of the reform of China's criminal forensic system. J State Prosecutor's Acad 4:95–115

Wang X (2016) Problems and judicial responses in environmental damage forensic identification. China Forensic Identif 1:1–8

Yu F, Tian C, Zhang YS (2015) A preliminary study on the judicial appraisal system of environmental damage in China. China Judicial Appraisal 5:125–129

Yu Y (2018) A study on the Jury system for environmental public interest litigation - centering on the relevant provisions of the People's Jurors law. Rule Law Stud 6:75–81

Zhang C (2016), Amicus curiae: the value of public participation and its operation, Tianfu Xinzhu, No 2

Zhang Q et al (2021) A study on the development history and problems of ecological environmental damage forensic identification in China. China Forensic Identif 4:1–9

Zhang S (2021) Research on the legal system of environmental damage appraisal and assessment, Dissertation. Lanzhou University of Technology

Zhang X, Peng Y (2016) Research on intellectual support for environmental civil public interest litigation. J Shanxi Agric Univ (Soc Sci Edn) 10:729–734

Zheng F (2019) On the "four-dimensional model" of judicial specialization in China. Polit Legal Forum 3:67–77

Zhu J (2019) Research on the management of judicial appraisal of environmental damage and the formation and admissibility of appraisal opinions, Dissertation. Nanjing Normal University

Chapter 13
Judicial Application of Ecological Remediation Liability in EPIL: A Commentary to the Guiding Cases, Gazettes Cases and Model Cases of the SPC in China

Zhiyu Huang

Abstract Starting from the Guiding Cases, Gazettes Cases and Model Cases issued by the SPC on EPIL in China, this chapter analyses the innovative approaches of the judiciary when considering ecological remediation liability. At the same time, there are problems in the judicial approach to ecological remediation liability, such as poor implementation of the principle of the priority of ecological remediation, insufficient supervision of the assessment and management of ecological restoration, compensation for damages, and serious divergences in some innovative practices. These issues and differences of opinion between members of the judiciary do not, on occasion, reflect the principle of "Like Cases be Treated Alike", nor are they conducive to the defendants and the public being satisfied with the outcome of a trial. Consequently, it is argued that there is an urgent need to clearly define the relevant rules, to establish operating guidelines which address the issue of which parties are to be responsible for any restoration programmes and which parties are to supervise such a programme. The overall intent should reflect the goal of ensuring the public that the result of any decision was fair, legitimate, and democratic in terms of securing public participation.

Keywords Ecological remediation liability · Judicial application · Environmental civil public interest litigation

13.1 Introduction

Ecological remediation liability has undergone a series of developments in China which are reflective of different interpretations of passages in the Civil Code. Consequently, the allocation of responsibilities, although relatively comprehensive

Z. Huang (✉)
Nanchang University, Nanchang, China
e-mail: huangzhiyu@ncu.edu.cn

X. Wang et al. (eds.), *Environmental Public Interest Litigation in China*,
https://doi.org/10.1007/978-3-031-26526-6_13

at the practical level, is nevertheless diverse from judgment to judgment. Additionally, many academic and judicial departments continue to debate over the allocation of responsibility for ecological environment restoration. In recent years, the SPC has released hundreds of Guiding Cases, Gazettes Cases and Model Cases on EPIL, which enable the public to analyse the innovative approaches of the judiciary when considering ecological remediation liability. Problems of the judicial approach to ecological remediation liability have been observed, such as poor implementation of the principle of the priority of ecological remediation, insufficient supervision of the assessment and management of ecological restoration, compensation for damages, and serious divergences in some innovative practices. These issues and differences of opinion between members of the judiciary do not, on occasion, reflect the principle of "Like Cases be Treated Alike", nor are they conducive to the defendants and the public being satisfied with the outcome of a trial. Analysis of these issues and problems constitutes the major theme of this chapter.

13.2 The Development of the Rules for Ecological Remediation Liability Under EPIL in China

13.2.1 The Development of Policies and Legal Rules

In 2007, the first environmental court in China was established in Qingzhen City, Guizhou Province. This essentially represented the legislative acceptance of environmental public interest litigation within China. Subsequently, only one month later, Guiyang's Administration of "Two Lakes and One Reservoir" brought a lawsuit to the Environmental Court of Qingzhen and won. In 2009, the first environmental public interest lawsuit was initiated by the Procuratorial Organs and won a similar lawsuit in Foshan, Guangdong Province. In the same year, the Xishan District People's Court of Wuxi City also accepted the first environmental public interest action in Jiangsu Province. In this case the court ruled that the defendant had an environmental restoration responsibility. These cases ran against the very Chinese principle that practice should precede legislation which had previously made it difficult to find a court that would accept an action based on the broad principle of an environmental public interest.

Moreover, due to the lack of a clear legal basis, the legality of environmental courts and environmental public interest litigation itself was widely questioned. Until the enactment of Article 55 of the *Civil Procedure Law of the PRC* (2012 revision),[1] it was clear that "For conduct that pollutes the environment, infringes upon the lawful rights and interests of consumers or otherwise damages the public interest, an authority or relevant organization as prescribed by law may institute an action in a people's court", This was the first time that some of the rules of

[1]Order No. 106 of the President of the People's Republic of China, 2012.

environmental civil public interest litigation at the legislative level in China had been clarified. Additionally, Article 58 of the *Environmental Protection Law of the PRC* (2014 revision)[2] was amended to further clarify that "relevant organizations" were those social organizations that "have been registered with the civil affairs departments of people's governments at or above the level of a district city" and "had specifically engaged in environmental protection for the public good for five consecutive years or more without any recorded violation of law".

Notwithstanding, in judicial practice, there was considerable controversy over what constituted "specializing in environmental protection public interest activities" and what constituted "without any recorded violation of law", These issues were dealt with by *The Interpretation of the Supreme People's Court (SPC) on Several Issues concerning the Application of Law in the Conduct of Environmental Civil Public Interest Litigations*[3] in 2015. These rules have opened a new era of environmental civil public interest litigation in China. As a result, the long-term existing phenomenon of "enterprises bring pollution, the public suffers and the government pays the bill" had been theoretically ended.

In order to protect the ecological environment more comprehensively, more and more local courts have begun to tackle the often next question as to which party or group of parties would be obliged to take responsibility for "ecological restoration". After the establishment of the Environmental and Resources Trial Division of the SPC, the obligation to enforce an obligation to undertake environmental restoration was again restated and subsequently confirmed by *Interpretation of the SPC on Several Issues concerning the Application of Law in the Conduct of Environmental Civil Public Interest Litigations* in 2015. Article 18 of the 2015 Interpretation established a liability on the basis of a civil liability for "restitution". Article 20 of the 2015 interpretation also stated that "where the plaintiff requests the restoration to the original environmental state, the people's court may render a judgment in accordance with law that the defendant shall restore the ecological environment to the state and functions before the damage occurred. If complete restoration is impossible, the people's court may permit the adoption of alternative restoration methods. The people's court may, when rendering a judgment that the defendant shall restore the ecological environment and determine the expenses of restoring the ecological environment that shall be borne by the defendant. However, if the defendant failed to perform the restoration obligation, it may render a judgment that the defendant is obliged to pay the cost of the work undertaken by another party. This has been reflected in 2021, China's" *Civil Code*"[4] Article 1234, which stipulates that "when a violation of the provisions issued by the state causes harm to the ecology and environment, and the ecology and environment are capable of remediation, the authority specified by the state or the organization specified by law shall have the right to require the tortfeasor to assume the liability for remediation

[2]Order No. 9 of the President of the People's Republic of China, 2014.

[3]Interpretation No. 1 [2015] of the Supreme People's Court.

[4]Order No. 45 of the President of the People's Republic of China, 2021.

within a reasonable time limitation. If the tortfeasor fails to do so, the authority specified by the state or the organization specified by law may conduct remediation, by itself or through any other party authorized by it, at the expense of the tortfeasor." This provision also establishes ecological restoration as a priority issue over and above other issues.

13.2.2 Assessment Approaches and Technical Standards of Eco-Environmental Damage

Before environmental public interest litigation was formally established, the demand for ecological environmental damage assessment and remediation liability was obvious in practice. For this reason, the former Ministry of Environmental Protection (former MEP, The Ministry of Ecology and Environment since 2018) issued the "*Recommended Methods for Calculating Environmental Pollution Damages (Edition I)*"[5] in 2011. The identification and evaluation of environmental damage in China has been officially launched to provide technical support for the fulfillment of environmental restoration. According to the existing problems and shortcomings of *The Recommended Methods* (*Edition I*), the Ministry of Environmental Protection focused on revising the environmental damage identification and evaluation methods and renaming it the "*Recommended Methods for Authentication and Assessment of Environmental Damage (Edition II)*"[6] in October 2014. The cause of damage from environmental pollution in *Edition I* has been expanded to ecological destruction in *Edition II*, and ecological destruction also includes environmental emergencies. On June 29, 2016, the former MEP issued the "*Technical Guidelines for Identification and Assessment of Eco-environmental Damage: General Program*"[7] and the phrase "environmental damage"[8] has been newly defined as "ecological environmental damage", which emphasizes that the ecological environment is an organic whole that not only pays attention to environmental pollution and ecological damage, but also addresses the impact of the damage on indirect ecological service functions.[9] Consequently, ecological environment remediation now refers to the adoption of necessary and reasonable measures to restore the ecological environment and its ecosystem services to the baseline level before the occurrence of ecological

[5] Notice No.60 of the former Ministry of Environmental Protection of the People's Republic of China, 2011.

[6] Notice No.90 of the General Office of the Ministry of Environmental Protection of the People's Republic of China, 2014.

[7] Notice No.67 of the General Office of the Ministry of Environmental Protection of the People's Republic of China, 2016.

[8] Notice No.67 of the General Office of the Ministry of Environmental Protection of the People's Republic of China, 2016.

[9] See Qiang et al. (2021), pp. 1–9.

environmental damage, and at the same time compensate for the damage during the period until restoration is completed. Depending on the objectives and stages of restoration, ecological and environmental restoration may include basic restoration, compensatory restoration and supplementary restoration".

To further standardize the appraisal and evaluation of ecological environmental damage, the Ministry of Ecology and Environment and State Administration for Market Regulation of RPC jointly published the "*Technical guidelines for identification and assessment of environmental damage-General principles and key components-Part 1:General principles*",[10] "*Technical guidelines for identification and assessment of environmental damage-General principles and key components-Part 2:Damage Investigation*",[11] "*Technical guideline for identification and assessment of environmental damage-Environmental elements-Part 1:Soil and groundwater*",[12] "*Technical guideline for identification and assessment of environmental damage-Environmental elements-Part 2:Surface water and sediment*",[13] "*Technical guidelines for identification and assessment of environmental damage-Principal methods-Part 1:imputed abatement cost for air pollution*"[14] and the "*Technical guidelines for identification and assessment of environmental damage-Principal methods-Part 2:imputed abatement cost for water pollution*".[15] The relevant definitions, applicable scenarios and calculation methods of various evaluation methods are further refined and standardized, which promotes the application of evaluation methods to be more scientific, standardized and refined.

13.3 Relevant Precedents

13.3.1 Brief Introduction to Guiding Cases, Gazettes Cases and Model Cases of the SPC

Follow the continental legal traditions, the Chinese legal system is relatively abstract, macroscopic and stable; as a result, different judges have different understandings of legal provisions. Therefore, the phenomenon of "different judgments in

---

[10]Ministry of Ecology and Environment and State Administration for Market Regulation of RPC, GB/T 39791.1-2020.

[11]Ministry of Ecology and Environment and State Administration for Market Regulation of RPC, GB/T 39791.2-2020.

[12]Ministry of Ecology and Environment and State Administration for Market Regulation of RPC, GB/T 39792.1-2020.

[13]Ministry of Ecology and Environment and State Administration for Market Regulation of RPC, GB/T 39792.2-2020.

[14]Ministry of Ecology and Environment and State Administration for Market Regulation of RPC, GB/T 39793.1-2020.

[15]Ministry of Ecology and Environment and State Administration for Market Regulation of RPC, GB/T 39793.2-2020.

the similar cases" has always been controversial. Although there are objective reasons for different judgments in the same case, such as the professionalism of the same judgment, the complexity of the case, the inconsistent standards of proof, and the differences in the social effects of case handling,[16] the public still has great doubts about the different judgments in the same case, coupled with the adverse effects of external intervention in some cases, resulting in a lack of judicial credibility. In order to unify the application of laws and the standards of adjudication for promoting judicial justice and improving judicial efficiency, the SPC has been attempting to establish and strengthen the case guidance system which has become one of the most important aspects of Chinese judicial reform. In 2010, the SPC issued the "*Provisions on Case Guidance*",[17] which pointed out that the guiding cases issued by the SPC should be referred to similar cases by the people's courts at all levels. In 2011, the SPC announced the first batch of guiding cases. In 2015, the SPC promulgated the "*Detailed Rules for the Implementation of the Provisions of the SPC on Case Guidance*",[18] which further clarified the standards for the uniform application of guiding cases. In 2020, the SPC issued the "*Guiding Opinions on Unifying the Application of Laws to Strengthen the Retrieval of Similar Cases (for Trial Implementation)*",[19] which requires the People's Courts to conduct type retrieval when handling cases.

Guiding cases of the SPC refer to cases that have been discussed and decided by the Judicial Committee of the SPC, and then are uniformly issued by the SPC. Guiding cases are specific cases clearly stipulated by law that have the status of "quasi-judicial interpretation". Gazette cases of the SPC refer to cases that are published in the Gazette of the SPC, but that have not been discussed and decided by the Judicial Committee of the SPC. Gazette cases are cases that courts could refer to, but they are not compulsory. Model cases refer to cases of legal disputes with strong typical significance and greater social impact. When handling similar cases in the region, model cases could be used to provide a reference, or as a sample or example for legal research. These cases serve as an instruction and guide when handling similar cases. Therefore, this paper mainly selects the Guiding cases, Gazette cases and Model cases promulgated by the SPC as samples, and then sorts out the judicial innovation and further improvement of eco-environmental remediation responsibility.

[16] See Jianping (2012), pp. 98–107.

[17] No. 51 [2010] of the Supreme People's Court.

[18] No.130 [2015] of the Supreme People's Court.

[19] No. 24 [2020] of the Supreme People's Court.

13.3.2 *Scope of the Case Study*

There are forty-six environmental civil public interest litigation cases studied in this chapter, including eleven Guiding cases, two Gazette cases and thirty-three Model cases, all of which were issued by the SPC. The span of time is relatively large, from 2012 to 2020. In terms of the type of dispute, there are twenty-six environmental pollution cases, mainly involving water pollution, approximately thirteen cases, five cases of air pollution, five cases of hazardous waste and solid waste pollution, and the remaining three cases of soil pollution. It is worth noting that water pollution, soil pollution and hazardous waste pollution often simultaneously exist in the same case. To avoid repeated calculations, the classification is based on the most important pollution. The remaining twenty cases are in the category of ecological damage, among which there are thirteen cases of wildlife protection, three cases of forestland and mineral resources respectively, and one case of a geological relic. Geographically, cases are distributed in 14 provinces across the country, such as Jiangsu, Zhejiang, Yunnan, and Chongqing.

Apart from two preventive environmental public interest litigation cases,[20] the responsibility for eco-environmental remediation has become the core issue of environmental civil public interest litigation in China. However, those cases show different characters and innovations in the performance of eco-environmental remediation responsibilities. Based on this, this paper believes that it is necessary to make a comprehensive comparative review.

13.4 The Application of Eco-Environmental Remediation Liability in EPIL Cases

In the EPIL cases announced by the Supreme Court, the ecological environment remediation liability as a new type of liability has evidenced some new approaches to the identification of relevant parties and how to impose responsibility on defendants. These innovative cases not only help the local court hear the case, but also assist in the judgement being accepted more generally by the public. To some degree they may result in achieving a unity of social and legal affects.

[20] Guiding Case No. 173 of the SPC: *Chaoyang District Friends of Nature Environmental Research Institute of Beijing Municipality v Hydrochina Corporation Xinping Development Co., Ltd* (2020), (2020) *Minzhong* No.824 High People's Court of Yunnan Province and POWERCHINA Kunming Engineering Corporation Limited (civil public interest action for ecological and environmental protection) and Guiding Case No. 174 of the SPC: *China Biodiversity Conservation and Green Development Foundation v Yalong River Hydropower Development Co., Ltd.* (civil public interest action for ecological and environmental protection) (2015), (2015), *Minchu* No.45 Ganzi Tibetan Autonomous Prefecture Intermediate People's Court of Sichuan Province. In these two cases, lawsuits were filed because the development behaviour caused huge risks to the living environment of endangered species, and no damage had been caused, so there was no problem of restoration.

13.4.1 Responsible Subjects

It can be seen from the cases that there are a diverse range of parties who become responsible for ecological environment restoration. Among them, there are twenty-six cases where the defendants are state-owned and private companies, cooperatives, factories and other corporate entities. There are twenty cases where the defendants are farmers, fishermen or self-employed individuals. This diversity flows from Article 6, which states that all relevant partiers have "(an) 'obligation to protect the environment" under the Environmental Protection Law.

In addition, some cases manage to break through the rules and restrictions on general civil and commercial liability matters. For example, in the *Jiangsu Province Nanjing Gulou District People's Procuratorate v Nanjing Sembcorp Water Co., Ltd.* was an incidental civil public interest litigation case.[21] Here Sembcorp, as a polluter, caused damage amounting to about 470 million RMB. According to Article 3 of the Company Law of the People's Republic of China, "a company is an enterprise legal person which has independent legal person property and enjoys the right to legal person property. The company shall bear the liabilities for its debts with all its property". The shareholders of a limited liability company are liable to the company to the extent of their subscribed capital contributions. "According to the "Company Law", Sembcorp's structure meant that all of its assets would be chargeable in the event of a default, but Sembcorp (China) Investment Co., Ltd. as a shareholder was only limited to its subscribed capital contribution ($17.556 million) (approximately 112.2 million RMB based on the 2021 China-US exchange rate). However, in this case, Sembcorp (China) Investment Co., Ltd., was actively joined as a "third party" for restoration responsibilities in two aspects. First, Sembcorp (China) Investment Co., Ltd assumed joint and several liability of affording 237 million RMB restoration expenses which effectively abrogated the limited liability rule. Second, the Investment Co., Ltd. took on 233 million RMB of the alternative remediation liability that far exceeded its subscribed capital. Based on the active application of the "third party" as the subject of compensation, the public interest litigation plaintiff submitted that the public interest must be considered a priority. This was also the opinion of a previous mediation. In this case, it is innovative that the court creatively agreed to add a third party (Sembcorp Investments) as the compensation obligor. Of course, the participation of this third party can only be voluntary and must not be forced or forced in disguised form.

[21] Case of incidental civil public interest litigation of *People's Procuratorate of Gulou District, Nanjing City, Jiangsu Province v Nanjing Sem bcorp Water Co.,Ltd., Zheng Qiaogeng, and 11 other persons for environmental pollution* (2019), (2019),Su 01 *xingzhong* No.525 Nanjing City Intermediate People's Court of Jiangsu Province.

13.4.2 Ways of Assuming Ecological Environment Remediation Responsibility

Except for two preventive lawsuits that did not involve ecological environment remediation responsibility, the methods of the other forty-four cases involved a variety of ecological environment remediation liability. There were three major categories of responsibility: direct restoration, compensatory restoration and alternative restoration, which were often realistically combined in such cases.

13.4.2.1 Direct Restoration

Direct restoration means that the responsible person takes effective measures to restore the ecological environment of the original area and the original body to the state before the damage occurred. The identification of direct restoration needs to meet three requirements. First, the main body of restoration must be the responsible person, i.e., the party responsible for the damage. Second, the restoration object must be the original area, or the original body of the ecological environment destroyed by the responsible person. Third, the restoration target must be the previous ecological state. Therefore, it can be seen that direct restoration has strict requirements on the subject of restoration, the object of restoration and the restoration method. This kind of liability is also a liability method closest to the restoration of the original state in traditional civil law. There are six cases using direct restoration by means of replanting, regreening, proliferation and replenishment, etc.[22]

[22]In Guiding Case No. 172 of "*People v Qin Jiaxue (criminal case of deforestation with an incidental civil public interest action)*" (2018), (2018), Xiang 3125 *Xingchu* No.5 Baojing County People's Court of Hunan Province, the Baojing County Court entered Incidental Civil Judgement that defendant Qin Jianxue should plant 5050 one-year-old Chinese fir seedings at Tudiao in the Hunan Baiyun Mountain National and Nature Reserve within two years after the judgement came into effect, at a survival rate of not less than 90%. After the judgement was pronounced, neither defendant nor the prosecution applealed, and the first instance judgement became effective. Pursuant to the judgment, defendant planted 5050 one-year-old Chinese fir seedings in the original forest cutting area and other areas, at a survival rate reaching 100%. In the Model case of "*All-China Environment Federation v Wuxi Lihu Huishan Scenic Spots Management Committee for Ecological Environment Tort*" (2012), (2012) *Minchu* No.0002 Binhu District People's Court of Wuxi City, Jiangsu Province, the defendant invested nearly 800,000 yuan to the Yangwan plot and reclaimed and planted the green land of 7000 m^2. In the Model case of "*People's Procuratorate of Lianyun District, Lianyungang City v Yin Baoshan et al.* (Civil action incidental to criminal proceedings concerning illegal fishing of aquatic products)" (2016), (2016) Su 07 *Xingzhong* No. 99 Lianyungang City Intermediate People's Court of Jiangsu Province, the people's court ordered the six persons to restore the marine ecological environment damaged by their criminal activities by multiplication and release of 13.65 million Chinese prawn seedings. In the Model case of "*Environment Public Interest Protection Association of Zhenjiang City, Jiangsu Province v Jiangsu Youli Optical Spectacles Co., Ltd. Civil Public Interest Litigation regarding pollution of solid wastes*" (2015), (2015) *Zhen Mingong* No.00003 Zhenjiang City Intermediate People's Court of Jiangsu Province,the court of first instance ordered that Youli Company should under the

The cases with direct restoration have the following commonalities: first, the case is relatively simple. The scope of ecological damage is relatively narrow, or the damage value is easy to assess. Second, the judgment is specific and clear. Most of them have restoration plans or restoration opinions from professional institutions that laid down specific requirements for restoration time, place, number, survival rate and so on. Finally, the efficiency of restoration is high. Nearly all of them were completed during the time of judgment or within the specified time after the judgment.

13.4.2.2 Compensatory Restoration

Compensatory restoration refers to a way to fulfil the responsibility by paying a certain amount of fees for ecological environment restoration. Generally speaking, the court can decide that the responsible person bears the compensatory restoration liability in the following two situations. When the damaged ecological environment cannot be restored or cannot be completely restored, and the other situation that the defendant directly rejects performing or fails to fulfil the direct restoration liability.

In the research sample, there are thirty-four cases involved compensatory repairs, accounting for 74% of the total cases. Those cases can be subdivided into the following types: (1) there are fourteen cases in which the judgment only mentions the bearing of compensatory expenses without further mentioning restoration, accounting for 40%. Among them, one case of compensatory expenses was delivered to the government's special account,[23] two cases were delivered to the account of court,[24] and the judgment of the remaining eleven cases did not mention the

---

supervision of the Environmental Protection Bureau of Danyang City, legally dispose of the wastes involved as general wastes. In the Model case of "*People v Li Xugen (Criminal case of Illegally Fishing Aquatic Products with Incidental Civil Public Interest Action)*" (2019), (2019) Su 0682 *Xingchu* No.605 Gao City People's Court of Jiangsu Province, the Jiangdu District People's Procuratorate of Yangzhou City, Jiangsu Province and Li Xugen reached a settlement agreement on ecological and environment remediation that Li should reintroduce fish fry worth 25,000 yuan within ten days from the date of signing this settlement agreement (already preformed) and further reintroduce fish fry worth 22,500 yuan within two years from the date of signing this settlement agreement. The Model case of "*Civil public interest litigation incidental to criminal proceedings of illegally felling trees by Defendant Jia Bozhou*" (2018), (2018) Chuan 3232 *Xingchu* No.25 Zoigê County People's Court of Sichuan Province, the court of first instance entered a judgment that Jiaozhou should replanted 390 spruces within six months after the judgment took effect.

[23] Guiding Case No. 135 of the Supreme People's Court: *People's Procuratorate of Xuzhou City, Jiangsu Province v Suzhou Qi'an Arts Co., Ltd. et al* (2018), (2018) Su 03 *Minchu* No.256 Xuzhou City Intermediate People's Court of Jiangsu Province, The Intermediate People's Court of Xuzhou City, Jiangsu Province entered a Civil Judgment that the defendant should pay 4,835,267 yuan in compensation for the ecological and environmental restoration cost caused by dumping three barrels of waste sulfuric acid to the special savings account the Environmental Protection Public Interest Fund of Xuzhou city within 30 days after the judgment took effect.

[24] *Henan Environment Federation v Liaocheng Dongran Chemicals Co., Ltd.* (2016), (2016) Yu 09 *Minchu* No.185 Puyang City Intermediate People's Court of Henan Province, both parties

account number or other information of the payment. (2) Seventeen cases were used for further restoration by specific departments on the basis of compensatory expenses, of which eight cases were delivered to the local government special account or the account of the Finance Bureau. In addition, four cases were delivered to the court account, one case set up the trust fund, one case was delivered to the national treasury, and the remaining three cases were not specified in the judgment. (3) There are three cases in which the responsible person should bear the responsibility for direct restoration but fails to perform timely or qualified restoration, and then it needs to accept compensatory restoration. Among these cases, two clearly stated that the compensation fee should be paid to a special account or designated account, and in one case, it was not specified. The key point of compensatory restoration lies in the management and use of restoration funds. To this end, special fund accounts for environmental protection have been established in various regions. In addition, courts and local finance bureaus have become the main recipients of compensation funds. At the same time, through the establishment of funds or the management of trust companies, it has innovated a market method that is different from the management of public power.

It is particularly worth mentioning that there are some detailed innovations in compensatory restoration. In the case of *Taizhou Environment Federation v Taixing Jinhui Industrial Co., Ltd,* the restoration costs were paid by instalments and deduction of technical transformation.[25] On the one hand, the installment payment ensures that the company can continue to operate normally. On the other hand, the restoration funds are deducted through technological transformation by encouraging enterprises to move towards a green path of development. In the case of

reached the folling mediation agreement that Dongran Chemicals Company should pay the environmental restoration and treatment cost of CNY3 million. The first installment of CNY3 million has been paid and the second installment of CNY3 million should be paid to the designated account of the Intermediate People's Court of Puyang city prior to January 1, 2019. *Public Welfare Case of Water Pollution between All-China Environment Federation and Yichun Zhong'an Industrial Co., Ltd. et al.*, (2018), (2018) *Gan Minzhong* No.189 High People's Court of Jiangxi Province, the Intermediat People's Court of Xinyu City held that Zhong'an Company should pay the environmental restoration and treatment cost to the accout of Environmental Public Interest Litigation Special Fund of Xinyu Intermediate People's Court.

[25] SPC Gazette's case, *Taizhou Environment Federation v Taixing Jinhui Industrial Co.,Ltd.* (2015), (2015) *Minshen* No.1366 The Supreme People's Court, Beijing. The court held that if the parties apply and can provide effective guarantee within 30 days from the effective date of this judgment, the 40% of the ecological restoration compensation fee of 160,666,745 yuan can be postponed to one year. At the same time, within one year from the effective date of the judgment, six companies such as Changlong can recycle by-product acid through technological transformation, significantly reducing environmental risks, and if they have not been punished for environmental violations within one year. The cost of transformation can be applied to the Taizhou People's Court for the 10% of the deferred payment based on the statement of the company's environmental compliance issued by the environmental protection administrative department, the environmental protection acceptance opinions on the completion of the project, and the audit report on the investment in technological transformation issued by the intermediary agency with legal qualifications. deductible within the amount.

Environmental Protection Association of Ganyu District, Lianyungang City v Wang Shengjie,[26] the court considered that it was difficult for the defendant to pay the full compensation due to the defendant's living difficulties, so the compensation was deducted through working for environmental public welfare. These innovations seek to unify the legal, social, and ecological effects of judgments.

Therefore, it can be seen from the above that the proportion of compensatory restoration in environmental civil public interest litigation is very high. However, only nineteen cases have clearly clarified the whereabouts of the ecological restoration costs, and in the remaining seventeen cases, it has not been specified in the judgment, especially the management of compensation costs in the above-mentioned first category of cases. These issues are worthy of vigilance and will be discussed later in this chapter.

13.4.2.3 Alternative Restoration

Interpretation of the SPC on Several Issues concerning the Application of Law in the Conduct of Environmental Civil Public Interest Litigations stipulates that if the damaged eco-environment "cannot be completely restored, alternative restoration methods may be allowed", which means that alternative restoration is only applicable to "the situation where the damaged ecological environment cannot be partially or fully restored". However, in judicial practice, alternative restoration is rarely applied alone. It is generally used in combination with direct restoration and compensatory restoration in various forms. The judgments illustrate flexibility. The relevant government agency itself may replace the responsible party or the responsible party may be replaced if it is not repaired properly, or the responsible party may pay compensation and those fees used to employ a third-party contractor.

The above types are mainly to replace the defendant to implement the liability of restoration. In addition, the replacement aspects are also reflected in the replacement of the object. For example, the object of restoration refers to the original damage that can no longer be repaired or is not necessary, so the compensation fee is used for the restoration of the entire area or watershed. For instance, Guiding Case No. 132 of *China Biodiversity Conservation and Green Development Foundation v Qinhuangdao Fangyuan Packaging Glass Co., Ltd* is an air pollution case, but the compensation fee was used for the restoration of the entire environment in Qinhuangdao City.

---

[26] *Environmental Protection Association of Ganyu District, Lianyungang City v Wang Shengjie* (2014), (2014) Lianhuanggong *Minchu* No.00002 Lianyungang City Intermediate People's Court of Jiangsu Province, the court held that Defendant Wang Shengjie should, within two years after the judgment came into force, provide a total of 960 hoursof environmental voluntary labour (no less than six hours per time and at least six times per month) to make up his insufficient compensation for environmental damages, and the Environmental Protection Bureau of Ganyu District, Lianyungang City should be responsible for supervising and managing the execution of the voluntary labor.

13.4.3 Professional Identification and Assessment of Damages

Whether the cost, method and effect of ecological environment restoration are reasonable and scientific is one of the most important issues in environmental civil public interest litigation. With the advancement of environmental public interest litigation, the Ministry of Ecology and Environment has successively issued normative documents for the identification and assessment of eco-environmental damage, which clarify highly professional rules. Therefore, in the research sample, there are thirty cases in which professional appraisal agencies and technical experts have participated to assess remediation expenses or formulate restoration plans. In addition, some cases even invited professionals as witnesses[27] and used academic articles as evidence.[28]

In the evaluation method of ecological environment restoration costs, the most commonly used method is the virtual treatment costs method with fifteen cases, followed by the actual restoration costs method,[29] the market value method,[30] and the willingness value method.[31] There are also cases that simultaneously use the market value method, the fictitious governance cost method, the shadow engineering method and the alternative market method to comprehensively estimate the different damages involved in the case.[32]

[27] *Chaoyang District Friends of Nature Environmental Research Institute of Beijing Municipality v Hydrochina Corporation Xinping Development Co., Ltd. and Powerchina Kunming Engineering Corporation Limited* (2020), (2020) Yun *Minzhong* No.824 High People's Court of Yunnan Province, during the trial, the plaintiff invites professional wildlife photographers to testify in court to prove the living conditions of green peafowl.

[28] *China Biodiversity Conservation and Green Development Foundation v Yalong River Hydropower Development Co., Ltd.* (2015), (2015), *Minchu* No.45 Ganzi Tibetan Autonomous Prefecture Intermediate People's Court of Sichuan Province. The plaintiff provided a series of papers to prove that the living environment of the precious and endangered wild plant Maple chinensis was seriously threatened. After cross-examination, the court supported it.

[29] It is mainly used in the case of third-party alternative restoration, the actual cost of third-party alternative restoration shall prevail, such as "*All-China Environment Federation v Tan Yaohong and Fang Yunshuang Environmental Pollution Civil Public Interest Litigation Case*" (2015), (2015) Sui Zhongfa *MinyiZhong* No.3804 Guangzhou City Intermediate People's Court of Guangdong Province.

[30] It mainly calculates the loss of wild animals and forestland resources. For example, *People's Procuratorate of Nanjing City, Jiangsu Province v Wang Yulin* (civil public interest litigation case of ecological damage) (2020), (2020) Su 01 *Minchu* No.798 Nanjing City Intermediate People's Court of Jiangsu Province, the unit price of the forestland damage price is based on the Chinese forestland compensation standard.

[31] *People's Procuratorate of Shangrao City, Jiangxi Province v Zhang Yongming, Mao Weiming, and Zhang Lu* (2020), (2020) Gan *Minzhong* No.317 High People's Court of Jiangxi Province.

[32] *People's Procuratorate of Guangzhou City, Guangdong Province v Weijie Integrated Waste Treatment Plant in Huadu District, Guangzhou City and Li Yongqiang* (2018), (2018) Yue 01 *Minchu* No.724 Guangzhou City Intermediate People's Court of Guangdong Province.

13.5 Problems Existing in the Application of Eco-Environmental Remediation Liability

13.5.1 Ineffective Implementation of Ecological Restoration Priority Principle

According to the current legal system, priority has been given to ecological restoration and monetary compensation can only be a supplement, which means that only when the ecological environment cannot be restored, monetary compensation will be implemented.[33] This was established in the "*Plan for the Reform of the System of Damages for Harm to Ecology and Environment*"[34] issued by the General Office of the CPC Central Committee and the General Office of the State Council in 2017. Unlike a traditional tort claim, the damage caused by an environmental tort has dual characteristics viz. that on the one hand, it causes human damage, and on the other hand, it causes ecological environmental damage. Therefore, the degradation of environmental elements and ecological functions should be restored in time to prevent further damage to human health and property.[35]

From the cases released by the Supreme People's Court, the priority of ecological restoration has occasionally not been fully reflected, mainly in the following aspects. First, in the majority of cases, the final judgment is made to compensate for ecological restoration, but direct ecological restoration is not able to be attained. Of course, there are many reasonable grounds for such a consequence. In some cases, the damage no longer exists because the self-regulating function of the ecosystem plays a role,[36] so the actual damage can only be assessed and compensated by the virtual governance cost method and other methods. Meanwhile, it may also be that the responsible person does not have the ability to restore the ecological environment. For example, in a criminal incident with an attendant civil lawsuit, the responsible person may be imprisoned, so paying the restoration fee is his only way to discharge his penalty. In such a situation it is better to hand it over to a professional organization or relevant government authority to carry out the restoration plan. Second, another manifestation of the poor implementation of the priority of restoration is that there are few judgements attached to ecological restoration plans. Among the evaluated cases, only eight cases explicitly involved restoration plans, but some of them only mentioned the need to formulate ecological restoration plans in the judgment or mediation agreement, rather than already

[33] See Dou (2019), pp. 136–154.

[34] The General Office of the CPC Central Committee and the General Office of the State Council, Plan for the Reform of the System of Damages for Harm to Ecology and Environment, 2017.

[35] See Zhongmei (2016), pp. 244–264.

[36] The Gazette case "*Taizhou Environment Federation v Taixing Jinhui Industrial Co.,Ltd.*" (2015), (2015) *Minshen* No.1366 The Supreme People's Court, Beijing.

formulated ecological restoration plans.[37] However, it is unknown whether a remediation plan has been formulated in the later stage and whether the plan is reasonable and scientific. In addition, the restoration methods in the ecological environment restoration plan are relatively simple, mostly replanting and 'regreening', and lack precise design and differentiated treatment. For example, the final ecological restoration plot of environmental pollution crimes in criminal incidental civil public interest lawsuits is to actively replant and restore greenery, without considering the different restoration needs of different environmental ecosystems. Judicial pragmatism may lead to the formalization of restorative liability.[38] Finally, there are only six subsequent judgments that clarify the supervision and effect evaluation of the performance of ecological restoration responsibilities.[39] If the most direct manifestation of ecological restoration priority in form is direct restoration, then the substantive manifestation must be that restoration costs are actually used for ecological restoration. Then, whether there is a scientific and reasonable restoration plan and whether the restoration plan is actually implemented is the key point.

[37] Guiding Case No.134: *Chongqing Green Volunteer Federation v Jianshi Huangchangping Mining Co.,Ltd.* (2016), (2016) Yu 02 *Minzhong* No.77 Intermediate People's Court of Chongqing Municipality. Of Enshi Autonomous Prefecture, the court entered a Civil Judgment that Jianshi Huangchangping Mining Co., Ltd. Of Enshi Autonomous Prefecture should,within 180 days this judgment took effect, develop a restoration plan and conduct ecological restoration with respect to the soil in the depression in the beneficiation plant in Guojiatang State-owned Gaoyanzi Forest Farm.

[38] See Xiaoxue and Zhanwei (2020), pp. 33–45.

[39] Among the research samples, Guiding Case No. 172 "*People v Qin Jiaxue* (see footnote 11)" requires replanting saplings with a survival rate of over 90%. It is clear that the court participates in the supervision and acceptance of the whole process and urges the scenic spot management committee to carry out the maintenance of trees for the defect period (one year) in the Model case of "*All-China Environmental Protection Federation v Wuxi Lihu Huishan Scenic Spots Management Committee for Ecological Environment Tort*". The Model case "*China Environmental Protection Fund v New Spring Water (Yangzhou) Co., Ltd.*" (see footnote 11) specifies that the restoration results must be reviewed and approved by the Yangzhou environmental protection department and reported to the Yangzhou Intermediate People's Court for the record. The restoration plan should be implemented within one year after the review and confirmation of Process and effect. In the case of "*Jiangsu Zhenjiang City Ecological Environment Public Welfare Protection Association v Jiangsu Youli Optical Spectacles Co., Ltd.*" (see footnote 11) clarifies that Youli Company disposed the waste involved in accordance with the law under the supervision of Danyang Environmental Protection Bureau. The case of "*China Environmental Protection Confederation v Jiangsu Jiangyin Changjing Liangping Cooperative and other breeding pollution civil public interest litigation case*", Wuxi City Intermediate People's Court of Jiangsu Province,the Chinese court entrusted the local environmental protection authority for acceptance. "*Tongren City People's Procuratorate v Guizhou Yuping Xiangsheng Chemical Co., Ltd., Guangdong Shaoguan Woxin Trading Co., Ltd* (2016), (2016) Qian 03 *Minchu* No.520 Zunyi City Intermediate People's Court of Guizhou Province.The Soil Pollution Liability has established a dual supervision of "responsible person restoration, government supervision, enforcement by the people's court and supervision by the people's procuratorate".

13.5.2 *Supervision of the Assessment and Management of Eco-Environmental Remediation Fees*

According to the existing published cases, the proportion of compensatory restoration cases is relatively high. Whether the ecological environment restoration funds are truly and effectively used for ecological environment restoration determines the success or failure of the environmental public interest litigation.[40] The assessment and management of ecological environment restoration fees has become a controversial issue because it is generally not subject to effective supervision.

13.5.2.1 Assessment of Ecological Environment Remediation Costs

Due to the high professional and technical requirements for damage identification, the assessment of ecological environment restoration costs is generally issued by one of the parties (the plaintiffs are the majority, and a small number of courts also commission relevant professional institutions or experts). The proceedings are highly dependent on expert opinion.[41] As a result, the independence of the courts has sometimes been questioned. In any event, the judicial organs have adopted a large number of expert arguments and opinions to assist in the determination of facts in environmental litigation, and the relevant judicial interpretations also clarify that expert opinion "can be used as a basis for determining the facts of the case after cross-examination by the parties". From a specific point of view, the virtual treatment cost method is the most commonly used evaluation method, but "the application scenario is relatively simple, and the key links such as pollutant emissions and unit treatment costs involved have greater uncertainty in some cases and are often prone to controversy".[42]

13.5.2.2 Management of Ecological Environment Restoration Costs

There are two criteria for assessing whether ecological remediation funds are well managed, that is, the security of funds and their effectiveness use.[43] The management mode of ecological environment restoration costs varies from case to case. Some are managed by the official authority, such as the court, the competent department, the national treasury or local finance departments of local authorities, while the remainder operate under a trust. At present, the management of these funds is relatively chaotic, and there is a lack of a long-term supervision mechanism to

[40] See Shekun and Yijiu (2019), pp. 44–53.

[41] See Le (2020), pp 37–42.

[42] See Mai et al. (2022), pp. 9–16.

[43] See Shekun and Yijiu (2019), pp. 44–53.

clarify the process, scope, and specific methods of using funds. Therefore, it is unknown whether the funds have been misappropriated or whether they are truly used to restore the ecological environment. For example, in Guiding Case No. 132 "-*China Biodiversity Conservation and Green Development Foundation v Qinhuangdao Fangyuan Packaging Glass Co., Ltd. Air Pollution Liability Civil Public Interest Litigation Case*", the appellant believed that the loss part paid to the special fund account of Qinhuangdao City in the first-instance judgment, which had no legal basis. The appellant neither knew whether the account existed or the rules for the use of funds in this account, nor did it judge the appellant to have the right to supervise the management and use of the special funds. Conversely, the appellant proposed that the establishment of an environmental charitable trust had a clear basis in Trust Law and the Charity Law.

13.5.3 Courts' Innovative Practices

In the implementation of compensation fees, a series of innovative practices have been introduced in various regions. For example, in order to achieve a balance between enterprise development, basic living security of citizens and environmental protection, the court innovatively adopted methods such as "instalment payment", "technical transformation deduction" or "labour service deduction". Alternatively, those methods of assuming responsibility have obvious defects, so that there are still great uncertainties in the theoretical study and judicial practice, resulting in difficulty in applying them uniformly. These problems include the following aspects: insufficient legal basis, unclear legal nature and poor practical operability. At present, only instalment payments can refer to *Article 1187 of the Civil Code* as the legal basis, but whether instalment payments can be directly applied to the compensation for a specific ecological or environmental damage is more controversial. According to relevant judicial interpretations and judicial practice, the instalment payment specified in this clause is mainly applicable to personal injury compensation.[44]

However, technical improvement deductions and labour deductions have no clear legal basis, which leads scholars to argue endlessly about their legal nature. For example, there are three different theoretical viewpoints on the issue of deduction for technological transformation, "the way of assuming responsibility", "the way of reducing liability" and "the way of compensation for losses".[45] Different viewpoints will directly or indirectly affect the legitimacy of value and practical operation. In fact, aside from disputes over legal basis and nature, methods such as technical improvement deductions and labor service deductions are not perfect operationally in judicial practice. Because the scope of application, conditions, rules and procedures are not clear, coupled with the complexity of ecological and environmental

[44] See Lixin (2021), pp. 1–15.

[45] See Gang (2021), pp 163–172.

cases, it is difficult to apply them directly. Therefore, courts are often suspected of arbitrarily handing down judgements that are contrary to judicial precedents.

13.6 Concluding Remarks

In view of the unclear sequence of various ecological restoration methods, the lack of ecological restoration schemes resulting in poor enforceability of judgements, and the lack of supervision in the evaluation and management of ecological restoration costs in the Chinese EPIL, this chapter proposes some targeted countermeasures to improve the judicial application of eco-environmental restoration responsibility.

13.6.1 Clarification of Ecological Environment Restoration Methods in the Legal Rules

In 2021, China's "*Civil Code*" came into effect and clarified the "priority principle of restoration", but it did not involve specific adjudication rules. Therefore, it is recommended to further clarify the order of ecological environment restoration methods when the judicial interpretation on Civil Code is promulgated and the *Interpretation of the Supreme People's Court on Several Issues concerning the Application of Law in the Conduct of Civil Environmental Public Interest Litigations*[46] is revised. The best order is that direct restoration priority; only when the person refuses to restore or cannot be restored due to objective reasons will a professional third-party institutions or competent authorities carry out alternative restoration. When ecological restoration is impossible or infeasible, compensatory restoration is applicable. Whether it is direct restoration, alternative restoration or compensatory restoration, a detailed restoration plan should be attached in the judgment, which should be reviewed by the competent authority of the damaged eco-environment and recognized by the court.

13.6.2 Governance of Ecological Restoration Programs

For EPIL cases, winning the case is not the end. To truly implement the judgment and take the judgment as a part of popularizing the knowledge of the rule of law on environmental protection, multi-participation is very important. Ecological restoration programs are not only scientific and professional but also often involve the prevention and control of human health risks and the overall planning of social

[46]Interpretation No. 20 [2020] of the Supreme People's Court.

factors. Therefore, it is necessary to build a multi-subject governance system of courts, administrative agencies, professional institutions and individuals, news media, those responsible for restoration, and ordinary people.

13.6.3 Supervision of Ecological Environment Restoration Costs

Although *the Management Measures for Compensation Funds for Ecological Environmental Damage (for Trial Implementation)*[47] jointly issued by the Ministry of Finance and nine other departments stipulate the management of compensation funds for ecological and environmental public interest litigation, it is different from environmental public interest litigation and cannot fully cover the latter. Whether it can be applied by reference is debatable. In addition, due to the lack of detailed regulations on the approval procedures and supervision methods for the use of compensation funds, the efficiency of fund use is low, and the proportion of accumulated funds is relatively high. Therefore, it is recommended to form a unified management method as soon as possible and clarify the management, use and supervision rules of funds.

13.6.4 Clarification of the Legal Basis for Innovation

The application of instalment payments in compensation for ecological and environmental damage should be further defined in judicial interpretation. At the same time, judicial interpretation should be used to regulate the types, scope, conditions and procedures of the application of technical improvement deductions and labour service deductions.

References

Dou H (2019) Response to environmental damage events: limitation of tort damage theory and construction of environmental damage theory. Law Soc Dev 25(02):136–154

Gang J (2021) Jurisprudence and application of "Technical Reform Deduction" in civil liability for eco-environmental damage. Law Rev 39(04):163–172

Jianping C (2012) Reasons and countermeasures for different judgments in the same case. Orient Law 4:98–107

Le X (2020) Empirical analysis and improvement of appraisal opinion acceptance in environmental civil litigation. J Hebei Univ Environ Eng 30(05):37–42

[47]Notice No.6 of the Ministry of Finance of the People's Republic of China, 2020.

Lixin Y (2021) The lack and specific application of the rules of periodic payment for tort damages. Legal Forum 36(05):121–129

Mai Y, Yixin S et al (2022) Research on the application of virtual governance cost method in forensic appraisal of environmental damage. Chinese J Forensic Sci 1:9–16

Qiang Z, Junxiong C et al (2021) Research on the development and problem of forensic appraisal of eco-environmental damage in China. Chinese J Forensic Sci 4:1–9

Shekun W, Yijiu W (2019) Comparison and selection of management modes of ecological environment restoration funds. J Nanjing Tech Univ 18(01):44–53

Xiaoxue Z, Zhanwei Y (2020) Judicial application of ecological environment restoration liability: a commentary on 347 Judgments. Shandong Judges Train Inst J 36(03):33–45

Zhongmei L (2016) Judicial rationality on environmental protection should not be overridden by the pricey compensation: an analysis on environmental public interest lawsuit in Taizhou City. China Legal Sci 3:244–264

Chapter 14
Looking Backwards, Looking Forwards: Environmental Public Interest Litigation in Soil Pollution Law

Huanhuan Wang and Zhenglian Zhang

Abstract This chapter reviews the historical developments of environmental public interest litigation within the context of soil pollution in China to identify what roles it has played in soil pollution prevention and remediation. Due to the complexity of land remediation, there is a heavy reliance on scientific evidence. A deficiency of EPIL exists especially in confirming compensation amounts, setting clean-up criteria and monitoring remediation process. With the commencement of the *Law on Soil Pollution Prevention and Control* in 2019, liability for land contamination is clearly stipulated as public liability, directly imposed by environmental agencies. This liability may include undertaking a site investigation, site risk assessment, risk control, remediation, evaluation of site risk control and remediation, and long-term stewardship. In view of the dominance of administrative power in soil pollution prevention and control, environmental public interest litigation should be reshaped by focusing on the preventive function, supplementing the responsibility of soil remediation with civil public interest litigation, and strengthening the supervision of administrative power with administrative public interest litigation.

Keywords Soil contamination · Chinese law on the prevention and control of soil pollution · Environmental public interest litigation · Environmental remediation · Supervision of administrative power

H. Wang (✉)
School of Law, East China Normal University, Shanghai, China
e-mail: hhwang@law.ecnu.edu.cn

Z. Zhang
School of Law, East China Normal University, Shanghai, China

X. Wang et al. (eds.), *Environmental Public Interest Litigation in China*,
https://doi.org/10.1007/978-3-031-26526-6_14

14.1 Introduction

In May 2016, the State Council issued the *Action Plan on Prevention and Control of Soil Pollution*, which clearly states: 'Encourage public interest litigation against environmental violations such as soil pollution according to law'.[1] This not only characterizes the state's affirmation of previous environmental public interest litigation in environmental pollution control, but also shows the state's expectation of environmental public interest litigation in soil pollution prevention and control. China's Environment Public Interest Litigation (EPIL) can be divided into Civil Environmental Public Interest Litigation (civil EPIL) and Administrative Environmental Public Interest Litigation (administrative EPIL). They have different functions and roles to play in environmental pollution management. While civil EPIL focuses on the prevention and direct relief of environmental pollution, administrative EPIL urges better fulfillment of regulatory duties of environmental administrative departments in combating pollution.

The Law on the Prevention and Control of Soil Pollution (PCSPL) of China came into effect on 1 January 2019. It establishes a soil pollution prevention and control system dominated by administrative power, strengthening the public law character of the responsibility for soil pollution prevention and control. It provides procedures, responsibilities, and investigation and remediation (where appropriate) of land that is contaminated, to eliminate damages caused by contaminated sites. In this context, the judicial relief by EPIL and the public law relief by administrative supervision and law enforcement will probably conflict with each other in the sense of protection, object and function. Therefore, the question about how to coordinate and connect these two types of relief becomes an important challenge for the soil pollution legal regime, the legal terms and provision of which are still too broad and vague. There is therefore a need for clarification by detailed regulations, standards, and other enforcement mechanisms.

There are several items that need clarification before providing our analysis. Soil pollution commonly refers to conditions brought by contaminants that cause, or have a risk of causing, damage to human health or the natural environment.[2] Liability for soil pollution could be as broad as the responsibility to repair all the damage caused by pollution in the soil. More specifically, it may include liability for soil

---

[1] State Council (2016) Action Plan on Prevention and Control of Soil Pollution. Available at: http://www.gov.cn/zhengce/content/2016-05/31/content_5078377.htm.

[2] See, for example, Article 2 of the Federal Soil Protection Act of Germany which defines 'contaminated sites' ('Altlasten' in German) to include closed-down waste management installations. . .that cause harmful soil changes or other hazards for individuals or the general public. See also, Section 78A(2) of Part 2A of Environmental Protection Act of 1990 from the United Kingdom which provides that 'contaminated land' is 'Any land which appears to the local authority in whose area it is situated to be in such a condition, by reason of substances in, on or under the land that: (a) significant harm is being caused or there is a significant possibility of such harm being caused; or (b) significant pollution of controlled waters is being caused, or there is a significant possibility of such pollution being caused.'

contamination cleanup, long-term damage to the environment or natural resources, and even liability for personal injury or private property damage.[3] However, soil pollution liability generally, and in this chapter too, refers to the responsibility to clean up contaminated land to its unpolluted status, or to meet a target set by law such that the land is suitable for a particular use - by either cleaning up pollutants or cutting off the transference pathway between soil contaminants and receptors. Globally, duties imposed by land contamination laws include emergency response, contaminated sites investigation, risk assessment, promulgation of a remediation plan, site cleanup, long-term monitoring and institutional controls.

This chapter takes the function of EPIL as a logical starting point and reviews the performance of EPIL prior to the implementation of PCSPL, to investigate its proactive functions and limitations in the prevention and control of soil pollution. By looking at the latest development of PCSPL and civil law in China, this chapter also explores the future avenue of EPIL to find out how the litigation system might be reshaped to further facilitate soil pollution prevention and control.

14.2 The Role of EPIL in Soil Pollution Prevention and Control Before PCSPL

In the absence of unified legislation on soil pollution prevention and control, EPIL in China was expected to effectively bridge the gap between substantive laws, such as civil law and environmental law, and procedural law, such as civil procedures and administrative procedures, to provide judicial remedies for soil pollution prevention and control. Broadly speaking, the functions of EPIL are mainly threefold as follows.

14.2.1 Enhance Soil Pollution Prevention

In China, soil pollution was notorious for the irreversibility of damage, as well as the difficulty, high costs and long period of soil remediation and cleanup. As a result, China has always taken prevention as the primary principle of its environmental laws and regulations. The concept of prevention has also been clearly incorporated into statutory provisions relating to EPIL in China. For example, Article 1 of the Supreme Peoples' Court (SPC) Interpretation on Application of Law on Environmental Civil Public Interest Litigation (2015) (hereinafter referred to as the civil EPIL

[3] CERCLA in the U.S. stopped short of imposing liability for personal injury or economic losses due to land contamination. However, many states in the U. S. have enacted laws that go further than CERCLA. For example, in Minnesota, statutes impose liability for personal injuries and economic losses. Damages that can be recovered for the release of hazardous substances include a variety of economic losses, death, and personal injury or disease. See Minn. Stat. Ch. 115B (2018).

Interpretation) states that: 'Where an authority or relevant organization as prescribed by law files a lawsuit against any conduct that pollutes the environment and damages the ecology, which has damaged the public interest or has the major risk of damaging the public interest, in accordance with the provisions of Article 55 of the Civil Procedure Law, Article 58 of the Environmental Protection Law, and other laws, if the provisions of item (2), (3) or (4) of Article 119 of the Civil Procedure Law are complied with, the people's court shall accept the lawsuit.'[4] This provision pierces through the requirement of 'direct interest' under the theory of traditional civil litigation by enabling eligible organizations to claim an interest on behalf of other individuals or groups, or the public in general. A firm normative basis for filing preventive civil EPIL is thus established.[5] Moreover, compared with private litigation, the cause of action for EPIL does not necessarily require the occurrence of damage, as long as there is reasonable suspicion of environmental interests at risk.[6]

Currently, eligible environmental non-profit organizations (ENGOs) and the people's procuratorates both can file preventive environmental civil public interest litigation in China. According to Article 55 of the Civil Procedure Law of the People's Republic of China (as amended in 2012) and Article 58 of the Environmental Protection Law of the People's Republic of China (as amended in 2014), social organizations that meet certain conditions can file civil EPIL lawsuits.[7] From 2015 to 2016, China started to pilot programs under which public interest litigation cases could be brought by the people's procuratorates. In 2017, China amended its Civil Procedure Law, which formally stipulated that the people's procuratorates are able to bring civil EPIL. Accordingly, in the context of soil pollution prevention and control, eligible ENGOs and the people's procuratorates can file preventive lawsuits against actions that are deteriorating soil environment or putting it at risk, so as to curb the expansion or occurrence of soil pollution damage in a timely manner. In judicial practice, although preventive environmental public interest litigation is not

[4]The document for the Supreme Peoples' Court (SPC) Judicial Interpretation on Application of Law on Environmental Civil Public Interest Litigation could be found at: https://www.court.gov.cn/fabu-xiangqing-13025.html.

[5]Liu (2021b), p. 40.

[6]Wang (2018a), p. 299.

[7]Article 55 of the Civil Procedure Law of the People's Republic of China (amended in 2012) stipulates: 'for acts that pollute the environment, infringe on the legitimate rights and interest of many consumers and damage the social and public interest, the organs and relevant organizations prescribed by law may bring a lawsuit to the people's court.' Later, the legislation further defines the 'organizations'. Article 58 of the Environmental Protection Law of the People's Republic of China (amended in 2014)provides that Chinese social organizations that meet the following two requirement requirements can bring suit on behalf of the public interest in cases involving pollution or ecological damage: (1) the organization has registered with the civil affairs departments at or above the municipal level within the district;(2) the organization has specialized in environmental protection public interest activities for five years and has no record of violating the law. The amended rule thus allowed broader standing in some respects than is allowed by U.S. courts under Article III of the U.S. Constitution. Chinese NGOs need not have members with a direct tie to the site of the pollution. See Sun and Jack (2017), p. 10499.

common in the field of soil pollution in China, it has been adopted in other public interest litigation practices, such as the 'Yunnan Green Peacock Case', which has been broadly acknowledged as a classic and innovative use of preventive civil EPIL.[8] Therefore, implanting the concept of preventive public interest litigation in soil pollution law and exploring the practice of preventive public interest litigation can strengthen the prevention of soil pollution. Moreover, as the soil is a finite resource and only able to regenerate to a limited degree, prevention is of great importance and makes more sense.[9]

14.2.2 Secure Ecological Restoration

Prior to the introduction of China's Soil Pollution Prevention and Control Law, both the Opinion on Strengthening the Work on Soil Pollution Prevention and Control (2008) by former MEP (before 2018) and the State Council's Action Plan on Prevention and Control of Soil Pollution (2016) clearly stipulated the responsibilities of polluters (organisations or individuals) for soil remediation and cleanup. However, the two above-mentioned normative documents at the central level neither clarified the way to impose the responsibility for soil pollution, nor had the mandatory effect of laws. This makes it difficult to implement soil pollution remediation, which is the fundamental solution for soil pollution abatement, by restoring the soil

[8] In March 2017, the Beijing Chaoyang District Friends of Nature Environmental Research Institute filed an environmental public interest lawsuit with the Kunming intermediate people's Court of Yunnan Province, claiming that the inundated area of the Jiasajiang I Hydropower Station on the mainstream of the Red River (Yuanjiang) is the habitat of the national first-class protected animal and the endangered species Green Peacock. Once the hydropower station is impounded, it will lead to the possibility of extinction of green peacocks in this area. The supporting project of the hydropower station will destroy the local precious seasonal rainforest ecosystem in the dry and hot valley. After hearing the case, the Kunming intermediate people's court held that the case was a preventive environmental public interest litigation. The plaintiff Beijing Chaoyang District Friends of Nature Environmental Research Institute has provided evidence to prove that the inundation area of the hydropower station involved in the case is an area where green peacocks are frequently active and constitutes its biological habitat. Once inundated, it is likely to cause serious damage to the survival of green peacocks. At the same time, the environmental impact report of the Jiasajiang I Hydropower Station does not involve the protection of rare rainforest plants such as Cycas chenii. If the hydropower station continues to be constructed, the survival of rare animals and plants in the area will face major risks. The Kunming intermediate people's court then ordered the defendant to immediately stop the construction project of the Jiasajiang I Hydropower Station Based on the existing environmental impact assessment, not to intercept and store water, and not to cut down the vegetation in the flooded area of the hydropower station. After completing the post environmental impact assessment and filing according to the requirements of the Ministry of ecology and environment, the relevant competent administrative department shall make a decision in conformity with legal provisions according to the specific situation. The court of second instance upheld the original judgment. See Cha and Yang (2022), pp. 1–2.

[9] Malte (2006), p. 256.

to an acceptable condition.[10] Considering that soil remediation is a long-term, costly process, the incapability of the administrative authorities in allocating the costs of remediating soil makes civil remedy an alternative way to seek to recover these costs.

Article 20 of the civil EPIL Interpretation provides that: 'Whenever the plaintiff requests the restoration to the original state, the people's court may, in accordance with the law, rule that the defendant shall restore the function of ecological environment to the state before the damage occurred. If the damage cannot be fully remedied, alternative restoration methods should be provided. The People's Courts may determine the ecological restoration costs to be borne by the defendant in the event that the defendant does not fulfill its restoration obligations, or may directly sentence the defendant to bear the ecological restoration costs. The costs of ecological restoration include, among others, the costs of preparing and implementing the restoration plan and the costs of monitoring and supervision'.

Given that 'restoration of ecological environment' has not been explicitly listed as a type of civil remedy provided by the former tort law of China, this interpretation provides the first step to recognizing that compensation may be ordered for environmental damage per se.[11] In addition, in order to ensure full recovery of the damaged ecological environment, Article 21 of the civil EPIL Interpretation also provides that the defendant is obligated to compensate for the loss of service from the time of the incident until restoration is completed. It can be seen that China has incorporated ecological restoration into the tort liability system of civil law by means of judicial interpretation, which provides a legal basis for the implementation of restoration liability.

Based on the recently enacted Civil Code and the urgent need for ecological restoration, in December 2020, the Supreme People's Court amended the civil EPIL Interpretation, further clarifying the provisions of civil EPIL regarding ecological restoration.[12] According to the amendment, the public interest litigation plaintiff can

[10] Shen (2021), p. 33.

[11] Hu and Cui (2019), p. 14.

[12] On December 29, 2020, the Supreme People's Court announced 19 revision decisions on judicial interpretations of civil litigation, including the civil EPIL Interpretation. The highlight of the revision of the judicial interpretation is reflected in recognized claim from the damaged parties, that is, 'Request for restoration of the ecological environment' replaces the expression of 'Reinstatement' (recovery to the initial status) in Article 20 and Article 21 of the civil EPIL Interpretation (formerly enacted in 2015). Its significance lies in, on the one hand, making it clear that the effect of 'restoration of ecological environment' is different from 'Reinstatement' in traditional civil law, while on the other, providing a clear basis for the comprehensive restoration of ecological environment. Specifically, Article 20 is amended as: 'If the plaintiff requests to repair the ecological environment, the people's court may, in accordance with the law, judge the defendant to repair the ecological environment to the state and function before the damage occurs. If it cannot be completely repaired, it may permit the use of alternative repair methods. The people's court may, while judging the defendant to repair the ecological environment, determine the ecological environment repair costs that the defendant should bear if the defendant fails to perform the repair obligations; it may also directly judge the defendant to bear the ecological environment repair costs.

directly bring a lawsuit requesting the restoration of the ecological environment, without relying on the controversial term 'restoration to the original state of the rights' in its claim.[13] This substantially benefits civil EPIL by extending the scope of civil remedies to incorporate the claim of ecological restoration. Furthermore, the newly amended provisions also include such losses caused by permanent damage to ecological and environmental functions in the scope of compensation for the defendant. Obviously, the civil EPIL provides a clearer path to a judicial remedy for soil pollution prevention and control in terms of both ecological restoration and long-term ecological loss.

14.2.3 *Promote Fulfillment of Administrative Agencies' Responsibilities*

According to Article 25(4) of China's Administrative Litigation Law (as amended in 2017)[14] and Article 21 of the SPC and SPP Interpretation on the Application of Laws in Procuratorial Public Interest Litigation (2018)[15]: 'the People's Procuratorates shall take the responsibility to supervise the authorized administrative agencies in fulfilling their duties in the field of ecological environment and resource protection. Procuratorates shall firstly alert and direct focal agencies to proactively execute laws whenever they behave illegally or inactively, which result in the infringement of national interests or social public interests. If the agencies refuse to correct their misconduct, Procuratorates, when necessary, have recourse to lawsuits against those

The cost of ecological environment restoration includes the cost of formulating and implementing the restoration plan, the cost of monitoring and supervision during the restoration, the cost of acceptance after the restoration, the cost of post restoration evaluation, etc.; Article 21 is amended as: 'if the plaintiff requests the defendant to compensate for the loss caused by the loss of service function during the period from the damage to the ecological environment to the completion of restoration, and the loss caused by the permanent damage to the ecological environment function, the people's court may support it according to law.' See Supreme People's Court (2020) Decision of the Supreme People's Court on Amending the Provisions of the Supreme People's Court on Several Issues Concerning the Civil Mediation Work of the People's Courts and Nineteen Other Judicial Interpretations of Civil Litigation Available at: https://www.court.gov.cn/fabu-xiangqing-282651.html.

[13] 'Restoring the ecological environment' is different from 'reinstatement'. From the perspective of environmental law, 'Restoring the ecological environment' is more applicable to the relief of public interests of the ecological environment, it cannot be simply equivalent to the 'reinstatement' of private interests. Judicial practice proves that although 'restoration of the ecological environment' and 'reinstatement' are seemingly identical, the differences between the two in terms of relief objects, restoration standards and relief methods are still evidential. Lv and Dou (2017), p. 141.

[14] The document for China's Administrative Litigation Law could be found at: https://flk.npc.gov.cn/detail2.html?MmM5MDlmZGQ2NzhiZjE3OTAxNjc4YmY4NTg1NTBhMGY%3.

[15] The document for the SPC and SPP Interpretation on the Application of Laws in Procuratorial Public Interest Litigation could be found at: https://www.spp.gov.cn/zdgz/201803/t20180302_368570.shtml.

administrative agencies.' The administrative EPIL exerting pressure on relevant environmental regulatory agencies to perform their duties also indirectly facilitates prevention and remediation of soil pollution actively and effectively.

14.3 A Practical Review of EPIL Before China's Soil Pollution Law

Before the enactment of China's PCSPL, EPIL has played an important role in the prevention and control of soil pollution, and to a certain extent made up for the lack of specific legislation on soil pollution prevention and control. However, case studies on EPIL cases in the field of soil pollution shows that EPIL also has various limitations. For example, the varied perception from the court as to what end should public interest litigation pursue substantially affected the realization of the public interest. The more compensatory attribute of public interest damages hindered the development of preventive public interest litigation. Moreover, due to the technical and complex characteristics of soil pollution prevention and control, the courts also faced many professional challenges in trial adjudication. The fact that plaintiffs are opted to use civil EPIL rather than on administrative EPIL also constrained the supervisory role of administrative EPIL.

14.3.1 The Narrow Purpose of Litigation

Practically, the purpose of EPIL remains vague with two conflicting views coexisting. The first view, which is broadly accepted by judges, suggests that the only aim of EPIL is remediation of soil pollution. In the 'Changzhou Toxic Land Case',[16] the plaintiffs argued that the three defendant companies caused severe land contamination and injuries to students at the adjacent high school during the land remediation process. In their submission the defendants should be liable for environmental pollution risk control and remediation. However, the court pointed out that since the local government had already carried out risk control and cleanup of the land in question, there was no need to order the defendant companies to

[16] In May 2016, Beijing Chaoyang District Friends of Nature Environmental Research Institute and China Biodiversity Conservation and Green Development Foundation, as plaintiffs, sued Jiangsu Changlong Chemical Company Limited, Changzhou Changyu Chemical Company Limited and Jiangsu Huada Chemical Group Company Limited as defendants to Changzhou Intermediate People's Court. The plaintiffs proposed that the three defendant companies were all processing enterprises engaged in chemical products, which seriously polluted the land in question and the surrounding environment during their production and operation and the management of hazardous waste, and failed to remediate the polluted land after moving out, resulting many students in Changzhou foreign language school suffered health damage.

remediate the land, as a result the claim was dismissed. The court further pointed out that letting the defendant companies recover any land contamination costs experienced by local government was beyond the scope of the case.[17] The court's position regarding the purpose of EPIL generated considerable public and academic controversy. Some scholars argued that the monism of 'purpose' cut the inherent nexus between restoration liability and restoration cost. According to Hongkun Sun (2021), since environmental public interest litigation aims to safeguard environmental public interest, including the loss of interests (economic interests and ecological interests) caused by environmental damage, remediating the contaminated soil in this case is clearly a form of "restoration of ecology and environment".[18] Therefore, in this case, the plaintiff has the right to request the three polluters to pay the cost of environmental restoration, which is not beyond the scope of environmental public interest litigation. Let's get back to the case. According to the adjudication of the Court of Appeal, plaintiffs' claiming for remediation of contaminated sites was again dismissed. The court further points out that, as for the cost in remediation already spent by the Xinbei District Government of Changzhou City, it may pursue compensation from the polluters though, however, a separate proceeding. This might cast a shadow of uncertainty on the implementation of the restoration costs. The Court of Appeal should therefore focus on enhancing clarity about the nature of environmental remediation costs, rather than merely determining whether pollution remediation has been or is being undertaken; Otherwise, there will be a narrow understanding of the purpose of environmental public interest litigation.

As another viewpoint, Xu and Zhou (2017) contend that the first instance of the case actually has a narrow and wrong interpretation of the purpose of civil EPIL, and further points out that EPIL should have a dual purpose: (1) to prevent and control environmental risks that harm the public interest of society and to solve environmental problems; and (2) to justly allocate the social costs and burden associated with environmental problem solving.[19] Obviously, the dualism of the purpose of environmental public interest litigation is more in line with 'polluter pays principle' (PPP) for soil pollution management. If public interest litigation only reaches the first layer of purpose, it may lead to 'the government to pay for the polluter', which is essentially detrimental to the public interest and national interest. Because endeavors for finding the real polluters are all too often costly and time consuming, governments are not strongly motivated to pursue compensation from the responsible parties.[20] So when looking back to the 'Changzhou Toxic Land Case', urging the government who completed the land restoration to file a separate case to recover its costs would probably be a waste of judicial resources.

---

[17] Higher People's Court of Jiangsu Province, Su Civil trial No. 232 (2017).

[18] Sun (2021), p. 169.

[19] Xu and Zhou (2017), p. 98.

[20] Sun (2021), p. 170.

14.3.2 Failure of the Prevention Function

The preventive effect of EPIL results in the effective prevention of an occurrence and expansion of.

the content environmental damage. However, in China, almost all the lawsuits using EPIL in relation to soil pollution prevention and control were filed after the occurrence of soil pollution damage; and hardly any cases were filed for the prevention of soil pollution. This damage-oriented pattern makes it difficult to fulfill the preventive judicial remedy function of EPIL.[21] The 'Changzhou Toxic Land Incident', the 'Tengri Desert Pollution Incident', and the 'Seepage Pit Incidents in Hebei and Tianjin' in recent years are typical examples which reflect the inefficiency of prevention orders in EPIL resulting in a situation where irreversible damage to the environment has been caused.[22]

There are a number of potential reasons for this. First, the effort of inlaying environmental public interest into a traditional tort liability system has proved problematic, especially when judges were not ready to get rid of the 'no damage, no remedy' doctrine, and attached great importance to the 'consequences of damage' when approaching tort liability. In the case of *China Biodiversity Conservation and Green Development Foundation v Beijing Jinyu Mangrove Environmental Protection Technology Co*., Ltd., for example, the court held that, 'For either a civil EPIL or an environmental civil private interest litigation, whenever the plaintiff claims that the defendant bears an environmental tort liability, the burden of proof is on the plaintiff to prove the existence of tortious activity, damage, as well as the causal relationship between acts and environmental damage'.[23] Second, the terms of initiating preventive EPIL are relatively abstract, which leads to difficulties in judicial practice. Article 1 and Article 18 of civil EPIL Interpretation state: 'the authorities and relevant organizations as provided by law may file lawsuits against acts that have harmed the public interest of the society or have a significant risk of harming the public interest of the society.' This provision is more vague, especially the concept of 'significant risk', causing difficulty and divergence in understanding and application and resulting in a very low caseload.[24] Third, some EPIL soil pollution damage disputes took months or even years to be adjudicated, especially

[21] Tang (2019), p. 29.

[22] Dong (2019), p. 121.

[23] Fourth Medium People's Court of Beijing City, Jing 04 Civil trial No.89 (2016).

[24] For example, 'Yunnan Oil Refinery Case' is the first case of preventive Civil EPIL against a large construction project in China. The main focus of the case is whether the defendant's unapproved construction of the project poses a significant risk of damaging the public interest. According to the ruling of the second trial, the primary reason why the first and second trials were not accepted was that the court held that the plaintiff failed to prove that the defendant's behavior had caused damage to the social and public interest or had a significant risk of damaging the social and public interest. The court's determination of 'significant risk' is very different from the plaintiff's understanding of 'significant risk', which was the first legal reason for the inadmissibility of this case. See Zhang and Wu (2020), p. 138.

for EPIL cases brought by the people's procuratorates, which require at least a 30-day notice period, not to mention the time required awaiting court decisions.[25] Due to cumulative effects of soil pollution as well as its potentiality to disperse to surrounding surface and underground water and vegetation, even if the public interest litigation could timely find the persons liable, the environmental damage could have continued throughout the whole litigation process.

14.3.3 Limited Professional Competence

EPIL is essentially a way of relying on judicial proceedings to identify persons who are liable for environmental pollution and allocate liabilities when there is more than one contributor to be brought into the cleanup process. However, the involvement of science and technology issues further complicate the court's decision process. The court is required to clarify the costs of soil cleanup, cleanup procedures, remediation method selection, remediation target determination, and remediation effectiveness assessment.

The costs of restoration, for example, usually requires a comprehensive decision by the judge with professional knowledge in recognizing the costs and benefits of different cleanup and restoration methods, how to cut down a pollutants' migration routes to restore contaminated land into a 'clean' status consistent with relevant legal standards and, finally, avoiding the expansion of harm or risk due to degraded land conditions. However, due to the existence of a professional knowledge threshold, some judges are only able to conceptualize the problems using jurisprudential concepts, rather than using a scientific analysis. For instance, in the case of *Tongren Municipal People's Procuratorate v Guizhou Yuping Xiangsheng Chemical Co., Ltd. and Guangdong Shaoguan WoXin Trade Co., Ltd*, the loss of service function of agricultural arable land needed to be calculated based on the number of years of remediation. According to the damage assessment report recommended by the agency, the expected restoration period would be 4–6 years by using the 'chemical passivation and ecological recovery' restoration method, and the judge chose a timeframe of 6 years. However, taking account into the iterative nature of the restoration and the uncertainty of the final restoration, from the perspective of ensuring the absolute safety of agricultural products after the restoration of agricultural land use, the judge decided to increase the time limit by 4 years from a punitive perspective and determined that the final timeframe was 10 years.[26] From this reasoning, the judge ultimately determined the restoration years mainly based on 'absolute safety', 'punitive' and other subjective judgments, and not from the environmental science or other professional interpretations and analysis. The results

[25] Medium People's Court of Xuzhou City, Su 03 Civil trial No.338 (2019); Medium People's Court of Xian City, Shan 71 Civil trial No.4 (2017), etc.

[26] Medium People's Court of Zunyi City, Qian 03 Civil trial No.520 (2016).

may achieve the purpose of restoration, but they may not be reasonable, and suspiciously increase the burden on the defendants in terms of liability. In fact, when determining the criteria for remediation, a doctrine based on the risk analysis of future utilization of the polluted land is much more commonly accepted, rather than 'return to zero pollution' status, to provide a site capable of sustaining the future use by ecological receptors.[27]

In addition, if the judges have a bias in their understanding of remediation methods, it can also affect their perception of environmental damage and harm. For example, 'restoration' in civil law and 'environmental remediation' in environmental science are two different concepts that should not be intermixed. In the 'Changzhou Toxic Land Case',[28] the main method of remediation was to 'remove and cover', that is to dig out the contaminated soil, send it to appropriate places for treatment, then cover the place with clean soil moved in from other places. The mulching method is unable to repair the soil, groundwater and underground ecosystem of the site involved thoroughly. If the judges mistakenly believed that the polluted ecological environment can be 'restored to its original state' or that the effect of environmental restoration can achieve the original appearance by this method, it is probably too lenient on the polluter's' illegal environmental obligations.[29]

Moreover, the identification and assessment of ecological environment damage usually involves the application of technical norms. These technical norms have become an important standard and effective reference for the court to decide whether the ecological environment damage exists and the severity of the damage. As a type of scientific evidence, this environmental forensic involves the application of scientific principles and technical means that are beyond the common sense of our daily life. As a result, judges are often unable to actively examine evidence 'by simply applying logical reasoning and the rules of thumb as they do with other evidence'.[30] For example, in the case of *Liaocheng City People's Procuratorate v Jixin Wang and Fengzhong Li* a soil pollution liability dispute, the judge confirmed that the environmental damage assessment report was made and signed by a qualified judicial forensic center, and the defendant failed to submit sufficient evidence to refute it.[31] In addition,in another case of *Xian City People's Procuratorate v Kaiqiang Liu, Gaopan Sui* a soil pollution liability dispute, as the costs of formal judicial forensic was extremely high and the experts who issued the opinion were senior engineers with relevant professional competence, the judge determined that

[27] Wang (2016a), pp. 57–59.

[28] Above note 17.

[29] Sun (2021), p. 175.

[30] Jin (2021), p. 155.

[31] Medium People's Court of Liaocheng City, Lu 15 Civil trial No.486 (2018).

expert opinions formed the ground of liability allocation.[32] This shows that the judges are also more passive in making a judgement in soil pollution cases and tend to rely extensively on the professional opinions by forensic organs or expert testimony.[33]

14.3.4 Neglected Supervision Role of Administrative EPIL

Among the selected soil pollution public interest litigation cases, environmental public interest litigation is mainly brought about in the form of civil public interest litigation, rather than administrative public interest litigation.[34] The reason for this phenomenon may lie in that before the implementation of China's Soil Pollution Law, there was no substantive law that governed soil pollution prevention and control, which made it difficult for the people's procuratorates to determine whether the relevant administrative agencies had fulfilled their responsibilities and thus to initiate administrative EPIL. In practice, it may be relatively easy to file a civil EPIL for specific ecological and environmental damage. However, after the implementation of the PCSPL, under which the responsibility of administrative agencies for soil pollution prevention and control has become clearer, the people's procuratorates have now a supervisory function within administrative EPIL.

---

[32] Medium People's Court of Xian City, Shan 71 Civil trial No.4 (2017).

[33] In fact, the SPC may have taken into account the difficulty of the professionalism of environmental cases and issued a special judicial interpretation for this purpose, which provides a system tool for judges to adjudicate in environmental cases with the help of professional opinions. See Article 10 of SPC Interpretation on the Application of Laws in Adjudicating Environmental Tort Cases (2015) stipulates: 'The investigation report, inspection report, test report, evaluation report or monitoring data of environmental pollution events issued by the Department responsible for the supervision and administration of environmental protection or its entrusted institution can be used as the basis for determining the facts of the case after cross examination by the parties.' Article 23 of SPC Interpretation on Several Issues Concerning the Application of Law in Environmental Civil Public Interest Litigation (2015) stipulates: 'If it is difficult to determine the cost of ecological environment restoration or the identification cost required to determine the specific amount is obviously too high, the people's court may, in combination with factors such as environmental pollution, the scope and degree of ecological damage, the scarcity of ecological environment, the difficulty of ecological environment restoration, the operation cost of pollution prevention and control equipment, the benefits obtained by the defendant due to infringement and the degree of fault, it can also be reasonably determined by referring to the opinions and expert opinions of the Department responsible for environmental protection supervision and management.'

[34] Taking 'administrative public interest litigation' and 'soil pollution' as the key words, a comprehensive search has been carried out in the legal database Lawinfochina.com of Peking University, and no results have been found.

14.4 Development of EPIL After Soil Pollution Law

China, the world's second largest economy, is undergoing high-speed industrialization along with serious environmental degradation in the past decades. Land contamination in China can be traced back to the 'Great Leap Forward' in the 1950s (even before the establishment of the People's Republic of China in 1949).[35] From 2005 to 2013, China conducted a National Land Contamination Survey. It took samples from survey units across 6.3 million square kilometers (approximately two-thirds of the country's land area). Of the investigated units, 16.1% showed contamination exceeding the permitted level; 11.2%, 2.3%, 1.5%, and 1.1% of the units were identified as subtly, lightly, moderately, and heavily polluted, respectively. As for highly suspected sites such as high-pollution-density industrial sites and landfill sites, around 30% of the surveyed sites were contaminated.[36] Faced with long-lasting contentions, the National People's Congress spent 12 years drafting national legislation and finally enacted the Law on Prevention and Control of Soil Pollution in 2018. This law went into effect on January 1, 2019. The provisions of the law are still general and require detailed regulations, standards, and other enforcement mechanisms to facilitate its implementation.

14.4.1 Improving the Feasibility of EPIL's Preventive Function

The PCSPL emphasizes adherence to prevention and the strengthening of risk control. Improving the preventive function is now an important direction for the future development of EPIL. The prerequisite for prevention lies in the ability to detect acts that may result in soil contamination in a timely manner and to take effective measures to address them, thereby to avoid the occurrence or expansion of damage. In this regard, the following two points need to be specially addressed.

14.4.1.1 Active Participation of ENGOs

The current law provides that the people's procuratorate needs to carry out pre-litigation procedures when filing civil EPIL, and the court will accept the lawsuit

[35] Xie and Li (2010), p. 3. See Overview of the current situation on brownfield remediation and redevelopment in China. World Bank. Available at: https://documents.worldbank.org/en/publication/documents-reports/documentdetail/450251468024319815/overview-of-the-current-situation-on-brownfield-remediation-and-redevelopment-in-china.

[36] Former Ministry of Environmental Protection (MEP), Ministry of Land Resources, Report on National Land Contamination Investigation Survey, 2014, Available at: http://www.gov.cn/foot/site1/20140417/782bcb888408l4ba158d01.pdf.

only if no relevant organs and social organizations file the lawsuit within the 30-day notice period. In judicial practice, some people' courts and people's procuratorates have a quite narrow understanding of this provision. This restrained the scope of ENGOs entitled to sue. This may easily cause delays in prevention of soil pollution, especially when local organizations are hesitating to challenge the potentially liable parties. Therefore, the people's courts and the people's procuratorate should reach a consensus that pre-litigation notice procedures should be applicable for eligible social organizations nationwide, rather than be limited to the those within the same local jurisdiction. Moreover, sufficient information disclosure which would facilitate an ENGOs' involvement in EPIL should also be secured.[37]

It should also be noted that, in order to improve the judicial relief mechanism for soil pollution prevention and control, some scholars proposed an expansion of the scope of the persons having standing to sue in civil EPIL, including individual citizens.[38] However, currently the latter proposal is far from practical. On the one hand, the burden to file an EPIL brings forth many difficulties for individual citizens, such as time-consuming proceedings, high litigation costs and requirement for special expert knowledge. This is especially true because soil contamination, compared with air and water pollution, is less prominent and more difficult to be detected, quantified or remediated. Consequently ENGOs have not played a major role, in EPIL in China,[39] and individual citizens are not financially able to protect the public interest. In view of this, two solutions can be suggested. Firstly, further lower the threshold requirements for ENGOs to initiate environmental public interest litigation, and incentivize participatory enthusiasm of ENGOs nationwide through facilitative proceeding arrangements, economic incentives and honorary encouragements. Secondly, currently, the weak enforcement of environmental laws and regulations is challenging China's rural environmental governance.[40] Therefore, the recognition of rural autonomous organizations, who are more familiar with the situation of agricultural land, to access EPIL could be considered as an important complementary factor which may assist the public powers while at the same time it may stimulate ordinary citizens to participate in environmental affairs.[41] Here, it should also be noted that due to the special structure and weak regulatory competence of rural autonomous organizations in villages, their participation is better in the form of being the initial whistleblower.

---

[37] Zhu (2018), p. 21.

[38] Dong and Wang (2022), p. 118.

[39] According to the survey data, at present, only 4% of the more than 700 social organizations that meet the legal prosecution conditions in China say that they are capable and willing to undertake this function. Therefore, the role of social organizations in environmental public interest litigation has not been maximized, which needs attention and support. See Zhou (2020), p. 50.

[40] Liu (2012), p. 237.

[41] Li and Wu (2021), pp. 53–58.

14.4.1.2 Judicial Application of Preventive EPIL Provisions

According to Article 18 of the civil EPIL Interpretation, the authorities and relevant organizations as stipulated in the law may file lawsuits against 'acts of pollution of the environment and ecological damage that have a significant risk of damaging the social public interest'. The understanding and application of this provision is the key to determining whether EPIL can play a preventive role. The future promotion of the application of preventive EPIL in soil pollution prevention and control requires clarification of the following issues.

Firstly, in the judicial concept, judges should break the traditional thinking of 'there is damage before there is relief' and establish a sense of risk prevention, not simply rejecting lawsuits on the grounds that there is no damage.

Secondly, whether the condition of 'significant risk' is met should be decided by the court, and the experts, scholars, administrative departments, and the public have the right to participate in the process of determination. If the determination of 'significant risk' is left to the administrative departments, once the departments are 'captured' by economic interests and other needs, it would be impossible to guarantee the scientific, objective and impartial results of the determination. In addition, the determination of 'significant risk' involves technical requirements, the balance of interests, the court must consider not only the actual impacts on the ecological environment, but also the possible social impacts of the result.

Thirdly, appropriately grasp the criteria for determining 'significant risk'. In judicial practice, the inherent uncertainty and irreversibility of significant environmental risks has led to unsatisfactory preventive protection of environment public interests.[42] How to determine whether there is a 'significant risk' has become a difficult issue in the practice of preventive public interest litigation. In this regard, there are views about the standard of evidence for determining 'significant risk'.[43] However, from the above analysis, basically it is still in the state of abstract discussion and does not serve as a good guide. We believe that the determination of the criteria for determining 'significant risk' should take into account the tolerance needs of different legislative norms, i.e., the acceptable level of risk, and needs to be judged in the context of specific legislative provisions. Taking the prevention requirements of PCSPL for soil pollution risk control as an example. China's Soil Pollution Law adopts the principle of prevention, but it does not provide that preventive and control measures should only be taken if there is a 'significant risk'. Instead, there is a 'possibility' to take preventive measures in a timely manner. For example, Article 18 of the PCSPL stipulates that the environmental impact assessment document shall include the possible adverse effects on the soil and the

---

[42] Yu and Mu (2019), p. 29.

[43] As for the standards of proof of 'significant risk', China's academic circles mainly have three views: 'high degree of probability', 'general probability' and 'probability' in the prosecution stage, 'high degree of probability' in the court trial stage. See respectively Zhang and Wu (2020), p. 141; Yu and Mu (2019), p. 31; Song (2021), p. 70.

corresponding preventive measures to be taken; Article 44 of the PCSPL stipulates that providing for the occurrence of emergencies that may cause soil contamination, the relevant entities shall immediately take emergency measures to prevent soil contamination. This also indicates that China's PCSPL places extremely high demands on soil pollution prevention. Therefore, as long as there is the possibility of pollution, the judicial application can extend the interpretation of the existence of 'significant risk'.

Fourthly, at the legislative level it should promote the study of environmental background values of soil pollution and environmental benchmarks, to provide a scientific and reasonable standard for soil environmental pollution damage and to assist judicial decisions.

14.4.2 Complementary Role of Civil EPIL in Soil Pollution Law

China has established a comprehensive public liability system pertaining to soil pollution control and remediation under PCSPL. In this context, it is necessary to clarify how to coordinate and promote the implementation of EPIL and to develop appropriate systems to achieve better management of polluted soil.

14.4.2.1 Public Liability Under PCSPL

PCSPL requires site investigation, site risk assessment, risk control, remediation, evaluation of site risk control and remediation, and long-term stewardship.[44] When the presence of contaminants at a site emerges, site investigation is required to obtain data on the physical characteristics of the site and the hazardous substances, including, among others, quantity, toxicity, concentration, and mobility of those substances.[45] If the contaminant exceeds the threshold level in the national standard on risk control of land contamination,[46] risk assessment should be commenced to evaluate risk, as well as to determine whether responsive actions are needed to mitigate impacts on the safety of agricultural products, public health, and the environment. Targets value for remediation and basic requirements for remedial

[44] Article 35, Law on Prevention and Control of Soil Pollution of the P.R.C. For a more detailed discussion about risk control and remediation procedures in China's soil pollution law, see Li et al. (2019), p. 556.

[45] Article 36, Law on Prevention and Control of Soil Pollution of the P.R.C.

[46] Soil Environmental Quality—Standard on Risk Control of Soil Pollution in Agricultural Land (GB 15618-2018) and Soil Environmental Quality - Standard on Risk Control of Soil Pollution in Constructional Land (GB- 36600-2018).

actions should also be included in the risk assessment report.[47] Responsible parties are obliged to take remedial actions through risk management (i.e., cutting off or blocking the potential pathway of contamination from hazardous substances by, for example, fencing, alternative agricultural planation, or restrictions on land usage)[48] or remediation (i.e., cleaning up or removing the hazardous substances, or lowering the concentration or toxicity of contaminants in soil, by which to reduce the environmental risk to an acceptable level).

The parties who should be liable for land contamination, according to the law,[49] include owners of usufructuary rights,[50] and local government in a general sequence. Under the Polluter Pays Principle stipulated in Article 3 of the law, parties liable for land contamination are responsible for remedial actions including site investigation, site risk assessment, risk control, remediation, evaluation of site risk control and remediation, and long-term stewardship.[51] Owners of usufructuary rights become responsible parties if 'parties liable for land contaminations', or more accurately, polluters, are unable to be located or identified.[52] If the polluters and owners of usufructuary rights cannot be identified, the local government will undertake the obligations under the law; the government's obligations are to be covered by funds for soil pollution prevention and control made available by the central and provincial government.[53]

---

[47] Article 37, Law on Prevention and Control of Soil Pollution of the P.R.C.

[48] This assembles institutional control in, for instance, The Land Recycling and Environmental Remediation Standards Act in Pennsylvania of the U.S. which defines that, 'institutional controls' are 'measure undertaken to limit or prohibit certain activities that may interfere with the integrity of a remedial action or result in exposure to regulated substances at a site. This includes, but not limited to, fencing or restrictions on the future use of the site.' The National Action Plan on Land Contamination Prevention and Control enacted by the State Council in 2016 already clarified that the principle of risk management should be adopted to abate China's land contamination.

[49] The law, as a matter of fact, provides no explanation on the scope of parties liable to land contaminations. According to the legislative document, the legislators intend to refer it as parties who caused the land contamination.

[50] China has a special land right regime of its own. Fundamentally, lands in urban area are largely owned by the State and those in rural areas are largely owned by the collective. Initially, use of state-owned land was directly granted by the state government, which was concretely implemented by governments at local levels above county at no cost. Since the 1980s, China has developed a real property right system with three main types: ownership rights, usufructuary rights and security rights. Amongst others, the owner of usufructuary rights, which might be entities like enterprises and individuals, has the right to possess, utilize and obtain profits from the land. Compared to ownership rights. Usufructuary rights are non-absolute rights, for example, right holders have no right to sell the land. As for collective-owned land, the right to land contractual management which allows contractor of agricultural land to possess, utilize and gain profits is created, correspond to usufructuary right of state-owned land.

[51] Article 45 and Article 46, Law on Prevention and Control of Soil Pollution of the P.R.C.

[52] The law immediately empowered the administrative agencies of the local government to identify parties liable for land contaminations through administrative procedures in Article 48.

[53] Article 71, Law on Prevention and Control of Soil Pollution of the P.R.C.

14.4.2.2 Advantages of Administrative Power in Soil Pollution Risk Control and Remediation

Accordingly, we may find that letting administrative power to manage the soil remediation process has several advantages: First, the government has established an advanced soil environmental monitoring system under the requirements of relevant laws, regulations and policies on soil pollution prevention and control. Compared with the passivity of judicial power, administrative power can detect problems earlier. The second is about the timeliness and emergent needs of prevention and remediation. As soon as illegal acts that caused damage to the soil environment are discovered, the administrative agencies can immediately issue an administrative order requiring correction of offenders' misconduct. There is no need to wait for the facts and consequences of the violation completely clear and responsibility can be determined as long as the fact that violation or failure of fulfilling legal obligations can be established. A prompt order to cease the violation may eliminate harm and the expansion of damage.[54] Third, in relation to judicial organs, administrative agencies are better equipped with professionals in the field of soil pollution prevention, through the introduction of third-party forces and the establishment of expert review mechanisms. As a local authority, may have more information on contaminated land and polluting enterprises, and are able to supervise remediation plans. Last, the administrative-led process is more likely to gain the trust of the public in the sense that judicial process is always lengthy and costly, especially considering the more sufficient public participation in administration.[55]

14.4.2.3 Shortcomings of Administrative Power in Soil Pollution Risk Control and Remediation

Deterrence of statutory penalties are limited. Since the effect of administrative power is highly dependent on whether the offenders correct their polluting behaviors and further execute cleanup work. Once they refuse to do that, two types of fines could be imposed: lump-sum fines imposed with certain limit and fines calculated on daily basis. According to the relevant provisions of the PCSPL, the fines in China's soil pollution prevention and control are capped. E.g., Article 86 provides that 'a fine of more than 200,000 yuan RMB (approximately 27,700 US dollars) and less than two million yuan RMB (approximately 277,000 US dollars) shall be imposed if the listed circumstances were met, and serious consequences resulted. This approach, however, could also been read as constraints on administrative power. For example, after the Songhua River incident in China, the former Ministry of Environmental Protection (MEP) imposed a fine of one million RMB on the enterprises involved, but the fine was insufficient compared to the huge amount of ecological and environmental

---

[54] Li (2020), p. 191.

[55] Wang (2020b), p. 188.

remediation costs. In the end, the cost of ecological and environmental damage was actually undertaken by the government.[56] Consequently, fines rarely function as a meaningful deterrence to noncompliance of environmental laws.[57] Consequently, ecological and environmental restoration often still requires the use of public funds.

For any responsible person who failed to fulfill the remediation duties, the administrative agency can designate a third party to remediate the contaminated land as a replacement then claim reimbursement from the responsible parties. This institutional arrangement is likely to remedy damaged public interest and enable the regulatory authorities to fully control the remediation process and ensure the quality of remediation. However, the relevant provisions of the PCSPL are still too vague without clearly providing for the compensable items as well as the compensating procedures.[58] This leads to difficulties in the practice of the relief system.

Similarly, confiscation of illegally obtained assets is relatively limited for soil contamination remediation due to unclear legal terms. For example, there are four Articles in PCSPL, which stipulate that 'if there is financial gain resulting from the illegal activities, the gains obtained shall be confiscated'. Even though, when searching court decisions with the keywords 'Soil Pollution Prevention and Control Law', 'illegal obtained gains' and 'administrative' in legal databases, no relevant results can be found.[59] It can be seen that confiscation proceedings as a remedy under the soil pollution law have been rarely used and thus are incapable of providing support for environmental remediation.

It should also be noted that environmental administrative enforcement is not able to impose penalties on those who did not violate the law but have caused soil pollution. Administrative penalties are intended to punish violations of the law, which are concretely indicated by sanctions on violators.[60]

14.4.2.4 Supplementing Soil Remediation Public Liabilities with EPIL

According to the civil EPIL system in China, eligible ENGOs can sue polluters directly without notifying the administrative authorities in advance. Clearly, this generates a competition between EPIL and public liability imposed by administrative authorities according to PCSPL. Based on our above analysis, as a daily regulatory approach, the public liability determined by the administrative agencies should be prioritized over EPIL through judicial process. Nevertheless, it is still necessary to use civil EPIL as one of the preferable civil remedies to plug the loopholes. There are two potential ways here.

---

[56] Wang (2018b), p. 108.

[57] Juan (2018), p. 511.

[58] Gong (2018), p. 61.

[59] Liu (2021a), p. 40.

[60] Zou (2022), p.50.

Firstly, when soil pollution damage occurs, administrative agencies do not perform their duties in accordance with the law, eligible ENGOs and the people's procuratorates should be empowered to bring civil EPIL against responsible parties (polluters or land use rights holders) to pursue responsibility. Moreover, if the land has, according to Article 68 of the PCSPL, been expropriated and the government gains full ownership of the polluted land, it is the government who should undertake the remediation. Here, the government should therefore acquire the compensatory right to claim reimbursement from those who caused land contamination. If the government was reluctant to do so, eligible ENGOs and the people's procuratorates could claim the compensation from the polluters through civil EPIL; otherwise, the public finance will be hampered.

Secondly, plaintiffs of EPIL are able to claim for residual damages to public interest even though the risk control and remediation have been well completed. Article 97 of the PCSPL stipulates that for contaminated soil that damages national interests and social public interests, relevant organs and organizations may file lawsuits in the people's courts in accordance with China's 'Environmental Protection Law', 'Civil Procedure Law', 'Administrative Procedure Law' and other relevant laws. This provision provides a comprehensive legal basis for damages covered by the PCSPL. Eligible ENGOs and the people's procuratorates can bring civil EPIL to claim damages have not been covered by realization of public liability compensation, for example, the loss of service function caused by damage to the ecological environment in the course of restoration and the loss caused by permanent damage to the function of the ecological environment. In addition, even though the remediation liability has been allocated by the administrative agency, whether the requirements have been strictly adhered to during the entire 'site investigation-risk assessment-remediation long-term stewardship' process should also fall into the scope of civil EPIL.

14.4.3 Ecological Environment Damage Litigation

In 2017, CCP Central Committee and State Council Reform Plan on the Compensation Regime for Eco-environmental Damage formally enacted,[61] established a new litigation approach targeting the compensation for ecological and environmental damage. The Judicial Interpretation of the Ecological Environment Damage Compensation Litigation of the Supreme Court in 2019 further provide for concrete

[61] The general office of the CPC Central Committee and the general office of the State Council (2017) CCP Central Committee and State Council Reform Plan on the Compensation Regime for Eco-environmental Damage. Available at: http://www.gov.cn/zhengce/2017-12/17/content_5247952.htm.

application provisions in a judicial verdict.[62] Later in the Civil Code, the scope of compensation for ecological environment damage is confirmed. Local government at or above the municipal level as well as the agencies designated may initiate the litigation. It can only be commenced when: (1) major environmental emergencies occurred; (2) environmental pollution and ecological destruction occurred in key ecological functional areas and prohibited development areas designated in the national and provincial main functional planned area; (3) other incidents that could seriously affect the ecological environment. The relief objective of this kind of litigation is basically the same as public interest litigation after the promulgation and implementation of the Civil Code. However, for the contaminated site, the remedy standard of the ecological environment damage compensation system is to restore the soil environment to the 'original' baseline level rather than the acceptable risk level - the baseline level is usually higher than the acceptable risk level which is applied in the public liability regime. Similar to the judicial path, it can also remedy the interim damage and permanent ecological service function damage caused by the soil pollution.

The operation mechanism of an ecological damage compensation system can be summarized as investigation and evaluation of pollution; determination of responsible person for administrative negotiation and filing of ecological damage compensation litigation. It is worth noting that the litigation will be initiated only when the administrative negotiation does not work out. In fact, the government has no administrative coercive force in the process of claim. The enforcement of 'administrative agreement' or 'litigation result' both depends on judicial power (i.e., court), which is much the same as the judicial relief path.

Then how to distinguish the roles of administrative agencies in regulating land contamination processes as well as in claiming for ecological and environmental damage? To answer this question, we need to, first, consider the feasibility of remediating ecological environment damage. For the remediable contaminated sites, the administrative relief path shall be prioritized, and whether *the Technical Guide for Identification and Assessment of Ecological Environment Damage - Soil and Groundwater* should be used as the regulatory basis for the specific remediation scheme of the contaminated sites (i.e. remediation to the baseline or acceptable risk level line) shall be determined by the degree of soil pollution damage and the area where the contaminated sites are located. The criteria provided by *the Technical Guide for Identification and Assessment of Ecological Environment Damage - Soil and Groundwater* are as follows: (1) the average concentration of pollutants in soil exceeds the baseline level by more than 20%; (2) there are significant differences in the number, density, structure, community composition and other indicators of biological species in soil compared with the baseline; (3) the soil has undergone qualitative change and no longer has the original ecological function. If one of the

[62]Provisions of the Supreme People's Court on Trial of Cases of Compensation for Ecological Environment Damage (for Trial) (2019) could be found at: https://www.court.gov.cn/fabu-xiangqing-162322.html.

above conditions was met, it can be considered that the ecological environment of the site has been damaged. Who should actually undertake the remediation - the responsible person or a third party under the 'acting performance' system in the administrative relief path, should be determined by the actual situation.

For sites which are impossible to be restored, the ecological damage compensation litigation system is adopted to recover from the responsible person - that is, to claim compensation for the loss during the service function period and permanent loss. To achieve that objective, it is necessary to bring the compensation for eco-environmental damage into PCSPL, build the system of compensation for eco-environmental damage from the administrative law level, and clarify the scope of application, starting conditions, constituent elements, reasons for exemption, mode of undertaking, fund management, identification and evaluation, restoration scheme, scope of compensation, implementation procedures, implementation supervision effect evaluation, information disclosure, public participation and even compulsory insurance, repair fund and other specific operation modes. At the same time, a series of supporting mechanisms such as investigation and evidence collection, evaluation and identification of damage should be designed.

14.4.4 Strengthening the Supervisory Role of Administrative EPIL on Administrative Power

According to China's practical experience, it is an inevitable trend to build a soil pollution prevention and control system dominated by administrative power. At present, public interest litigation, as a thorough mechanism to safeguard the interests of the public, can protect the supervision right of the public in participating in the relief system of soil pollution damage. As mentioned in Sect. 14.2, administrative EPIL plays a supervisory role in the administrative power and can better assist the administrative agencies to undertake both supervision and law enforcement. Consequently, it is necessary to highlight the supervisory role of administrative EPIL on administrative power.

14.4.4.1 Reasons for Strengthening the Supervision of Administrative Power

From the current judicial adjudication of soil pollution public interest litigation, administrative EPIL has not received sufficient attention in soil pollution prevention and control, and has not given full play to the role of public interest litigation in urging administrative agencies to fulfill their responsibilities. With the establishment of the dominant position of administrative power in soil pollution prevention and control, the administrative power has been further expanded, so it is necessary to strengthen the supervision of the administrative power.

Firstly, research has showed that it is unrealistic to expect the political design of the administration to be completely accurate and in accordance with the law.[63] In addition, because administrative power covers a wide area and it has great influence and natural expansiveness, if not supervised and restrained appropriately, it is most likely to be abused. As a result, strengthening the supervision of administrative power has been the choice for countries under the rule of law.[64]

Secondly, negative 'campaign-style enforcement' is easy to slacken environmental supervision and enforcement. Due to the weak position of environmental interests, the weakness of environmental legislation and the limitations of the environmental management system, 'campaign-style enforcement' has a natural priority and inevitability.[65] One of the consequences of 'campaign-style law enforcement' is 'selective law enforcement', specifically the focus of law enforcement tends to be key regulatory units, enterprises and areas, while often ignoring small and medium-sized enterprises, individuals and non-essential regulatory areas. In fact, a number of soil contamination incidents have occurred in rural areas and have also resulted in more serious consequences of contamination damage. Therefore, soil pollution prevention and control should not be biased and should be strengthened with comprehensive supervision. Soil pollution prevention and control relies on strong government supervision, and once lacking supervision, it will directly put the environmental public interest at risk. At present, China still has the problems of insufficient environmental supervision and enforcement capacity,[66] which need to be supplemented by supervision by the public at large. In addition, restoration standards determined based on risk assessment and management methods may trigger subsequent risk control responsibilities that should be borne by those responsible for restoration in the long term or even permanently. In this regard, the administrative authorities should also follow up on their regular or irregular regulatory responsibilities on an ongoing basis.[67]

Thirdly, soil pollution prevention and control is highly technical, time-consuming, and costly, so administrative authorities may not act proactively due to considerations of costs and other factors. In practice, the cost of restoration, appraisal and evaluation are usually very costly and need to be paid by the administrative organ in advance, which is usually difficult for the administrative authorities to bear, and thus it is often not motivated to make administrative orders (such as ordering restoration, etc.).[68] Although administrative authorities can take the initiative to file lawsuits for ecological and environmental damages, given the cost concern, they can also be passive and wait for the people's procuratorates and eligible ENGOs to file the civil EPIL. Especially, in the case of small site contamination, the remediation

[63] Xi (2004), p. 81.
[64] Research group of Guangdong Provincial People's Procuratorate and Huang (2019), p. 66.
[65] He (2016), p. 26.
[66] Wang (2020a), p. 2.
[67] Wang (2016a), p. 57.
[68] Zhang (2021), p. 47.

costs may not be high, but the resulting appraisal and assessment fees may not be low, in which case the administrative agency may indeed abandon the administrative remedy route in the face of high upfront costs.

14.4.4.2 Improvement of Administrative EPIL Supervision Mechanism

Like many other environmental laws in China, the PCSPL still relies on criminal punishment and bureaucratic accountability to enforce the law.[69] Thus, strengthening the implementation of the responsibilities of administrative departments remains the focus. EPIL is not only a supplement to administrative power in soil pollution prevention and control, but also a kind of effective tool of supervising administrative power. At present, China mainly supervises administrative authorities to perform their duties in accordance with the law and act positively through administrative EPIL initiated by the people's procuratorates, but the supervision mechanism in force is not perfect, which still needs further optimization.

Firstly, the pre-litigation procedure deserves further improvement. According to China's EPIL system, the people's procuratorate has to fulfill its pre-litigation notification obligation before filing administrative EPIL,and urge the administrative authorities to correct illegal administrative acts or actively perform their duties by means of procuratorial suggestions. If the administrative authorities still fail to act, the people's procuratorate can directly file administrative EPIL with the people's court to realize the supervision demands through judicial orders. Specifically, the administrative authorities must perform their duties in accordance with the law within 2 months from the date of receiving the procuratorial suggestions and reply to the people's procuratorate in writing. In the event of emergent situations such as damage to national interests or social public interests continue to expand, the administrative authorities must respond in writing within 15 days. However, the absence of the important role of the 'government' in the 'Changzhou Toxic Land Case'[70] shows that China's pre-litigation procedure is not perfect and need further improvement.

---

[69] Yue (2018), p. 133.

[70] In 2011, the government of Xinbei District of Changzhou planned to use the plot involved in the case for commercial residential development, but found that the pollution of soil and groundwater in the plot involved in the case was serious, so Changzhou Institute of Environmental Sciences was entrusted to prepare the technical scheme for remediation of soil and groundwater in the contaminated site of the plot of the original site of Changlong (Huada, Changyu) Company for remediation. Later, the secondary air odor during the repair process affected the teachers and students of Changzhou foreign language school and other sensitive people around, causing serious health damage, and the repair project was completely stopped. After the incident, the Education Steering Committee of the State Council sent a special steering group, the former Ministry of environmental protection (MEP) and the Jiangsu Provincial People's Government jointly established an environmental investigation group, and the national health and Family Planning Commission and the Jiangsu Provincial Health and family planning commission established a medical and health expert group to carry out supervision, investigation, analysis and research in Changzhou. The investigation

Nevertheless, we can refer to the practice of the United States of America. In the U.S., 'citizen litigation' must go through two important pre-litigation procedures before being filed, namely 'notification' and 'diligent law enforcement'. The plaintiff must inform the polluter and government regulatory authorities within 60 days before the prosecution, giving relevant government departments the opportunity to deal with problems and make up for mistakes, so as to ensure that the administration takes the lead in the process of soil pollution control. In view of this, China should improve the pre-litigation notice procedure of administrative EPIL. That is, before filing administrative EPIL, the procuratorial authorities should first notify the administrative authorities to handle the case, and determine whether the administrative authorities have taken substantive measures to perform their duties, such as issuing administrative orders, imposing administrative penalties, organizing and implementing restoration, etc.; if the administrative departments are dilatory in handling or mishandling the case, and then can file the case. At the same time, the administrative authorities should also actively participate in the litigation process, especially in the implementation process after the adjudication. As the supervisor of soil pollution prevention and control in the region, the administrative authorities have a large amount of soil pollution information and need to cooperate with the court in the management of funds, assessment of remediation, acceptance of the project and subsequent management of the parcel of land involved in the case.

Secondly, allowing eligible ENGOs to bring administrative EPIL is necessary. At present, only the People's Procuratorates can initiate administrative EPIL in China. ENGOs can only file normal (non-EPIL) administrative lawsuits, based on their direct interests.[71] The supervision of administrative power by the people's procuratorate is the embodiment of its own functions, and if the supervision is limited only to the procuratorial authorities, the social supervision of administrative power will undoubtedly be neglected. In order to give full play to the supervisory role of social organizations, the eligibility of plaintiffs in administrative EPIL should be consistent with the relevant provisions of civil EPIL, i.e., social organizations meeting specific requirements should be allowed to file Administrative EPIL, rather than being limited to the procuratorial authorities.[72] Under this scenario, there is a consequent problem that eligible ENGOs and people's procuratorates can both file civil EPIL and administrative EPIL, thus creating the problem of subordinate articulation. As mentioned above, after the occurrence of soil pollution incidents,

team believes that the campus environment of Changzhou Foreign Language School is safe, but there are problems with the construction and supervision of the restoration work of the land involved in the case, requiring the Changzhou Municipal People's Government to urgently develop and implement a pollution prevention and control and restoration program for the land involved in the case. See Higher People's Court of Jiangsu Province, Su Civil trial No. 232 (2017).

[71] According to Article 25 of the Administrative Litigation Law of the PRC as mended in 2017, a plaintiff other than a procuratorate must show a direct interest in the administrative action at issue. As a result, only procuratorates can initiate Administrative EPIL. See Richard and Benoit (2017), pp. 209–210.

[72] Wang (2016b), p. 66.

it is necessary to comply with the principle of 'priority of administrative power'. That is, first file administrative EPIL and urge the administrative authorities to perform their responsibilities, and then file civil EPIL for accountability and supplement when the administrative authorities do not perform their responsibilities in place. We believe that there is no need to qualify the subordination between ENGOs and people's procuratorates in filing administrative EPIL, so as not to dampen the enthusiasm of ENGOs in filing public interest lawsuits. With the continuous improvement of information sharing mechanism and the increasing awareness of cooperation, ENGOs and people's procuratorates can also strengthen prior information communication, leaving the option to the related subjects to prosecute according to the specific circumstances of the case may be much better.

14.5 Conclusion

By reviewing the common institutions and practice in soil pollution prevention and control, this chapter analyzes the functions of EPIL especially before the enactment of China's soil pollution law. Despite of its vital importance in imposing liability of the potential polluters, several limitations have still been found pertaining to EPIL in this arena. Firstly, the preventive function of civil EPIL has not been fully utilized and cannot meet the requirements of soil pollution control at source. Secondly, in the whole process of soil pollution prevention and control, there are some defects in civil EPIL, such as lack of professionalism, low governance efficiency and weak implementation, which hinders its potential leading role in afterwards soil contamination.

With the enactment of China's PCSPL, administrative authorities are endowed with more powers and responsibilities, highlighting the dominant position of administrative power. It's closely related to the inherent advantages of administrative execution of law in technological expertise and an abundant regulatory toolbox. This also leaves room for the development of civil and administrative EPIL. Therefore, in order to better fit into the system, EPIL should be reshaped by emphasizing more the preventive nature of environmental public interest litigation and by strengthening the supervision of administrative power through administrative EPIL.

References

Cha Y, Yang F (2022) 'Yunnan Green Peacock Case': the first case of preventive environmental civil public interest litigation for the protection of rare Wild Fauna and Flora in China. Journal of People's Court Newspaper 2022-03-07 (003). https://doi.org/10.28650/n.cnki.nrmfy.2022.000758

China's Administrative Litigation Law. Available at: https://flk.npc.gov.cn/detail2.html?MmM5MDlmZGQ2NzhiZjE3OTAxNjc4YmY4NTg1NTBhMGY%3D

Dong B (2019) Optimization path of the applicable rules of preventive responsibility in environmental civil public interest litigation. J China Environ Manag 11(5):121–126. https://doi.org/10.16868/j.cnki.1674-6252.2019.05.121
Dong H, Wang Y (2022) Study on the legislation of prevention and control of cultivated land soil pollution in China. J Northwest A&F Univ (Soc Sci Edn) 22(2):114–119. https://doi.org/10.13968/j.cnki.1009-9107.2022.02.13
Fourth Medium People's Court of Beijing City, Jing 04 Civil trial No.89 (2016)
Gong G (2018) Soil remediation from the perspective of public law responsibility--analysis based on Soil Pollution Control Act. J Legal Sci 2018(10):52–64
He X (2016) Cracking the dilemma of deterrence-based environmental enforcement in China--analysis based on concept and mechanism. J Fashang Yanjiu 33(4):24–34. https://doi.org/10.16390/j.cnki.issn1672-0393.2016.04.004
Higher People's Court of Jiangsu Province, Su Civil trial No.232 (2017)
Hu J, Cui M (2019) Feasibility study on responsibility performance of ecological environment restoration under binary litigation mode. J China Univ Geosc (Soc Sci Edn) 19(6):13–28. https://doi.org/10.16493/j.cnki.42-1627/c.2019.06.002
Jin Z (2021) Identification and assessment of environmental damage as scientific evidence--investigation based on environmental justice cases. J Law Rev 39(5):150–163. https://doi.org/10.13415/j.cnki.fxpl.2021.05.014
Juan C (2018) Vindicating public environmental interest: defining the role of environmental public interest litigation in China. Ecol Law Quart 45(3):485–532. https://doi.org/10.15779/Z38599Z22F
Li M, Wu P (2021) On the plaintiff qualification of environmental public interest litigation of grass-roots autonomous organization. J Environ Protect 49(21):53–58. https://doi.org/10.14026/j.cnki.0253-9705.2021.21.013
Li T et al (2019) Soil pollution management in China: a brief introduction. Sustainability 11(3):556
Li Z (2020) Study on the mechanism of administrative order type ecological environment restoration. J Law Rev 38(3):184–196. https://doi.org/10.13415/j.cnki.fxpl.2020.03.015
Liu F (2021a) The reconstruction of China's environmental confiscation of illegal income system. J China Univ Geosci (Soc Sci Edn) 21(6):38–49. https://doi.org/10.16493/j.cnki.42-1627/c.2021.06.005
Liu J (2012) Environmental justice with Chinese characteristics: recent developments in using environmental public interest litigation to strengthen access to environmental justice. Florida Univ Law Rev 7(2):229–260
Liu M (2021b) The rationale and approach of preventive environmental public interest litigation: based on normative interpretation of environmental risk. J Nanjing Univ Technol (Soc Sci Edn) 20(2):37–48
Lv Z, Dou H (2017) Responsibility of 'restoration of ecological environment': an empirical analysis 39(3):125–142
Malte K (2006) German soil protection law. J Eur Environ Plan Law 3(3):250–257
Medium People's Court of Liaocheng City, Lu 15 Civil trial No.486 (2018)
Medium People's Court of Xian City, Shan 71 Civil trial No.4 (2017)
Medium People's Court of Xuzhou City, Su 03 Civil trial No.338 (2019)
Medium People's Court of Zunyi City, Qian 03 Civil trial No.520 (2016)
Ministry of Environmental Protection, Ministry of Land Resources, Report on National Land Contamination Investigation Survey, 2014., Available at: http://www.gov.cn/foot/site1/20140417/782bcb88840814ba158d01.pdf. Accessed 8 April 2022
Research Group of Guangdong Provincial People's Procuratorate and Wenai Huang (2019) Procuratorial supervision of administrative violations under the framework of public interest litigation system. J Fazhi Shehui 2019(2):62–71. https://doi.org/10.19350/j.cnki.fzsh.2019.02.007
Richard ZQ, Benoit M (2017) Public interest environmental litigation in China. Chinese J Environ Law 1(2):202–228. https://doi.org/10.1163/24686042-12340013

Shen Y (2021) Basic principle of soil pollution remediation and soil ecological protection strategy. J Res Conserv Environ Protect 2021(12):33–35. https://doi.org/10.16317/j.cnki.12-1377/x.2021.12.011
Song B (2021) Study on the identification of major risks in preventive environmental public interest litigation. J Hebei Radio TV Univ 26(04):67–72. https://doi.org/10.13559/j.cnki.hbgd.2021.04.013
State Council (2016) Action plan on prevention and control of soil pollution. http://wwwgovcn/zhengce/content/2016-05/31/content_5078377.htm. Accessed 2 Apr 2022
Sun H (2021) Public interest litigation, capacity barriers and relative rationality--Based on the legal thinking of the second instance judgment of Toxic Land Case in C city of a province. J Orient Law 2021(2):168–177. https://doi.org/10.19404/j.cnki.dffx.2021.02.013
Sun Q, Jack T (2017) An exploration of and reflection on China's system of environmental public interest litigation. Environ Law Rep News Anal 47(6):10497–10510
Supreme People's Court (2020) Decision of the Supreme People's Court on Amending the Provisions of the Supreme People's Court on Several Issues Concerning the Civil Mediation Work of the People's Courts and Nineteen Other Judicial Interpretations of Civil Litigation. https://www.court.gov.cn/fabu-xiangqing-282651.html. Accessed 6 April 2022
Tang H (2019) Value interpretation and realization of environmental public interest litigation in risk society from the perspective of preventive judicial relief. J SJTU (Philos Soc Sci) 27(03):29–37. https://doi.org/10.13806/j.cnki.issn1008-7095.2019.03.003
The General Office of the CPC Central Committee and the General Office of the State Council (2017) CCP Central Committee and State Council Reform Plan on the Compensation Regime for Eco-environmental Damage. http://wwwgovcn/zhengce/2017-12/17/content_5247952.htm. Accessed 12 April 2022
The SPC and SPP Interpretation on the Application of Laws in Procuratorial Public Interest Litigation. Available at: https://www.spp.gov.cn/zdgz/201803/t20180302_368570.shtml
The Supreme Peoples' Court (SPC) Interpretation on Application of Law on Environmental Civil Public Interest Litigation. Available at: https://www.court.gov.cn/fabu-xiangqing-13025.html
Wang B (2020a) The situation of soil pollution is still grim. J People's Daily, 2020-10-16(006). https://doi.org/10.28655/n.cnki.nrmrb.2020.010301
Wang H (2016a) Preliminary study on standard system of contaminated soil remediation. J Fashang Yanjiu 33(3):54–62. https://doi.org/10.16390/j.cnki.issn1672-0393.2016.03.009
Wang H (2020b) Study on the responsibility of soil pollution control. Fudan University Press, pp 176–193
Wang J (2018a) Environmental law, 4th edn. Peking University Press, pp 298–299
Wang L (2018b) Study on the eco-environmental damage relief mechanism. J Soc Sci 2018(6):104–111. https://doi.org/10.13644/j.cnki.cn31-1112.2018.06.010
Wang M (2016b) Study on development direction of environmental public interest litigation in China: analysis based on the theory of the relationship between administrative power and judicial power. J China Legal Sci 2016(1):49–68. https://doi.org/10.14111/j.cnki.zgfx.2016.01.003
Xi L (2004) Expansion and supervision of administrative power. J Theor Explor 2004(02):80–81
Xie J, Li F (2010) Overview of the current situation on brownfield remediation and redevelopment in China. World Bank. Available at: https://documents.worldbank.org/en/publication/documents-reports/documentdetail/450251468024319815/overview-of-the-current-situation-on-brownfield-remediation-and-redevelopment-in-China, at p. 3. Accessed 8 April 2022
Xu Y, Zhou X (2017) On the purpose of environmental civil public interest litigation and the interpretation application. China Popul Resour Environ 12:97–105. https://doi.org/10.12062/cpre.20170523
Yu W, Mu T (2019) Study on judicial determination of 'significant risk' in environmental civil litigation. J Law Appl 2019(14):25–32
Yue H (2018) Chinese superfund law: China's new soil pollution law an analysis. Int Chem Regul Law Rev 2018(3):131–135. https://doi.org/10.21552/icrl/2018/3/9

Zhang B (2021) Differentiation and integration of environmental public interest protection mechanisms in China. J Soc Sci Hunan Normal Univ 50(02):41–50. https://doi.org/10.19503/j.cnki.1000-2529.2021.02.004

Zhang Y, Wu A (2020) Judicial determination of 'significant risk' in precautionary environmental public interest civil litigation. J China Environ Manag 12(2):138–144. https://doi.org/10.16868/j.cnki.1674-6252.2020.02.138

Zhou Y (2020) Division and integration of compensation litigation for ecological environment damage and environmental public interest litigation: a perspective of functionalism. J Soc Sci Hunan Normal Univ 49(05):47–54. https://doi.org/10.19503/j.cnki.1000-2529.2020.05.006

Zhu X (2018) Explaining the law with environmental public interest litigation cases. People's University of China Press, pp 7–123

Zou Y (2022) Defining the punitiveness of administrative punishment. J Adm Law Rev 2022(02):44–55

Part III
Reform the EPIL System in China: Problems and Solutions

Chapter 15
The Erroneous Setting of the Essential Attribute of Environmental Public Interest Liability in China

Shuai Xin

Abstract The mainstream view of the law academia and several important pieces of legislation in China mistakenly set the essential attribute of environmental public interest liability as civil liability. But its correct, implicit, and pre-existing essential attribute should be administrative liability. The erroneous setting of the essential attribute of environmental public interest liability as civil legal liability has led to a series of jurisprudential and legal institutional consequences including the conflicts of two types of environmental public interest liabilities with different attributes, the conflicts between administrative enforcement procedures and civil procedures of realizing environmental public interest liability, the conflicts between social organizations' public authority and administrative agencies' public authority. The causes of this erroneous setting include the impact of the theory of civil remedies for environmental public interest damage and the theory of crystallization of environmental rights as civil rights, and conceptual misunderstandings in the transplantation of environmental legal regimes from a common law system. Amendments in legislation should be made to change this erroneous setting of the essential attribute of environmental public interest liability to administrative liability.

Keywords Environmental public interest liability · Environmental administrative liability · Environmental public interest civil liability · Ecological and environmental tortious liability · EPIL · Civil EPIL · Environmental law enforcement · Environmental enforcement litigation · Compensation regime for eco-environmental damage · China

---

S. Xin (✉)
Environmental Law, Law School of Qingdao University of Science and Technology, Qingdao, Shandong, China
e-mail: xinshuai@qust.edu.cn

X. Wang et al. (eds.), *Environmental Public Interest Litigation in China*,
https://doi.org/10.1007/978-3-031-26526-6_15

15.1 Introduction

Environmental public interest liability refers to the legal liability that should be borne for damage to the environmental public interest, which is pursued in environmental public interest litigation and other environmental public interest damage relief regimes. Today, the mainstream view of the law academia and some important legislation in China have suggested that the essential attribute of environmental public interest liability is civil liability (hereinafter referred to as civil liability). However, such an essential attribute setting is wrong. The correct essential attribute of environmental public interest liability should be administrative legal liability (hereinafter referred to as administrative liability).

Since environmental public interest liability is the core concept of an environmental public interest damage relief system, this erroneous setting has led to many consequences in theory and to regimes which are not yet recognized, discussed, and resolved. Therefore, this issue should be an important topic in current research on environmental public interest protection theory and regimes. This chapter is an effort in this direction. First, this chapter will describe the current erroneous essential attribute setting of environmental public interest liability in theory and legislation. Second, it will establish that the essential attribute of environmental public interest liability should be administrative liability. Third, it will analyze in detail the consequences in theory and environmental legal regimes caused by this erroneous setting. Fourth, it will analyze the causes giving rise to this erroneous setting. Finally, suggestions are made to amend the legislation to correct the essential attribute setting of environmental public interest liability.

15.2 The Current Erroneous Essential Attribute Setting in Theory and Legislation

It has become a stereotype in theory and legislation that the essential attribute of environmental public interest liability is civil liability. Civil liability has become the explicit attribute of environmental public interest liability.

First, the mainstream view in China's law academia insisted that environmental public interest liability is a new type of civil liability. As Professor Jin Ruilin mentioned, 'Traditional torts are generally harmful acts that directly act on the property and individual victim's body, which is relatively simple and direct; environmental torts are acting on people, public and private property as a direct consequence of an environmental damage or the threat of such damage. The objects of such infringement include various natural elements, public and private property, and persons, which are far more extensive and complex than ordinary infringements.'[1] According to this understanding, damage to a natural element, which is damage to

[1] Jin (1997), p. 194.

an environmental public interest, will lead to environmental tortious liability. Similarly, Professor Cao Mingde believes that the scope of environmental torts includes environmental pollution and ecological damage.[2] This means that an environmental public interest liability caused by environmental and ecological damage is included in the scope of an overall tortious liability.

Second, influenced by this mainstream theoretical view, environmental public interest liability was subsequently set as civil liability (environmental tortious liability) in legislation. Article 64 of the 2014 revision of the Environmental Protection Law (EPL) provides: 'In the case of damages caused by environmental pollution and ecological destruction, tortious liability shall be borne pursuant to the relevant provisions of the Tort Law of the People's Republic of China.'[3] Later, article 1229 of the PRC Civil Code also clearly stipulates: 'In the event of damage caused to others as a result of environmental pollution and ecological destruction, the tortfeasor shall bear tortious liability.'[4] Neither of those two articles uses the concept of 'environmental public interest', but they implicitly include this concept in ecological and environmental damage, since environmental and ecological damage includes public and private interests. Because environmental and ecological liability is defined by those two articles as environmental tortious liability, an environmental public interest liability is certainly defined as an environmental tortious liability as well.

15.3 The Correct Essential Attribute of Environmental Public Interest Liability

The correct and implicit essential attribute of environmental public interest liability should be administrative liability. The evidence is as follows.

15.3.1 Environmental Public Interest Protection Obligation Is Administrative Obligation

My first argument is that the attribute of legal obligation corresponding to environmental public interest liability is an administrative legal obligation. Environmental public interest obligation is established to protect an environmental public interest. Here, the concept of 'environmental public interest' shall be clarified. In some other backgrounds and texts, environmental public interest could be used in the sense that many individuals' personal interests are involved, just like class actions raised by

---

[2] Cao (2000), p. 18.

[3] Article 64 Environmental Protection Law (1989, revised 2014).

[4] Article 1229 Civil Code (2021).

numerous victims of industrial pollution could be regarded as environmental public interest cases. However, such a type of 'environmental public interest' is a collection of private interests affected by environmental harm and not the type of 'environmental public interest' discussed in this chapter. In this chapter, an environmental public interest (or to say pure environmental public interest) is the environmental interest owned by all the people together in public law. It is not a composition or collection of private interests related to environmental harm. One type of it cannot be divided and distinguished as specific private interests but can only be owned by all citizens collectively. For example, it is not only meaningless but also impossible to divide oxygen in the air of this country into specific private shares or to divide the value of the diversity of species on endangered species into specific private interests. The other type of it is owned by all the people together as well. However, there are some differences. It could be partially divided and privatized as personal interest in certain situations. Mine resources supply a good example. When those natural resources have been taken as personal property through mining permits and activities, they are no longer in the public interest. However, the rest of the mine is still a public interest because it is still owned by all the people.

Since the environmental public interest is owned by all the people or citizens, environmental public interest management and protection cannot be conducted by civil mechanisms (such as civil co-ownership). Given that the environmental public interest is owned by so many people in one jurisdiction or country, its management and protection must be conducted by decision-making mechanisms based on representative democracy in an attempt to crystallize the general will (volonté générale). Legislation is a means by which the general will can be expressed and legislated by laws and regulations.

Among all the laws made for environmental public interest management and protection, administrative law is the most important. Almost all important matters concerning daily management and protection of the environmental public interest are provided for in environmental administrative law. Environmental administrative law delineates the protection level of the environmental public interest, grants an administrative agency management and protection authority, and establishes environmental public interest protection obligations.

Environmental public interest protection obligations stipulated in environmental administrative law are not established per se for the private benefit of private parties, but for public welfare in one country. Environmental administrative agencies holding public trust have the authority to enforce such obligations. Parties performing or violating such obligations are in the administrative legal relationship with administrative agencies, not in a civil legal relationship with other equal private parties. Thus, the environmental public interest protection obligation is an administrative legal obligation. Environmental public interest liability is the adverse legal consequence of violating environmental public interest protection obligations and as such may give rise to administrative liability.

15.3.2 Fault as a Constituent Element of Environmental Public Interest Liability

The second argument is that fault (breach of duty of care) is a necessary constituent element of both environmental public interest liability and administrative liability. Here, fault means violation of a legal obligation. The judgment of fault cannot be separated from the objective judgment of actors' external behavior. By comparing an actors' external behavior with the actors' legal obligations, it can be judged whether such actors are at fault. In law enforcement and judicial decisions, whether there is a violation of legal obligations has become the criterion for judging whether there is fault. No breach of legal obligation means no fault. A breach of a legal obligation constitutes a fault.

Administrative liability adopts a fault liability principle and takes fault as a necessary constituent element. 'No penalty without law making it so' is a basic principle of administrative law. This principle requires that the establishment of administrative liabilities requires the premise and basis of clear legal obligations set by legal norms. If an administrative counterpart[5] does not violate the administrative legal obligations provided by law, he does not need bear administrative legal liabilities. That means that a party without fault does not have to bear liabilities. The fault should be a basic constituent element of an administrative liability.

Similar to administrative liability, environmental public interest liability shall have fault as its constituent element as well; thus, the beach of environmental public interest protection obligation is the basic requirement for environmental public interest liability to be proven. The reason behind this statement is that environmental public interest protection is a fixed concept with limits in public law. Environmental public law does not safeguard the objective totality and integrity of the environment but the bottom line of environmental protection delineated by environmental public decisions. From a practical point of view, as long as human society exists, environmental exploiting and damaging activity will not cease. Existing historical experience shows that the degree of human damage to the environment continues to deepen with social and economic development. Based on a multi-value balance, environmental protection policies and laws overall can only limit and mitigate the damage to the environmental public interest as an actual goal, rather than achieving zero damage to the environmental public interest. As Professor Lazarus, R. J. has put it, 'Environmental protection policy has been almost exclusively concerned with two basic issues during the last several decades: (1) what is an acceptable level of pollution; and (2) what kinds of legal rules would be best suited for reducing pollution to that level.'[6]

---

[5]Administrative counterpart refers to the opposite party to the administrative agencies in administrative management relationship, that is, the individual and organization whose rights and interests are affected by the administrative agency's decision and activities.

[6]Lazarus (1993), p. 787.

Since an environmental public interest protection level is a fixed concept provided in public law, public law defines environmental damage on the basis of those fixed levels. The environmental public interest damage that does not exceed those levels will not be recognized as legal damage by environmental public law, and only those that exceeded the statutory limits set by environmental administrative law will be recognized by public law as illegal damage. Even if the former is still damage to environmental quality in a scientific sense, it is no longer an illegal environmental public interest damage in an environmental public policy. In other words, in the field of environmental public interest protection, environmental public law allows legitimate environmental public interest damage, but prohibits and corrects illegal environmental damage beyond a certain point.

Since public law will not define environmental damage which does not exceed acceptable levels as illegal environmental public interest damage, it will spare those parties causing such damage and will not establish liabilities for them. If only environmental public law is abided by and no environmental public interest protection obligations are violated, there should not be any liability for the environmental public interest established.

In this respect, environmental public interest liability is quite different from and contradictory to traditional environmental tortious liability, in which the no-fault liability principle is applied and fault is not a necessary constituent element overall. The no-fault liability principle is used in environmental tortious liability to protect victims' private civil interests damaged by environmental pollution, but not environmental public interest. By balancing the interests of polluting enterprises and environmental victims under the uncertainty of environmental risks, the application of the no-fault liability principle will realize better environmental justice than by applying a fault liability principle. Out of the complexity of environmental science, most duties of care borne by discharging parties determined in legislation mainly come from public law norms, which are based on valuation and balance of environmental harm level and a societal development requirement level. Those duties of care do not guarantee that the discharging activity is totally safe for individual health and their property. When private interests are damaged by pollution, polluters can defend their innocence by proving that they have exercised due diligence by abiding by statutory duties of care and the fact that they cannot predict the consequence of damage to private interests beyond their statutory obligations. This challenges the justifiability of requiring fault as a necessary constituent element in environmental tort cases. The introducer of the environmental risks leaves risks to individuals whose personal health and property are unproportionally threatened by those environmental activities.

For example, in the Minamata disease case in Japan,[7] environmental legislation, administrative agencies and polluting companies did not foresee one chemical compound discharged from the factory entering the human body after undergoing complex and long-term biochemical and social processes. After the onset of

[7] Hiroki Kawamura (2018), pp. 9–11.

Minamata disease, it is no longer possible to define the legal discharging of pollutants many years ago as illegal by the latest standards, and if continuing to use fault as a constituent element of tortious liability, those victims will not be able to claim compensation. Therefore, to protect a victim's personal and property rights in such special circumstances, fault becomes an unnecessary element for environmental tortious liability. By doing so, the interest and cost balance between the introducers of the environmental risks and the victims could be made.

Thus, fault is a basic constituent element of environmental public interest liability as administrative liability but not a necessary element for environmental tortious liability, though environmental public interest liability and environmental tortious liability are so closely related in conceptual appearance. This argument is even supported by provisions on public interest liability in the Civil Code, although this liability is incorrectly included in the environmental tortious liability category. In the Civil Code, environmental public interest liability is provided in Chapter 7 Liability for Environmental Pollution and Ecological Damage of Part VII Liability for Tort. The Civil Code stipulates two liability attribution principles in Chapter 7. Article 1229 of the Civil Code provides: 'In the event of damage caused to others as a result of environmental pollution and ecological destruction, the tortfeasor shall bear tortious liability.'[8] This article shows that the legislature applies the no-fault liability principle to environmental tortious liability overall.[9] However, a fault liability principle is stipulated for environmental public interest liability as an exception to no-fault liability under the Civil Code. Article 1234 provides: 'Where damage to the ecological environment is caused in violation of State regulations and the ecological environment is able to be restored, the authorities stipulated by the State, or the organizations stipulated by the law shall have the right to request the tortfeasor to bear the liability for restoration...'. In addition, article 1235 provides: 'In the event of damage caused to the ecological environment in violation of State regulations, authorities stipulated by the State or organizations specified by the law shall have the right to request the tortfeasor to compensate for the following losses and costs...' Both articles emphasize that liability for damaging the ecological environment shall be established when damage is caused by actions which were in violation of State regulations. That means, if damage to the ecological environment is caused without violation of state regulations, there shall not be liability for that damage to be established. Obviously, the Civil Code distinguishes damage to the ecological environment from other types of environmental torts and applies a fault liability principle to environmental public interest liability. It can be concluded that the Civil Code stipulates a fault liability principle for environmental public interest liability and stipulates no-fault liability for most of the remaining types of environmental torts. It is incorrect for the Civil Code to put environmental public interest liability into the environmental tortious liability category and Part VII Liability for Tort.

[8] Article 1229 Civil Code (2021).

[9] Huang (2020), pp. 242–243.

However, it is correct for the Civil Code not to assign the no-fault liability principle to environmental public interest liability.

15.3.3 Environmental Public Interest Liability Is Pursued by Administrative Agencies

As mentioned above, the protection of environmental public interest needs to rely on public decision-making mechanisms under representative democracy. Under representative democracy, government agencies are authorized by the people to engage in environmental public interest decision-making. Among them, the legislature has the highest authorization. The level of authorization obtained by an administrative agency is lower than that of the legislature, and it can make abstract entrusted legislative decisions (regulation enactment) and specific administrative law enforcement decisions according to the authorization of the legislature. In fact, the government agencies closest to the daily management and protection of the environmental public interest are administrative agencies. After the occurrence of environmental public interest damage, the relief of environmental public interest should belong to the natural content of the administrative law enforcement authority and jurisdiction, which is authorization of the people granted to administrative agencies under representative democracy. Thus, administrative agencies should be the parties having the jurisdiction to enforce environmental public interest liabilities. This is another argument to support that the essential attribute of environmental public interest liability should be administrative liability.

15.3.4 Environmental Public Interest Liability Has the Same Burden of Proof as Administrative Liability

The fourth argument is that environmental public interest liability should have the same burden of proof as administrative liability. In administrative enforcement procedures, the burden of proof of the establishment of administrative liability is assigned to administrative agencies. Administrative agencies need to prove that an administrative counterparty has engaged in certain acts and violated provisions of relevant laws; in case of damage to the public interest rather than just the destruction of public order, it must also prove that there is a causal link between the act of the administrative counterparty and the damage caused. For example, in the case of serious damage to a downstream wetland environment caused by an upstream enterprise discharging excessive pollutants permitted by its environmental licence, the burden of proof for proving this enterprise's administrative liability should be on the administrative agencies, who must collect evidence of the polluting activity, the downstream environmental damage, the causal relationship between the excessive discharge of pollutants and the environmental damage of the downstream wetland, and the enterprise's fault (violation of environmental statutes and regulations).

For administrative liability enforcement, no matter what kind of procedure is applied and how such a procedure is described, civil or administrative or even another new concept, such a procedure is in essence a public law enforcement procedure. In such a procedure, public law is enforced, public authority and power are used, administrative agencies are involved as leading entities, environmental public interest liability is assumed by administrative counterparties against the country and its administrative agencies, and the status imbalance between administrative entities and potential liable parties exists. Thus, such a procedure should serve the value of public law to regulate power and protect private rights. In short, the burden of proof to establish environmental public interest liability shall be assigned to administrative agencies.

The reverse burden of proof for environmental tortious liability should not be applied for environmental public interest liability. In environmental tort law, after the prima facie evidence on causation between defendants' pollution and private interest damage is proposed by the plaintiff, it is up to defendants to disprove it. The reverse burden of proof is used to reduce the total social cost of a litigation proof caused by the asymmetry of litigating strength and information between environmental pollution defendants, given the fact that most enterprises own more capital and social resources and more information about their polluting activities than victims. However, for environmental public interest liability, a reverse burden of proof is not applied. Administrative agencies have institutional and technical strength in relevant procedures no less than polluting enterprises, and they even have the public power to collect evidence from polluting enterprises. There is no rationale to apply the reverse burden of proof for environmental public interest liability. If this happened, a potential liable party would bear a very strict burden of 'self-incrimination' to disprove his liability, meaning that he will lose the special procedural protection he deserves in public law.

The Civil Code does not clearly provide whether the reverse of the burden of proof shall be applied to environmental public interest civil liability in the Civil Code or not. However, in the 'Interpretation of Liability for Tort Part in the Civil Code of the People's Republic of China' written by the Civil Law Office in the Legislative Affairs Commission of the Standing Committee of the National People's Congress, which expounds the legislative background and reasons for designing Part VII Liability for Tort in the Civil Code, the legislature's opinion can be found that the reverse of the burden of proof shall only be applied to traditional environmental tortious liability, but not to ecological and environmental tortious liability (environmental public interest civil liability).[10] Similar to the fault issue, the legislature's

[10]Huang (2020), p. 202. Ecological and environmental tortious liability and environmental public interest civil liability are same concepts in environmental legislation of China. However, in Civil Code, the terminology of ecological and environmental tortious liability is more often used to refer to environmental public interest civil liability.

understanding of not assigning a reverse burden of proof to environmental public interest liability is correct, while it is incorrect to put environmental public interest liability into an environmental tortious liability category, and there is still ambivalence in real provisions of the relevant articles about whether a reverse burden of proof should be applied to ecological and environmental tortious liability(environmental public interest civil liability), which leaves a potential and controversial space in practice.

15.3.5 Administrative Liability Connotates the Concept of Environmental Public Interest Liability

In fact, the essential attribute of environmental public interest liability should not only be administrative liability but also has long been connotated in the conceptual scope of administrative liability. It is easy to overlook the fact that administrative law itself is a law of public interest and public order management and protection. The legislature crystalizes the general will for the management and protection of public interest and public order into public law norms, of which the largest number and the highest importance are administrative law norms. Administrative legal norms set public interest protection obligations for actors to protect the public interest and maintain public order, that is, administrative legal obligations. Administrative liability is the consequence of the breach of such an obligation. Thus, an administrative liability is the legal liability with respect to the public interest and public order. Specific to the field of environmental public interest protection, environmental administrative liability is the legal consequence of the relevant parties violating obligations to protect the environmental public interest and to maintain the environmental public management order. Environmental administrative liability itself is an environmental public interest liability, the name and expression of which can be used interchangeably. The erroneous definition of environmental public interest liability as civil liability clearly ignores the homogeneity between public interest liability and administrative liability.

15.4 The Consequences of the Erroneous Essential Attribute Setting

From the above discussion, it can be observed that the essential attribute of environmental public interest liability should be administrative liability. Even though environmental public interest liability has so many differences from traditional environmental tortious liability, it is stipulated by the Civil Code as a special type of environmental tortious liability, namely ecological and environmental tortious liability. However, at the same time, the Civil Code recognizes that there are several important differences between traditional environmental tortious liability and environmental public interest liability and crystallizes those differences by differentiated provisions. Should this legislative arrangement be an acceptable compromise for environmental public interest liability? The answer is no. Realizing environmental public interest liability that belongs to an administrative liability under the essential attribute of civil liability will inevitably lead to several serious consequences in theory and environmental legal regimes. This section analyzes these consequences.

15.4.1 The Conflicts Between Two Types of Environmental Public Interest Liabilities

As discussed above, environmental administrative liability subsumes environmental public interest liability. In an administrative law system, environmental public interest liability has already been manifested as and included in environmental administrative liability. After the Civil Code stipulated ecological and environmental tortious liability, environmental public interest liability had a second manifestation in the civil law system. Currently, environmental public interest liability has two kinds of essential attribute settings, namely, administrative liability and civil liability.

This setting is quite contrary to the role of civil liability in the whole liability system structure in China. Within this system, all liabilities can be first classified into public law liability and private law liability branches. Within the public law liability branch, there are constitutional liability, administrative liability, and criminal liability. Within the private law liability system branch, there is mainly civil liability. The roles of those different liabilities are distinguished from each other by the interests they protect. Administrative liability protects the public interest, and civil liability protects the private interest (and collective private interests). This means that the role of civil liability in environmental cases should be to redress private interest damage, but not public interest damage. By stipulating environmental public interest liability as civil liability, the civil liability system is burdened with a new task to redress public interest damage, which is contrary to its role in the whole legal liability system.

The double essential attribute setting leads to overlaps between the roles of those two types of environmental public interest liability, namely 'old' environmental administrative liability and 'new' ecological and environmental tortious liability. They have the same environmental public interest as their target interest to protect, and they play the same role.

With regard to remedial relief, the Civil Code provides two forms of remedial environmental public interest liability. One is the 'liability for restoration within a reasonable period' provided in Article 1234, and the other is the compensation liability for losses and costs provided in Article 1234 and Article 1235. However, the same forms of remedial liabilities exist in environmental administrative law prior to relevant provisions in the Civil Code. Article 12 of the Measures on Environmental Administrative Sanction provides that environmental agencies can issue an order against a party in a prescribed period, as a response to environmental pollution problems.[11] The form of liability in such an order has the same function as 'liability for restoration within a reasonable period' in the Civil Code. Article 50 of the Administrative Compulsion Law provides that: 'Where after an administrative agency lawfully makes an administrative decision requiring the party concerned to perform obligation such as removal of obstacles or restoration to the status quo ante, the party concerned fails to perform the obligation within the prescribed period and still fails to do so despite being reminded, the administrative agency may perform the obligation on behalf of the party concerned or entrust a third party without interest to perform the obligation on behalf of the party concerned ("acting performance") provided that such a failure of the party concerned has endangered or is likely to endanger traffic security or has caused or is likely to cause environmental pollution or natural resource damage.'[12] Following this, Article 51 provides: 'The expenses arising from acting instead of the counterparty shall be determined reasonably at cost and borne by the party concerned, unless otherwise stipulated by law.'[13] The form of liability in this Article is the compensation liability for costs of restoring the environment or a natural resource, which has the very same function as the compensation liability for costs in Article 1234 of the Civil Code, which provides 'Where damage to the ecological environment is caused in violation of State regulations and the ecological environment is able to be restored, the authorities stipulated by the State or the organizations stipulated by the law shall have the right to request the tortfeasor to bear the liability for restoration within a reasonable period. Where the tortfeasor fails to carry out the repair within the time limit, the authorities stipulated by the State or organizations as prescribed by the law may carry out the repair on their own or entrust others with the repair, and the expenses required shall be borne by the tortfeasor.'[14]

With regard to punitive liability, there is overlap between those two types of environmental public interest liabilities as well. Article 1232 of the Civil Code provides: 'Where a tortfeasor, in violation of laws and regulations, intentionally pollutes the environment or damages the ecology, thereby causing serious consequences, the infringed is entitled to claim appropriate punitive damages.'[15] The punitive compensation liability form in this article can be used for environmental public interest liability as well. However, in administrative liability systems, environmental administrative agencies can make an administrative fine decision. Environmental administrative fines are designed to punish liable parties for damaging the

[11] Article 12, Measures on Environmental Administrative Sanction (2010).

[12] Article 50, Administrative Compulsion Law (2011).

[13] Article 51, Administrative Compulsion Law (2011).

[14] Article 1234, Civil Code (2021).

[15] Article 1232, Civil Code (2021).

environmental public interest and public management order. This liability form is quite the same as the one in Article 1232 of the Civil Code.

It might be argued that the double essential attribute setting and the overlap of their roles might serve as a double security for environmental public interest to be restored. When one fails, the other one will be there to do the job. But this is not the case. The double essential attribute setting and the overlap between them is not designed, expected, and even recognized by the legislature. There are no reconciliation mechanisms to adjust their relationship in practice. If in the same environmental public interest damage case, two sets of procedures are initiated, administrative liability will arise under administrative law enforcement procedures, and civil liability will arise under environmental public interest civil relief procedures. The mutual exclusion and isolation of the attributes of the two legal liabilities will hinder administrative law enforcement agencies and courts facing relevant cases from examining, considering, and realizing the correspondence between the total amount of environmental public interest liability and the total amount of damage in a unified liability system. This makes the correspondence between the total amount of environmental liability and the total amount of damage difficult to effectively achieve in practice. For example, for the same environmental public interest case, if administrative liabilities have been established in administrative enforcement procedures, judges in a civil court hearing the same case might not realize that remedial or punitive compensation liabilities in their judicial decisions have the same essence and function as the previous administrative liability. This might lead to excessive liability. This is only one example for all possible situations of imbalance between damage and liability. The inner conflicts within the environmental public interest liability system caused by unintentional double essential attribute setting will inevitably reduce the efficiency of the system in practice.

15.4.2 The Conflicts Between Administrative Enforcement Procedures and Civil Realization Procedures

In China's legal system, which follows continental/European legal traditions, the concept of environmental public interest liability is connotated in the concept of environmental administrative liability. The administrative legal system provides administrative enforcement procedures for the realization of environmental public interest liability. After environmental public interest liability was incorrectly defined as civil liability, civil legal systems participated in the realization of public interest liability as well, forming a series of civil procedures for realizing environmental public interest damage. This has essentially changed the basic pattern of China's legal system where public law liability is defined by public law and realized by public law procedures, and private law liability is defined by private law and realized by private law procedures. Now, environmental public interest liability can be realized through administrative law enforcement procedures under the attribute setting of administrative liability, or through civil relief systems for environmental public interest damage under the attribute setting of civil liability. A series of procedural conflicts arise between administrative law enforcement procedures and civil procedures with respect to environmental public interest liability. Those issues that have not yet been fully recognized and discussed by the existing legislative and academic discussions will be discussed below.

To be clear in the discussion, it is necessary to first classify the civil procedures of realizing environmental public interest liability. In the first step, they can be divided into two categories, namely non-litigation procedures and litigation procedures, depending on whether litigation is involved. The non-litigation procedure refers to the compensation negotiation procedure in the Compensation Regime for Eco-environmental Damage.[16] There are two types of litigation procedures provided by laws and regulations: civil environmental public interest litigation and ecological environmental damage compensation litigation in the compensation regime for eco-environmental damage. In essence, both are environmental public interest litigation. For those litigations, there are two main types of plaintiffs. One type is administrative agencies, and the other type is social organizations. Civil environmental public interest litigation and ecological environmental damage compensation litigation filed by administrative agencies as plaintiffs are in essence public environmental enforcement litigation initiated by administrative agencies as a way of environmental enforcement. Civil environmental public interest litigation initiated by social organizations should have been environmental public interest supervision litigation by which social organizations can supervise administrative agencies' performance in environmental public interest protection. However, the Civil Procedure Law has incorrectly made it to be a private environmental enforcement litigation in which social organizations can realize environmental public interest liabilities against liable parties directly. Therefore, civil litigation procedures for realizing environmental public interest liability now consist of public environmental enforcement litigation and private environmental enforcement litigation. According to the above classification, the conflicts and their consequences between administrative law enforcement procedures and various civil realization procedures will be discussed.

15.4.2.1 Conflicts Between Administrative Law Enforcement Procedures and the Compensation Negotiation Procedure for Eco-Environmental Damage

The Reform Plan on the Compensation Regime for Eco-environmental Damage (2017) established the Compensation Regime for Eco-environmental Damage, and the compensation negotiation procedure for eco-environmental damage (hereinafter

[16] CCP Central Committee and State Council Reform Plan on the Compensation Regime for Eco-environmental Damage (2017).

referred to as compensation negotiation procedure) is part of it. [17] The compensation negotiation procedure should have been a special procedure in administrative law enforcement procedures—the negotiation enforcement procedure. The negotiation enforcement procedure refers to a type of non-litigation administrative law enforcement procedure in which settlement agreements exist between administrative agencies and administrative counterparties. The agreements are on the specific forms of administrative liability and specific performance details. They are also concluded on the premise that they do not conflict with the statutory authority of administrative agencies, do not infringe on the statutory public interest, do not need to be realized immediately, and then will be performed by administrative counterparties.

This kind of law enforcement procedure has not yet been established in China's Administrative Litigation Law, but it has achieved partial germination through the development of the Compensation Regime for Eco-environmental Damage. In the compensation negotiation procedure of this regime, the administrative agencies can 'negotiate with the obligor on specific issues such as the commencement time and time limit for remediation, the manner and time limit for bearing compensation liability, etc.. .. If a consensus is reached through negotiation, an agreement will be signed'.[18] By reaching consensus on allocating environmental public interest liabilities through negotiation, this procedure enables environmental administrative enforcement in a new way. This procedure is in essence an enforcement procedure via negotiation. However, the problem is that when the environmental public interest liability is incorrectly set as civil liability, the compensation regime for eco-environmental damage and the compensation negotiation procedure in it are established as civil legal procedures for administrative agencies to realize civil liability for eco-environmental damage. The compensation regime and the negotiation procedure are set up in the civil legal system and are not identified as administrative law enforcement procedures.

The consequence of incorporating the compensation negotiation procedure into civil legal procedures is that in environmental public interest relief cases, at a non-litigation enforcement stage, competition arises between the administrative law enforcement system and the compensation negotiation procedure. On one hand, environmental public interest liability can be imposed and realized under administrative liability attribute setting through administrative orders for correction and administrative penalties. On the other hand, it can be realized under the civil liability attribute setting through the civil compensation negotiation procedure. In the same environmental public interest damage case, the potential liable party to environmental public interest liability may bear administrative liability through traditional administrative law enforcement procedures, or it may bear civil liability in compensation negotiation procedures. Both procedures are aimed at protecting the

[17] CCP Central Committee and State Council Reform Plan on the Compensation Regime for Eco-environmental Damage (2017)

[18] VI. Negotiation on compensation, the Opinions on Several Specific Issues Concerning Promoting the Reform of the Compensation Regime for Eco-environmental Damage (2020).

environmental public interest and taking environmental administrative law as an important legal basis, but they belong to two distinguished legal procedures of different natures. However, now, there is a lack of rules to address their relationship, which will inevitably lead to overlap and conflict in practice.

At the same time, for potential liable parties, such overlap and conflict will give rise to indirect loss of procedural rights provided for them in administrative procedure law. In these two types of procedures, the procedural rights that potential liable parties can enjoy are different. In view of the asymmetric relationship between administrative agencies and administrative counterparties, administrative procedure law provides special procedural safeguards for administrative counterparties by imposing procedural norms and burdens on administrative authorities and supplying relief procedures against abuse of power. Potentially liable parties can enjoy abundant special procedural rights and protection. However, civil legal procedures are set up for the resolution of legal disputes between equal parties and do not assume the function of balancing power and private rights. When public interest liability is realized in civil procedures, administrative procedure law will be bypassed, and potential liable parties cannot enjoy special procedural rights and protection as in administrative enforcement procedures, which amounts to a de facto unintentional deprivation of procedural rights.

15.4.2.2 Conflicts Between Administrative Law Enforcement Procedures and Environmental Enforcement Litigations

Law enforcement litigation refers to a type of litigation initiated by administrative agencies directly against potential liable parties in administrative law with the purpose of accomplishing administrative law enforcement. This type of litigation has not yet been provided in administrative law legislation as a way of administrative enforcement in China. For enforcement, in the first phase, administrative agencies should make administrative decisions by ordering administrative counterparts to accept their administrative liabilities. If this order is not followed by administrative counterparts, the enforcement procedure will go to a second phase of compulsory execution. Administrative agencies may apply to a people's court for compulsory execution, or proceed with compulsory execution on their own by special authorization provided by law.[19]

Although law enforcement litigation has not yet been established in China's administrative law legislation as an enforcement method, this type of litigation has achieved breakthroughs in the development of environmental litigation. When the fact that litigation cannot be used as a way of administrative enforcement has become a stereotype about the administrative enforcement system, it is difficult to find that environmental public interest litigation and eco-environmental damage compensation litigation in the Compensation Regime for Eco-environmental

[19] Article 97 Administrative Litigation Law (1989, amended 2014 and 2017).

Damage filed by administrative agencies are essentially law enforcement litigation. They are not only new types of environmental litigation, but also new paths for administrative agencies to promote their administrative law enforcement activity and to realize environmental public interest liabilities in the environmental law field.[20]

However, due to the stereotype about the administrative enforcement system and the setting of essential attributes of environmental public interest liability as a civil liability, these litigation procedures are not recognized by the legislature as 'environmental law enforcement litigations'; instead, they are recognized and established as civil tort litigation procedures initiated by administrative agencies as plaintiffs.

Owing to the ignorance of the enforcement essence of those 'enforcement litigations', the legislature did not reconcile traditional administrative law enforcement procedures with those new 'environmental enforcement litigation' procedures from the enforcement perspective. Instead, the legislature inadvertently created a tension between these two procedures. In the same case, the administrative agency can use both paths. However, there is no adequate stipulation in legislation about whether they can be initiated at the same time or priority should be given to one of them. In addition, these two paths of enforcement belong to public law and private law procedures, respectively, which makes the competitive relationship more complex and entangled. This can be illustrated by an example. Supposing in an environmental public interest damage case, in traditional administrative law enforcement procedure, an administrative agency has issued an order of correction, requiring a liable party to clean up pollution and remediate the environment within a time limit. However, if the administrative counterpart insisted that there was no violation of any law or regulation, and initiated an administrative reconsideration procedure or administrative litigation (judicial review), could the administrative agency file an environmental public interest litigation or eco-environmental damage compensation litigation against the liable party at the same time? If the answer is Yes, how can the relationship between those procedures be reconciled? Unsolved issues like these are abundant. Traditional enforcement procedures and new 'environmental enforcement litigation' procedures will interact in a faulty way, not in a reconciled way.

In addition to those environmental enforcement litigations, the legislature has unintentionally created private environmental enforcement litigation. According to article 58 of the Environmental Protection Law[21] and article 55 of the Civil Procedure Law,[22] social organizations meeting legal requirements can sue potential liable parties for environmental public interest damage. In such litigation, social organizations become the parties who can realize environmental public interest

---

[20] Xin (2019), p. 86.

[21] Article 58 Environmental Protection Law (1989, revised 2014) authorizes social organizations meeting legal criteria the standing to file a lawsuit to a People's Court for environmental pollution and ecological destruction acts which harm public interest.

[22] Article 55 Civil Procedure Law (1991, amended 2007, 2012 and 2017) authorizes relevant organizations the standing to file a lawsuit to a People's Court for acts which harm public interest such as environmental pollution, infringement of the legitimate rights and interests of multiple consumers, etc.

liability directly, sue liable parties as the only type of defendants directly, and completely abandon the administrative agencies in the litigation. This indicates that the essence of such litigation is private environmental enforcement litigation initiated by social organizations.[23]

However, blindfolded by the veil of the wrong essential attribute setting of environmental public interest liability, this essence is not recognized by the legislature. The legislature's logic is simple. Since environmental public interest liability is civil liability, then the litigation realizing such liability should be tort litigation (civil litigation), and social organizations as civil entities shall have a legal standing to realize such liability in civil courts. Therefore, the legislature gives social organizations standing to pursue environmental public interest liability directly against liable parties.

The legislative confirmation of private environmental enforcement litigation creates competition and conflict between private enforcement procedures and public enforcement procedures, namely, administrative enforcement procedures initiated by administrative agencies. Although there are legal rules to coordinate them, those rules are far from sufficient, and procedural conflicts in practice are inevitable.[24] Furthermore, for such litigation, procedural conflict is only a small issue, and the more important issue will be discussed in the following section.

15.4.3 The Conflicts Between Social Organizations' Public Authority and Administrative Agencies' Public Authority

As discussed above, setting essential attributes of environmental public interest liability as civil liability leads to the creation of private enforcement litigation in which social organizations meeting legal requirements have standing to sue liable parties who caused environmental damage to pursue the latter's environmental public interest liability. For such type of litigation, in addition to its procedural conflict with public enforcement procedures, there is a more important conflict hidden under social organizations' standing to sue. Social organizations' standing in private environmental enforcement litigation equates to a public authority identity. Granting such type of standing to social organizations leads to a conflict between social organizations' public authority identity and administrative agencies' public authority identity.

Social organizations' standing in private environmental enforcement litigation is the result of setting essential attributes of environmental public interest liability as a civil liability as well. Since environmental public interest liability is defined as civil liability, and social organizations as civil parties can enjoy rights in the environment

[23] See Xin (2020), p. 80.

[24] See Xin (2020), pp. 84–85.

(in theory), infringement of environmental public interest will be transformed into infringement of environmental rights of citizens and social organizations, and it will become a logical consequence for citizens and social organizations to directly pursue environmental public interest liabilities as parties who own corresponding rights. Therefore, in the legislature's design, social organizations' standing in the civil environmental public interest litigation is just a right of action to pursue a special civil liability. The legislature did not realize that its design unintentionally granted public authority identity and public power to social organizations.

In such litigation, social organizations' standing (right of action) is essentially a kind of public power, and their plaintiff identity is essentially public authority identity. As discussed above, the environmental public interest is not a collection of private interests. It must be maintained, disposed of, and remedied by public decision-making mechanisms under representative democracy. The legal power of preservation, disposition and relief of such interest is not a private right, but a public power that serves general will. The entities that enjoy such public power are entities with public authority. After environmental public interest damage occurs, the right of action to pursue environmental public interest liability through litigation is a manifestation of public power. The identity of a plaintiff in such litigation is a manifestation of public authority. The creation of private environmental enforcement litigation means that social organizations are granted public power[25] and a public authority identity in the environmental public interest protection area by the legislature.

However, there is constitutional jurisprudential issue in such legislative authorization. Such authorization conflicts with existing administrative agencies' authority. Administrative agencies established according to the Constitution are the first public authority to manage and maintain public interest after legislation. It is the statutory authority of the environmental administrative agencies to pursue the liabilities of environmental public interest damage, and environmental administrative agencies should be the eligible plaintiffs who directly file environmental public interest lawsuits against defendants who caused environmental damage. In civil environmental public interest litigation, social organizations are given the right to directly sue defendants who caused environmental public interest damage. This results in coexistence two public authority entities in the relief of environmental public interest damage with overlapping and conflicting powers.[26]

It might be argued that granting administrative power to non-administrative entities is allowed in China's legal system. However, such authorization shall meet a basic premise that authorized administrative power shall not conflict with the existing administrative power of administrative agencies. Granting social organizations standing in private environmental enforcement litigation is contrary to this premise in many aspects. Administrative agencies have been authorized with public power to environmental interest protection by the Constitutional Law and

[25] Wang (2016), p. 112.

[26] Xin (2015), p. 138.

environmental law much long before the provisions of the legislation on social organizations' right of action in environmental public interest litigation. The right of action to pursue environmental public interest liability shall be the justifiable content of such authorization. The unintentional authorization to social organizations steps into authorization given to administrative agencies without justifiable rationale for such second and overlapping authorization. In fact, the legislature did not even realize that its provision of social organizations' standing in civil environmental public interest litigation is an authorization of public power. Thus, the legislature did not make this authorization with consideration of this premise at all.

It is difficult to prove the legitimacy of such authorization to social organizations when administrative agencies have already been given a comprehensive authorization of public power in environmental public interest management. However, this fact does not mean that social organizations shall not have legal standing in environmental public interest litigation. Social organizations' standing shall be justified by citizens' supervising right provided by the Constitution.[27] In practice, it shall be established on the fact that administrative agencies fail to exercise their public duties diligently. In such situations, citizens' right of supervision enables them to have standing to sue administrative agencies. Administrative agencies must be defendants in such litigation. Thus, environmental public interest litigation raised by social organizations shall not be private environmental enforcement litigation, but public interest supervising litigation.[28]

15.5 The Causes of the Erroneous Essential Attribute Setting

It is not accidental that the essential attribute of environmental public interest liability has been set as civil liability. It is attributable to two important factors.

15.5.1 The Impact of the Theory of Civil Remedies and the Theory of Crystallization of the Environmental Right as a Civil Right

The theory of civil remedies for environmental public interest damage is a general term for theories advocating using civil remedies to redress environmental public interest damage. Many scholars have argued that environmental tort law could be used to redress not only personal injury and property infringement caused by

[27] Article 41, the Constitution of China (1982, amended 1988, 1993, 1999, 2004 and 2018).
[28] See Xin (2020), pp. 86–89.

ecological and environmental damage, but also environmental public interest damage in ecological and environmental damage. This theory expanded the function of the traditional environmental tort law which is mainly for redressing private interest damage.

However, this theory faces a massive challenge. In the traditional civil rights system, the environmental public interest, which refers to the environmental public interest owned by all the people was not manifested by any right. Because environmental public interest damage does not qualify as a violation of any civil right, the civil tort claim to environmental public interest damage cannot be raised. Thus, the lack of corresponding rights makes it is impossible for the environmental public interest to be redressed by traditional environmental tort law. The theory of the crystallization of environmental rights as civil rights is proposed to resolve this challenge. This theory argues that every citizen should have an environmental right to the public and private environment, and this right should be crystallized as a type of civil right, which can be redressed by environmental tort law. Consequently, damage to a pure environmental public interest will become damage to citizens' environmental civil rights, and environmental public interest liability naturally becomes civil liability. Thus, it enables citizens to protect the environmental public interest through civil relief channels.

However, those two theories ignore public law theories and legal regimes on public affairs governance. Civil rights should be the manifestation of private interests that can be classified or distinguished as individual quotas. The management of those private interests relies on private parties' autonomy in the private law arena. However, the environmental public interest, like other types of public interest, is managed and protected by public decision-making mechanisms based on representative democracy to crystallize the general will. Citizens' rights to the environmental public interest, like other forms of public interest, are manifested as political rights in public law, rather than civil rights in the private law system. Thus, environmental public interest liability is not the consequence of violating civil rights, and it is not a type of civil liability.

15.5.2 The Conceptual Misunderstanding in the Transplantation of Environmental Legal Regimes from the Common Law System

Another important reason for setting public interest liability as civil liability is the misunderstanding of the corresponding position and essential attribute of the 'civil liability' concept in the transplantation of the environmental public interest protection system from the common law system to the legal system of China. In order to construct environmental public interest litigation system and a compensation regime for eco-environmental damage, the relevant research in China often takes the relevant legislation of the United States as an extraterritorial example. However, what the relevant research often overlooks is that the concept of civil liability in the US legal system is very different from that of China which partially inherits continental legal traditions. The concept of civil liability in the US legal system covers civil liability and administrative liability in the Chinese legal system. There is no independent administrative liability system away from private civil liability in the US legal system. Legal liability for the public interest and public order, which falls into the category of administrative liability in China, still falls into the category of civil liability in the US, where even the concept of administrative punishment uses the expression 'civil penalty'.[29] This overlooking is highly related to the misplacement of environmental public interest liabilities and their related systems in China's civil law legislation system.

15.6 Recommendations for Correcting the Essential Attribute Setting

From the above discussion, it can be observed that setting the essential attribute of environmental public interest liability as civil liability has caused many jurisprudential and legal institutional consequences. It is urgent for the Chinese legislature to correct the essential attribute of environmental public interest liability as administrative liability.

First, Article 64 of EPL should be amended. Its current provision: 'In the case of damages caused by environmental pollution and ecological destruction, tortious liability shall be borne pursuant to the relevant provisions of the Tort Law of the People's Republic of China.'[30] shall be changed to 'Whoever causes specific civil interest damage by his environmental and ecological behavior shall bear tortious liability in accordance with the relevant provisions of the Civil Code. Whoever causes harm to the environmental public interest within the jurisdiction of administrative agencies, administrative legal liabilities shall be borne in accordance with the provisions of administrative laws and regulations related to environmental management.'

Second, several articles in the Civil Code should be amended. Article 1229 of the Civil Code provides that 'In the event of damage caused to others as a result of environmental pollution and ecological destruction, the tortfeasor shall bear tortious liability.'[31] It should be changed to 'Where specific civil damage is caused by environment pollution, the polluter shall bear civil tortious liability in accordance with this Law. Where environmental public interest damage is caused by environment pollution, administrative legal liability shall be borne in accordance with administrative laws and regulations.'

[29] See Zhang et al. (2013), p. 46.

[30] Article 64, Environmental Protection Law (1989, revised 2014).

[31] Article 1229, Civil Code (2021).

At the same time, articles 1234 and 1235 of the Civil Code should be moved to the administrative law system. Those two articles are specific provisions about remedial forms of environmental public interest liability. Article 1234 stipulates the liability for restoration within a reasonable period and the liability of compensation for restoration costs. Article 1235 stipulates the compensation liability for losses and costs incurred for environmental public interest damage.

Article 1232 of the Civil Code should be amended as well. This article provides the following: 'Where a tortfeasor, in violation of laws and regulations, intentionally pollutes the environment or damages the ecology, thereby causing serious consequences, the infringed is entitled to claim appropriate punitive damages.' It stipulates the punitive liability for environmental damage but does not distinguish between private interest liability and public interest liability. Since environmental administrative punitive liability has already been provided in administrative law, this article should remove environmental public interest damage from its coverage and keep this liability only for environmental private interest damage. This article can accordingly be changed to 'Where a tortfeasor, in violation of laws and regulations, intentionally pollutes the environment or damages the ecology, thereby causing serious consequences to personal life, health or property, the infringed is entitled to claim appropriate punitive damages.'

Finally, like the above suggested changes, relevant provisions about environmental public interest liability in other laws, regulations, judicial interpretations, and other legal normative documents shall be adjusted to administrative liability as well.

15.7 Conclusion

The connection between the environmental public interest and the private law system needs serious examination. As a country that follows continental legal traditions, the public law system is the appropriate legal system for environmental public interest protection in China. From the perspective of public law, this chapter explains why environmental public interest liability should be an administrative liability, rather than a civil liability. As a central concept in the civil procedure of realizing environmental public interest liability, the correction of the essential attribute of environmental public interest liability will supply a foundation for those procedures to be adjusted to administrative law procedures.

References

Cao MD (2000) Environmental tort law. Law Press China, Beijing

Huang W (ed) (2020) Interpretation of liability for tort part in the civil code of the People's Republic of China. Law Press China, Beijing

Jin RL (1997) Environmental tort and civil remedies: problems in the environmental legislation. China Environ Sci 03:2–7

Kawamura H (2018) The relation between law and technology in Japan: liability for technology-related mass damage in the cases of Minamata disease, asbestos, and the Fukushima Daiichi nuclear disaster. Contemp Japan 30(1):3–27. https://doi.org/10.1080/18692729.2018.1423459

Lazarus RJ (1993) Pursuing "environmental justice": the distributional effects of environmental protection. Northwest Univ Law Rev 87(3):787–857

Wang X (2016) Legalizing litigations for public interest in environment in China. Tsinghua Univ Law J 10(06):101–114

Xin S (2015) Mission impossible: limits of civil remedies for environmental damage. China University of Political Science and Law Press, Beijing

Xin S (2019) Rectification of environmental enforcement litigation system in China. Study Explorat 03:86–94

Xin S (2020) On the attribute and procedure correction of environmental citizen litigation system in China. J Qingdao Univ Sci Technol (Soc Sci) 36:79–89. https://doi.org/10.16800/j.cnki.jqustss.2020.04.014

Zhang JY, Yan HF, Qin H (2013) A compendium of environmental enforcement cases in the United States. China Environment Press

Chapter 16
From the Ternary Model to the Binary Model: On the Reconstruction of Litigation Systems for Environmental Public Interest

Xiaobo Bo

Abstract China has created three litigation systems for environmental public welfare relief: civil environmental public interest litigation (civil EPIL), administrative environmental public interest litigation (administrative EPIL), and ecological and environmental damage compensation litigation (EEDCL). First, judicial power intervenes in environmental public welfare protection and performs its duties instead of the government, which will lead to a misalignment of judicial power and administrative power. After the EEDCL emerged, the overlap with the civil EPIL system exposed the shortcomings of the existing ternary litigation model. The basis of claims for ecological and environmental damage compensation is not the national ownership of natural resources or the responsibility of environmental supervision, but the environmental public interest. There is no significant difference between these two kinds of litigation in terms of value objective, system function, basis of claim, object of relief, object of litigation, etc. The coexistence is unnecessary. Administrative EPIL should be retained, and environmental protection organizations could be added as plaintiffs. The civil EPIL system should be removed, and "EEDCL" should be replaced with "ecological environmental damage public interest litigation". As the plaintiff, the government can request judicial relief, and the pre-consultation procedure should be retained. The plaintiff's proof rule should be adopted, and the introduction of a punitive damage compensation mechanism should be considered.

Keywords Civil EPIL · Administrative EPIL · Ecological and environmental damage compensation litigation (EEDCL) · Basis of the right of claim · Ecological environmental damage public interest litigation

---

X. Bo (✉)
Law School, Jiangnan University, Jiangsu, China

X. Wang et al. (eds.), *Environmental Public Interest Litigation in China*,
https://doi.org/10.1007/978-3-031-26526-6_16

16.1 Introduction

As ecological and environmental issues have received widespread public attention, legislators have paid increasing attention to the maintenance of ecosystem functions and the protection of environmental public welfare. When environmental problems arise, traditional civil law only provides remedies for private rights damages (i.e., damages to personal and property rights) suffered by the victims in environmental torts, and there are insufficient remedies for the damage to the ecological environment itself. In order to cope with this problem, China's environmental legislation has been actively reformed in recent years, and the civil EPIL system, administrative EPIL and ecological and environmental damage compensation systems have been established successively, which together constitute a ternary litigation model for remedying ecological environmental damage in China.

It is worth mentioning that Chapter VII of the "Tort Liability" of the Civil Code stipulates the issue of liability for restoration of ecological environmental damage, and clarifies the scope of compensation.[1] This is contrary to the functional orientation and relief object of tort law and will lead to confusion of legislative logic (this issue will be discussed in detail in the last part of this article). There are problems in the existing ternary litigation model, especially the self-defects exposed by the civil EPIL system and the institutional conflicts and overlaps between it and the ecological environment compensation system, which are worthy of serious reflection.

16.2 Insufficiency of the System: Analysis of the System of Civil EPIL

16.2.1 Misalignment of the Functions of Judicial Organs and Administrative Agencies in the Civil EPIL System

The number of civil EPIL cases in China is far more than that of administrative EPIL. Judicial practice relies more heavily on the former than the latter. Scholars have long questioned the sequence of these two types of lawsuits. Civil EPIL has become a way for environmental protection NGOs or Prosecutor's Offices to draw support from the judiciary to correct behaviors that damage environmental public welfare. Local governments or functional departments (such as ecological environment departments and forestry departments) who should undertake their responsibilities of protecting environmental public welfare have instead become bystanders or assistants who provide professional assistance to environmental NGOs in environmental civil public interest lawsuits, resulting in a distorted relationship between

[1] For details, please refer to Articles 1234 and 1235 of the Civil Code of the People's Republic of China.

administrative power and judicial power. The heavy responsibility of maintaining environmental public welfare, which should have been taken by environmental regulators using their law enforcement powers, is transferred to the judicial organs.[2] This leads to the judicial organs overstepping and acting in the place of administrative organs, which in turn leads to the misalignment of the functions of the judicial and administrative organs.[3]

When the civil EPIL system was established in China, it was mainly to avoid the "tragedy of the commons". The ecological environment is a public good. When it is damaged, there is often a lack of specific rights holders to protect it and request relief, resulting in the lack of such a right to request relief for public interests. As Justice Douglas stated in his dissenting judgement in *Sierra Club v Morton* 405 US 727 (1972):

> Environmental standing should be vested in the natural creatures themselves, although the poorly articulated members of the ecosystem cannot express their opinions, but those who are familiar with those whose values and miracles can speak for the entire ecosystem community.[4]

Giving those whose primary focus is on environmental protection the right to sue and ask them to take responsibility for those who cause damage to the ecological environment, is an important means to prevent the frequent occurrence of "tragedies of the commons." The Environmental Protection Law of the People's Republic of China identifies social organizations that meet the requirements of Article 58 as eligible plaintiffs. At the same time, the Pilot Program for Procuratorial Organs to Initiate Public Interest Litigation Reform, an enforceable document issued by the Supreme People's Procuratorate on July 2, 2015, adds the Prosecutor's Offices as another type of plaintiff in civil EPIL. Compared with environmental protection organizations, the Prosecutor's Offices, as the plaintiff, started late in the field of civil EPIL, but they have occupied a much larger proportion. From January 2015 to September 2018, courts across the country accepted a total of 2041 EPIL cases, of which 1836 were filed by Prosecutor's Offices and 205 were filed by environmental protection organizations.[5] When environmental protection organizations and Prosecutor's Offices target the perpetrators of environmental pollution or ecological damage, what can the government, that should be the trustee of environmental public welfare, do? The existing civil EPIL system has not given the government its due space to play its role.

---

[2]In China, the judiciary in a broad sense includes courts, procuratorates, public security organs, judicial administrative organs, and state security organs. Courts and procuratorates are judicial organs that specifically exercise adjudication and procuratorial powers. Although the latter three belong to the administrative organs, they also assume some judicial functions. The narrow sense of the judiciary refers specifically to the courts. The term "judicial organs" here refers specifically to the courts.

[3]Xi (2016), p. 103.

[4]Quoted from Houfu (2006), p.32.

[5]Bixin (2019), p. 50.

In China, the State enjoys ownership of most important natural resources. Governments at all levels are responsible for the quality of the environment and are the Environmental Protection Work Supervisors.[6] When the ecological environment is damaged, the government should show a more positive attitude than environmental protection organizations and procuratorates. However, in the current civil EPIL system, environmental protection organizations or Prosecutor's Offices, instead of the government, can directly sue the judicial organs, which evades the government's environmental management responsibilities and leaves the government with nothing to do.

16.2.2 Overlap Between Civil EPIL and Ecological and Environmental Damage Compensation Litigation (EEDCL)Systems

Since the advent of the ecological and environmental damage compensation system, civil EPIL has become even more problematic. After the ecological environment is damaged, environmental protection organizations or Prosecutor's Offices can initiate environmental civil public interest lawsuits, and at the same time, the government or its designated departments can also request compensation for ecological environmental damage. This may lead to the situation where two claims are made that relate to the same damage. By sorting out judicial precedents, it is not difficult to find that the facts, claims and defendants of these two lawsuits are basically the same except for the plaintiffs. The formation of excessive competition and cooperation is worthy of our reflection on whether it is necessary for these two systems to coexist.

Before the promulgation of the "Several Provisions of the Supreme People's Court on Hearing Cases of Compensation for Eco-environmental Damage (for Trial Implementation)" (hereinafter referred to as the "Several Provisions"), there has been judicial chaos, in which different local courts handled the two lawsuits in different ways. For example, in the Cangjing Pavilion case in Chongqing and the DyStar waste sulfuric acid disposal case in Jiangsu, etc., the environmental civil public interest lawsuit brought by the court against the environmental protection organization and the ecological environment damage compensation lawsuit brought by the government were tried together. However, in the Taizhou "high price compensation" case,[7] the Taizhou Environmental Protection Federation filed an

[6]Paragraph 2 of Article 6 of the EPL stipulates: "Local people's governments at all levels shall be responsible for the environmental quality of their respective administrative regions."

[7]Court decisions for these cases are available at the legal database 'China Judgements Online'. See https://wenshu.court.gov.cn/website/wenshu/181107ANFZ0BXSK4/index.html?docId=2fae6a88799248408c41a923009abfac; https://wenshu.court.gov.cn/website/wenshu/181107ANFZ0BXSK4/index.html?docId=904cf1840d84459eaa00a8b300b06484; https://wenshu.court.gov.cn/website/wenshu/181107ANFZ0BXSK4/index.html?docId=1689340450e24cb2bd99f83543735738.

environmental civil public interest lawsuit against six polluting companies, and the Taizhou Environmental Protection Bureau also filed a claim for compensation for ecological and environmental damage against the same defendants. The Taizhou Intermediate People's Court adopted an alternative approach, and finally made a judgment of "Huge compensation"[8] in civil EPIL. Although the "Several Provisions" have made relatively simple normative arrangements for the connection of the two types of litigation, if there is no significant difference between the two in terms of system functions, litigation objects, scope of appeals, and relief objects, the separation of the two systems will lead to a waste of judicial resources and hidden dangers of conflict between the two systems. Therefore, these two systems should be integrated.

From the perspective of institutional functions, the institutional function of civil EPIL is mainly to protect environmental public interests, and EEDCL is also committed to safeguarding "people's environmental rights and interests" in the form of "realizing ecological damage responsibility".[9] The EPIL initiated by the NGOs and the ecological environment damage compensation litigation initiated by the government are both important components of the ecological environment damage compensation system".[10] The institutional functions of the two are consistent.

From the perspective of the scope of the appeal, comparing Article 18 of the *Judicial Interpretation of Civil EPIL* with Article 11 of the Several Provisions,[11] it can be seen that, except for the slightly different expressions of "restoring the original state" in the former and "restoring the ecological environment" in the latter, and the order of various types of responsibilities is slightly adjusted, other responsibilities are assumed in exactly the same way. Article 20 of the *Judicial Interpretation of Civil EPIL* further states that "restitution to the original state" is to require the defendant to restore the ecological environment.[12] Therefore, in these two kinds of lawsuits, the way of assuming responsibility is the same, which means there is no substantial difference in the claims that the plaintiff can bring to the court in these two kinds of lawsuits. If the qualified plaintiffs file these two lawsuits separately,

---

[8]In this case, the court ordered the defendant to compensate the total cost of environmental remediation of 160,666,745.11 RMB and pay 100,000 RMB for appraisal and assessment costs.

[9]Ganjie (2017), p. 1.

[10]Jing (2015), p. 1.

[11]Article 18 of the "Environmental Civil Public Interest Litigation Judicial Interpretation" determines that the defendant's liability methods include "stopping infringement, removing obstacles, eliminating danger, restoring the original state, compensating for losses, and making apology, etc.". Article 11 of the "Several Provisions" determines that in such a lawsuit, the defendant can be sentenced to "undertake civil responsibilities such as restoring the ecological environment, compensating for losses, stopping infringement, removing obstacles, eliminating danger, and making an apology."

[12]Article 20 of the Judicial Interpretation of Environmental Civil Public Interest Litigation stipulates: "If the plaintiff requests restoration to the original state, the people's court may judge the defendant to restore the ecological environment to the state and function before the damage occurred. If it cannot be completely restored, alternative restoration may be allowed.

although the court can deal with the order and connection issues in accordance with the "Several Provisions", it will inevitably lead to an unnecessary waste of judicial resources. In the context of the lack of judicial resources in China and the increasing number of environmental cases, it is urgent to consider how to integrate and reconstruct the two types of litigation at the legislative level, optimize the structure of litigation subjects, and improve the overall efficiency of litigation.

Before deciding whether the two litigation systems should be unified and integrated, it is necessary to clarify the issue of the source of litigation rights, which is a decisive factor in determining the value, goal and nature of litigation. It is clear that the civil EPIL system focuses on the relief of environmental public interests, but the nature of the ecological and environmental damage compensation lawsuit and the basis of the right of claim are still controversial. Clarifying this issue is the key to deciding whether it can be integrated with the civil EPIL system.

16.3 The Source of Rights: The Basis of the Right of Claim in EEDCL Lawsuits

The concept of the basis of the right of claim was created by Windscheed and introduced to China by Dr. Wang Zejian, which states, "One can claim which rights against another, according to which legal norms".[13] He argues that the main task of solving cases is to find legal norms that support one party and have a claim against the other party. Such a legal norm that can support a party's claim to the other party is the basis for the norm of the request, referred to as the basis of the right of claim. Although "the basis of the right of claim" is a concept created in the study of civil law theory and has never appeared in Chinese statutes, as civil judicial trials pay increasing attention to legal methods and the application of legal thinking, the law application method centered on the basis of the right of claim is increasingly widely used in Chinese judicial practice. Finding an appropriate "basis of the right of claim" has become one of the core tasks of the adjudication of civil disputes. If the plaintiff wants to get the support from the court, "the claim must be filed within the legal effect of the basis of the right of claim, and the elements of the basis of the right of claim must be presented ... When the court judges a case, it must examine the plaintiff's claims and evidence in detail based on the basis of the claim, and examine the defenses raised by the defendant on the basis of the claim".[14] This means that the plaintiff must prove that the claim made by him or her to require the defendant to bear the responsibility has basic rights or legal basis.

Although the basis of the right of claim and its legal thinking and application methods were initially applied in civil trials, they are not limited to this field. It is also necessary to clarify the legal basis for the plaintiff's action in any other type of

[13] Zejian (2002), p. 35.

[14] Haiyong (2015), p. 53.

litigation. In other words, the procedural rights of the plaintiff in any kind of litigation should have its source in the substantive law. The plaintiff in an EPIL also needs to prove the basis of its claim right, i.e., the environmental public interest. The court should also review the claims and evidence of both the plaintiff and the defendant on this basis. Therefore, the "basis of the right of claim" and its thinking method[15] should also be applied to judicial remedies for environmental public welfare.

Regarding the new type of lawsuit for ecological damage compensation, there are various opinions on what is the basis of the right of claim in academic circles.

16.3.1 The Theory of National Natural Resource Ownership

There are three theories of national natural resource ownership that underpin the basis for lawsuits taken to protect the environment. First, the "national interest litigation theory", which provides that the lawsuit for ecological damage compensation is different from the public interest lawsuit brought by the procuratorate or social organization, and it is a lawsuit to protect the interests of the state. Since the state is the owner of natural resources, and the administrative agency exercises the ownership of the natural resources on behalf of the state, the administrative agency has the qualification of claimant. Second, the "Private Interest Litigation Theory" provides that the state empowers administrative organs to file lawsuits for damages to the ecological environment, which is actually to protect their own natural resources. The private nature of state ownership of natural resources means that ecological and environmental damage compensation lawsuits are private interest lawsuits.[16] On the basis of synthesizing the above two theories, some scholars put forwards a third "hybrid litigation theory", arguing that the state ownership of natural resources has the dual power (right) attributes of both private and public rights, and thus the EEDCL "is not simply a private interest lawsuit or a public interest litigation, but a special category of litigation in which public and private interests overlap".[17] The above theories approach the nature of national natural resource ownership from different perspectives, but it is unanimously recognized that state ownership of natural resources is the basis for claiming rights in the ecological environment damage compensation system.

In China, national natural resource ownership is exercised mainly through the following paths: First, the State Council exercises its natural resource ownership on

---

[15]The thinking method of the "basis of the right of claim" includes four steps: (1) Truly understand the facts of the case; (2)In response to the questions raised, shuttle back and forth between the facts of the case and the legal norms; (3)Concretize the problems, eliminate useless facts and select the basis of the claim for key facts of the cases; (4) Decompose and encompass elements of the basic of the right of claim.

[16]Shuyi and Huaqi (2018), p. 71.

[17]Song Lirong (2018), p. 8.

behalf of the state, and when such rights are damaged, the provincial and municipal governments are authorized by the State Council to seek compensation for ecological environmental damage. According to the official interpretation, the purpose of this system is to make up for the lack of claimants in the existing system for damages to state-owned natural resources. The state ownership of natural resources constitutes the basis of the government's right to claim compensation for ecological environmental damage. It seems quite reasonable in the existing legislative basis and interpretation path, but in-depth thinking will reveal that there are logical defects and institutional conflicts.

First, similar to other natural resources, state-owned natural resources have both an economic and an ecological value. Economic value has the characteristics of divisibility and transferability, and can become the object of property rights, while the ecological value is indivisible and shared by the public. It does not have the characteristics of "property" in law, so it can only become the object of environmental public welfare, rather than public ownership. The state, as the owner of natural resources, has the right to the economic value of the natural resources. Correspondingly, when the ownership is infringed upon, it can only claim relief for the damage to the economic value of the natural resources. Although some scholars have pointed out that the ownership of national natural resources is a pure public right,[18] there is no doubt that ownership of national natural resources has great economic value which allows governments at all levels to reap the extensive benefits of resource exploitation. Therefore, regardless of whether the national natural resource ownership contains public rights attributes, it does not affect the manifestation of its private rights in reality. Moreover, the concept of "ownership" was originally created and used in civil law, even if the interests of the state in the exercise of ownership do not belong to a private subject but to the entire state, it cannot deny the natural connection between the ownership and the economic value contained in the object of right. It can be deduced that if the national natural resource ownership is the basis of the right of claim, the claim for remedies for damage to these natural resources relates to their economic value. However, this is obviously contrary to the design intention of the ecological environment damage compensation system. In this system, the compensation right holder requests compensation not only for the economic value of natural resources, but also for the restoration of the ecological environment. This is to restore the damaged eco-environmental functions and the public interests contained therein, although such interests are not elements of ownership.

Secondly, the scope of the ecological environment in the compensation for ecological environmental damage is very broad, including not only minerals, land, water bodies, forests, grasslands, wasteland, tidal flats and other natural resources determined by the Constitution to be owned by the state, but also the whole ecosystem, which is composed of various environmental elements, such as air and sunlight if ecological damage occurs over state-owned natural resources, such as

---

[18] Gu (2014), p. 142.

deforestation, degradation of grass due to overgrazing, or pollution of water bodies. The owner of the damaged natural resources, i.e., the state, demands compensation from the perpetrator of the damage. However, how should damage to the environmental elements that are not natural resources be handled? If a company emits waste gas and causes air pollution, will it be impossible to file a lawsuit for ecological damage compensation because the air is not a state-owned natural resource? The answer is No. The circumstances listed in Article 1 of the "Several Provisions" on which a lawsuit for damages to the ecological environment can be brought[19] do not exclude environmental events that occur in non-natural resource environmental elements. Therefore, it is unclear why the government seeks judicial remedies for damage to other environmental elements or even ecosystem functions as a whole, rather than seeking ecological damage compensation.

Thirdly, natural resource ownership in China includes both the state and the collective. If the government initiates a lawsuit for ecological damage compensation because it exercises national natural resource ownership on behalf of the state, can collective natural resources be remedied through such a litigation system when they are damaged? Again, Article 2 of the Several Provisions does not exclude the damage to collective natural resources from the scope of litigation claims. If the ownership of natural resources is used as the source of litigation rights, the owner, which is the collective economic organization, should be the right holder of compensation when ecological environmental damage occurs to the collective natural resources. At present, the compensation right holders stipulated in the ecological and environmental damage compensation system are limited to the government and its designated departments and agencies, excluding collective economic organizations, and the institutional arrangements lack legitimacy and rationality. Even if the government complies with the boundary of rights and does not request relief for damages to collective natural resources, different remedies may be granted due to different ownership rights. This is against the spirit of fairness and justice.

---

[19] Article 1 of the "Several Provisions": Under any of the following circumstances, the provincial, municipal governments and their designated relevant agencies, or the departments entrusted by the State Council who exercises the ownership of natural resource assets owned by all Chinese people can file a lawsuit for compensation of ecological environment damage as plaintiffs, if they fail to reach an agreement or are unable to conduct negotiation with the legal persons or other organizations who have caused ecological damage:

1. In the event of a relatively large, major or particularly major environmental emergency;
2. Environmental pollution and ecological destruction events occur in key ecological functional areas and prohibited development areas designated in the national and provincial main functional area plans;
3. Other serious consequences affecting the ecological environment occur.

The municipal governments specified in the preceding paragraph include cities divided into districts, autonomous prefectures, leagues, and regions, prefecture-level cities not divided into districts, and district and county people's governments of municipalities directly under the Central Government.

Finally, if the ownership of natural resources is regarded as a traditional private right, and the government and even collective economic organizations are allowed to file lawsuits for compensation for infringement of their ownership rights, such lawsuits are not fundamentally different from environmental infringement lawsuits and can be included in the private interest litigation system. Thus, some scholars claim that ecological damage compensation litigation can be classified as environmental civil private interest litigation.[20] In this way, the establishment of a litigation system for ecological damage compensation will lose its independent value and function.

If the ownership of state-owned natural resources is interpreted as a public right (which is indeed a major feature that distinguishes it from other types of ownership), the mechanisms for exercising public and private rights are very different. When exercising the ownership of national natural resources from the perspective of public rights, more emphasis should be placed on the use of preventive public rights instruments, i.e., administrative management, rather than post-facto remedies (such as negotiation and litigation). Administration management and compensation for damages are applied in different ways and to solve different problems. Administrative management mainly uses preventive means and measures to prevent the occurrence of acts that endanger the legitimate rights and interests of others. Even if post-event measures are taken, administrative management is mainly concerned with the punishment of the offenders and the mandatory corrections for illegal acts, rather than the relief of the rights and interests of the infringed. In contrast, compensation for damages focuses on the relief of damaged rights and interests, mainly relying on third-party adjudication to clarify the right and wrong, and to divide the rights and interests of the parties to the dispute. The logical premise of the exercise of the right to claim compensation for damages is that the right holder and the obligor are equal in legal status, and are not in a relationship of command and obedience to each other. The basis of any right of claim for damage relief should not be public power. If the public right characteristic of national natural resource ownership is emphasized, the exercise of this right (power) should not be carried out in the form of litigation for compensation of ecological environmental damage.

There are many loopholes in the institutional system and logical derivation if the ownership of national natural resources is used as the basis of the right of claim for compensation for ecological environmental damage, and it is difficult to justify it.

16.3.2 The Theory of Environmental Supervision Responsibility

In accordance with the theory of environmental supervision responsibility, the government can act as the representative of the state to exercise the ownership of

[20] Jin (2018), p. 40.

national natural resources, and it is also the unified supervisor of the ecological and environmental protection work that needs to be responsible for the quality of the environment. From the perspective of public rights, the government should rely heavily on administrative means, such as administrative management, to protect the ecological environment. However, if the government has exhausted the means of public law and still cannot effectively protect the interests of the ecological environment, the purpose of ecological environmental protection and supervision could be achieved through judicial means. Rather than exercising the right to request ecological environment restoration as the state owner of natural resources, it is more likely to exercise the right to claim compensation for damage to the ecological environment based on its fulfillment of environmental protection duties.[21]

The author believes that the relationship between judicial power and administrative power must be considered when critiquing this theory. If the environmental supervision responsibility of the law enforcement department is transferred to the judicial authorities only because administrative law enforcement cannot achieve the goal of ecological environmental protection, it will fall into the pattern of excessively strengthening the role of judicial power in environmental public affairs. Judicial jurisdiction transcends its proper boundaries and replaces executive power, which leads to authoritarianism and utilitarianism in the judicial procedure.

When considering the relationship between executive power and judicial power, regard must be given to the rule of law. Although law enforcement is undertaken by both the executive and judiciary, there is an essential difference between the two: executive power relates to management, and judicial power relates to judgment, and the two should not be confused.[22] The functional characteristics of the two should be fundamentally distinguished, the separation of powers should be maintained, and exerting the judicial power to supervise executive power should be the focus.[23] The traditional view holds that the judiciary is passive and the law execution is active, which means that the law executive authorities must initiate the action, with the judicial authorities insisting on "no trial without complaint". In recent years, the concept of "active justice" has gradually emerged. Some scholars argue that the most ideal state of effective justice is the integration of legal effects and social effects. Judging a case should not only be based on legal norms, but the judiciary must fully consider what kind of social effects the judicial process and results will produce.[24] This repositioning of judicial power will lead to changes in the relationship between judicial power and executive power, and some scholars have suggested that it may undermine the strictness of rules and procedures, thus raising concerns about the elimination of the rule of law.[25]

---

[21] Yucheng (2019), p. 87.

[22] Xiaoxia (1998), p. 34.

[23] Lihong (2000), p. 56.

[24] Li (2011), p. 133.

[25] Jinzhao (2011), p. 111.

Lessons can be learned about problems experienced in western societies regarding the relationship between the executive and judiciary—in particular the 'night police state',[26] the 'welfare state'[27] and 'risk society'.[28] First, the expansion of executive power is a general development trend; Second, on the issue of public interest, the function of judicial power should be adjusted in accordance with the requirements of the administrative power. Third, to ensure the public's participation in the administrative process, judicial power must ensure this participation, but the public's participation in the judicial process cannot be replaced by its participation in the administrative process. It is concluded from this that executive power and judicial power should form a dynamic balanced relationship, in which, on the one hand, the two powers should respect each other; on the other hand, the executive power should be given priority. That is, we should first consider solving various public decision-making and management problems within the scope of administrative power. Only when the executive authorities cannot perform their duties properly, or fail to act in a timely manner, should judicial authorities intervene in the process. They should not bypass the executive authorities and directly intervene in public decision-making or regulatory affairs.[29] Even in the context of the modern Chinese active judiciary, it should still abide by the boundaries and play a role in supervising the executive power. Judicial power can correct the inaction or disorderly actions in the administrative litigation procedure, and can also show a positive attitude of active justice, put forwards judicial suggestions to the administrative agency, urge it to correct mistakes in a timely manner, and solve the problem before litigation.[30] However, the judiciary should not replace the administrative agencies role in handling affairs within the scope of administrative powers.

As the ecological environment manager, the government enjoys unified supervision and management responsibilities for China's ecological environment. Environmental infringement has the typical characteristics of complexity and uncertainty. Administrative power has the advantages of initiative, flexibility and diverse forms, which facilitate the effective implementation of pre-prevention and ex-post sanctions against environmental problems. Therefore, the government occupies a central position in the environmental legal system and should be the direct guardian of environmental public interests.[31] Even if the administrative counterpart does not cooperate or even opposes law enforcement, the government can also use administrative coercive measures or administrative penalties to enforce the law. Compared with litigation procedures, law execution is more flexible and efficient, and environmental law enforcers are more specialized in ecological and environmental issues.

[26] Stewart (2002), p. 21

[27] Pézier (2002), p. 15–17.

[28] Baker and Wilms (2001), p. 119.

[29] Mingyuan (2016), pp. 58–64.

[30] Tao (2016), p. 37.

[31] Bao (2017), p. 17.

There is no valid reason to request judicial intervention based on environmental supervision duties when the ecological environment is damaged.

16.3.3 *Environmental Public Interest Theory*

There are also scholars who argue that the ecological and environmental damage compensation lawsuits can be regarded as special environmental civil public interest lawsuits. This is mainly because the ecological environment damage compensation lawsuit is similar to the environmental civil public interest lawsuit in terms of its purpose, the reasons for undertaking the action and the claims that are made. There are however differences in the plaintiff's qualification, the scope of application and the pre-litigation procedures, so it is appropriate to characterize the former as a special civil EPIL.[32]

According to the environmental public trust theory, "environmental elements such as atmosphere, sunlight, and water are the common property of the public. Without the consent of all the common owners, no one may arbitrarily occupy, dominate or deplete it. In order to reasonably utilize and protect such public property, the public entrusts it to the government for management based on trust, and the state, as the trustee, has the responsibility to manage and protect the entrusted property for the realization of the public interest. When such public interests are damaged, compensation can naturally be claimed by the government".[33]

The theory of public trust originates from the common law. Although Chinese legislation and academic circles have not confirmed this theory, there is a similar theory of "representative trust". In this theory, the property ownership of the trust is indivisible, and all citizens entrust the environmental public interest to the government for supervision and management. Even though the law retains a system in which the public can supervise environmental issues by reporting, making suggestions, etc., it is impossible to directly sue for these public properties based on citizenship. The representative trust system provides that public authorities, including administrative agencies and Prosecutor's Offices, as representatives of the people have the right to request judicial relief in the name of the state when environmental public interests are infringed.[34] This theory is basically similar to the principle of public trust theory in the common law system, and reasonably explains why the government represents the environmental public welfare. In fact, the state's power to supervise the environment is an inevitable product of the development of the market economy, and it is needed to protect public interests.[35]

---

[32] Hao (2019), p. 60.

[33] Shouqiu (2013), p. 7.

[34] Tao (2007), p. 6.

[35] Zhongmei (2000), p. .92

It can be seen from the reform plan and judicial interpretation that the main goal of establishing an ecological and environmental damage compensation system is not to supplement the economic value, but to realize compensation for the ecological value of the ecosystem. "The logical starting point and ultimate goal of administration is to realize public interests".[36] The government's performance of environmental supervision duties is the process of continuously realizing environmental public welfare. When this process is blocked, the government can use administrative power to solve it, but administrative power is not able to solve all the problems, especially in the context of the relatively low cost of environmental violations in China (although the new "Environmental Protection Law" and the revision of the separate laws have solved this problem to a certain extent). It is often difficult to completely restore the ecological environment to its pre-damaged state by using only administrative coercion or administrative punishment to pursue offenders. Especially when the behavior is not illegal, for example, pollution behavior complying with pollutant discharge standards causes environmental pollution. This behavior is not reprehensible in public law, and administrative responsibility cannot be pursued. However, the environment may still be damaged. When administrative power does not work, the relief function of the judiciary should spare no effort to play its role.

Therefore, when the government represents the public to maintain environmental public welfare, on the one hand, it uses executive power to manage the ecological environment. On the other hand, when executive power is insufficient, an ecological environment damage compensation lawsuit could be filed to request judicial relief. The functions of these two processes are quite different. The former focuses on preventing environmental problems through administrative regulation and punishes offenders when environmental problems occur, while the latter focuses on restoring the damaged ecological environment and providing relief to the damaged environmental public welfare.

Although in the Chinese environmental protection system, environmental public interests remain at the theoretical level instead of being adopted by legislation, the public's requirement to live in a healthy and beautiful environment cannot be denied. Although the legislation has not yet clearly stipulated "environmental rights", a series of legal provisions on the basic state policy of environmental protection, the government's responsibility for environmental quality, and public participation fully reflect the country's respect and protection of citizens' environmental rights. Even if it is not legally named, it cannot be denied that such interests are embodied in legislation under other names. In addition, the development of socialized large-scale production has increased the integration of the political state and civil society. The changes in the relationship between public interests and individual interests have led to the socialization of legal functions and the publicization of legal interests. Environmental law has always been designed to protect the environment. The new system established for the purpose of maintaining public welfare is the product of the privatization of both public and private law as well as the product of breaking

[36] Yingfu and Xisheng (2018), p. 35.

through conceptual jurisprudence.[37] In the process of writing and implementing environmental laws, it is necessary to break away from traditional legal concepts; otherwise, it will create an obstacle to the function of environmental law to protect the environment. The EPIL system also has the environmental public interest as the object of relief.[38] On this basis, taking the environmental public interest as the source of rights of claim in the lawsuit for damages to the ecological environment would not be described as a 'castle in the air' lacking an institutional foundation.

To sum up, there are logical and institutional flaws in taking the ownership of natural resources or the responsibility for environmental supervision as the basis of the right of claim for damages to the ecological environment. Only by taking the environmental public interest as the basis of the claim can division and cooperation between the administrative and judicial powers in the field of environmental protection be reasonably explained. Although the responsibility of environmental supervision is also based on the government's fiduciary duty to maintain environmental public interests, the basis of the right of claim must be clarified and the most fundamental types of rights and interests that can be litigated and protected through litigation should be carefully identified. This is analogous to the situation in which a minor's rights are infringed upon and a guardian files a lawsuit on the minor's behalf. The basis of the right of claim is the personal rights of the minor, not the guardianship of the guardian. Guardianship is only a procedural right that enables guardians to sue on behalf of the minors, not a substantive right to remedy in a tort lawsuit. In the same way, although the government's environmental supervision right is inseparable from environmental public welfare, only the latter can be used as the basis for the right to claim compensation for ecological environmental damage.

16.4 System Reconstruction: Establishment of Binary Mode of Administrative EPIL and EEDCL

In each of the three types of lawsuits mentioned above, there are generally four types of subjects: the government, the prosecutor's office, the environmental protection organizations and actors who cause environmental pollution and/or ecological damage. Except for the government (and the relevant departments of environmental law enforcement), which may become the defendants of administrative EPIL, other lawsuits are directed at the perpetrators of ecological and environmental damage. The potential litigation relationship is shown in Fig. 16.1.

[37] Zhongmei (2000), p. 103.
[38] Xiaobo (2019), p. 40.

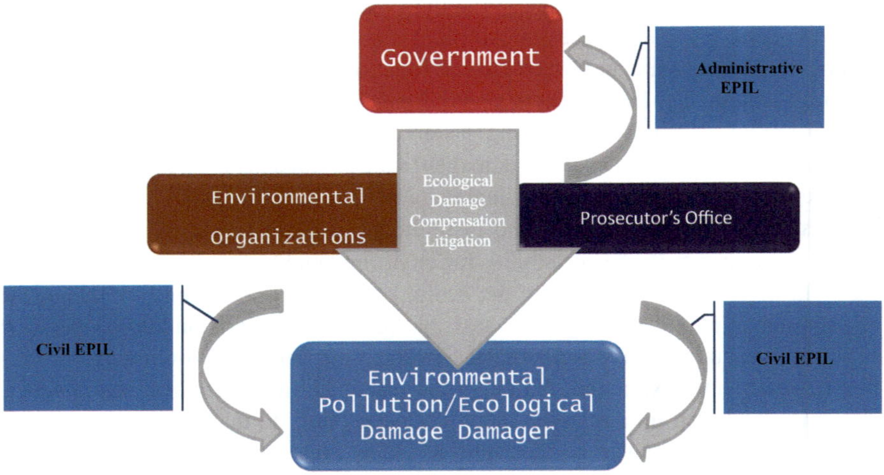

Fig. 16.1 Environmental Public Interest Relief Litigation of the Ternary Model (Source: by author)

16.4.1 The Value of Administrative EPIL System and the Expansion of Plaintiff Scope

16.4.1.1 The Value of the Administrative EPIL System

As the main body of environmental supervision, the government bears the legal obligation to protect environmental public welfare, and also has various resources, professional technologies and business capabilities to deal with ecological environment-related issues. The government's management of the environment has the advantages of normalization, systematicness and authoritativeness. However, many local governments often cause or aggravate environmental problems due to decision-making mistakes, administrative omissions or improper actions. Therefore, it is of great importance to urge the government to correctly perform its environmental supervision duties and thereby constrain enterprises or individuals to carry out various environmental development and utilization behaviors in accordance with the law. The cost of supervising the government is much lower than that of supervising a large number of scattered enterprises. Therefore, some scholars propose that administrative EPIL is of greatest importance in the EPIL system.[39]

From the perspective of comparative law, German environmental groups cannot directly file civil public interest lawsuits against enterprises and can only exercise administrative public interest litigation rights. The German EPIL system began in 1979. The state of Bremen stipulated that environmental groups could file public interest lawsuits when it revised the Nature Conservation Act, and other states have since followed suit. At that time, Germany had not yet established environmental

[39] Xi (2016), p. 114.

group lawsuits in federal law because most scholars believed that the regulation of environmental issues should be mainly completed by administrative agencies; otherwise, it would be inconsistent with the traditional German legal system and legislative model. In 2003, the European Union stipulated a group lawsuit system in the "Environmental Participation Directive". The purpose of this system is to allow the public to supervise administrative law enforcement. On this basis, Germany, a member of the European Union, also clarified that environmental groups should be granted the right to appeal in environmental administrative public interest matters at the federal level when revising the Federal Nature Protection Act. In the 2007 *Environmental Damage Law*, it is more clearly stipulated that when business activities cause environmental damage, environmental protection groups have no right to directly file a civil lawsuit against the person responsible for restoration or take appropriate measures to restore the environment to its original state, but can file a lawsuit against the government. The administrative lawsuit requires the government to order the operator to restore the environmental damage to make up for the damage caused by it.[40]

Representative citizen suits in the United States also mostly use law enforcement agencies as defendants. For example, in the *Sierra Club v Morton*, 405 U.S. 727 (1972), Morton was the Secretary of the Interior and in the Storm King case (*Scenic Hudson Preservation Conference v Federal Power Commission*, 354 F.2d 608 (2d Cir. 1965)), the defendant was the Federal Power Commission. It should be noted that US environmental citizen lawsuit defendants are not limited to government departments. For example, the *Clean Air Act* stipulates that there are two main types of citizen lawsuit defendants: the first is all those who violate the standards or restrictions stipulated in the law; the second is the Director of the Federal Environmental Protection Agency in the event of a violation of a non-discretionary act or duty.[41] In addition, the *Resource Conservation and Recycling Act* has a similar provision: "Defendants include everyone who violates any permit, regulation, provision, requirement, prohibition or order that has entered into force under this Act, as well as the directors of Environmental Protection Agencies who fail to perform acts or obligations under this Act that are not within the discretion of the EPA." It can be seen that the Environmental Protection Director is one of the key defendants in citizen lawsuits in the US environmental statutes, and that citizens who take the Environmental Protection Director as the defendant in litigation is similar in nature to administrative EPIL in China. In the system of separation of powers in the United States, judicial power influences the exercise of executive power to a large extent. Therefore, in the maintenance of environmental public welfare, the judicial power often changes its role, and even conflicts with the administrative power in terms of the division of authority and functional positioning,

---

[40] Shijun (2007), p. 9.

[41] Dong (2004), p. 90.

all of which are based on the independence of the judiciary.[42] Therefore, EPIL can be regarded as a social change in the environmental field driven by judicial power.

The human, financial, and material powers of China's judicial organs are all controlled by the local government. There are countless links between the local judicial organs and the government.[43] However, from the perspective of power allocation and functional positioning, judicial power has the efficiency of supervising administrative power. Therefore, in the field of ecological and environmental protection, the government (and its relevant environmental law enforcement departments) confers more qualified subjects the power on environmental supervision, as a way to avoid failure of performance in environmental supervision by itself. It is necessary to file a lawsuit to safeguard environmental public welfare. This system design follows the management mode in which the government controls environmental risks, while the judiciary mainly supervises the administration of government departments according to law through administrative EPIL, playing the role of environmental regulation checks and balances.[44] The Interpretation of Procuratorial Public Interest Litigation stipulates that the judgment outcomes of administrative EPIL include "confirming the violation of the law, ordering to take remedial measures and renewing the judgment, making administrative acts, performing within a time limit and changing of judgment", etc. In addition to the direct judgment change for the obvious misconduct of administrative punishment, in other judgment methods, the court has no right to directly issue instructions to the perpetrator of ecological environmental damage, but can only require the defendant to perform their duties according to law, to indirectly realize the purpose of environmental public welfare relief. The system has the theoretical support and institutional basis of cooperation among China's legislative, law enforcement and judicial organs, and its legitimacy and effectiveness are beyond doubt.

16.4.1.2 Incorporate Environmental Protection Organizations into the Scope of Plaintiffs in Administrative EPIL

According to China's current laws, judicial interpretations and normative documents, only the prosecutor's office can initiate administrative EPIL. The Constitution stipulates that "the prosecutor's office is the legal supervision organ of the state." Article 11 of the Administrative Procedure Law endows the prosecutor's office with the power of legal supervision over administrative proceedings. In essence, the prosecutor's office initiates administrative public interest litigation to supervise the exercise of administrative power with judicial power, so as to safeguard public interests. Fundamentally speaking, the people are the ultimate holders of state power. They hand over their individual rights to the common body of the

[42] Mingyuan (2016), p. 51.
[43] Shan (2017), p. 108.
[44] Jiang (2020), p. 145.

state, thus forming the power of the state and the government.[45] "The Theory of State Active Action" advocates restricting one type of power with another type of power in legal system construction. Administrative EPIL is a positive way of restricting administrative power with procuratorial supervision power.[46]

The reason for choosing the prosecutor's office as the only qualified subject of prosecution in administrative EPIL is as follows. On the one hand, the prosecutor's office as the plaintiff in administrative public interest litigation can coordinate and cooperate with legal supervision and judicial powers. Compared with other subjects, it can play a more effective supervisory role.[47] On the other hand, it is difficult for natural persons, legal persons or other organizations to compete with the power of administrative organs in litigation. Therefore, it is relatively difficult for them to maintain environmental public welfare by correcting the behavior of administrative organs. Finally, the prosecutor's office and administrative organs have no stake in the law enforcement behavior, so they can maintain relative independence, and can more objectively and impartially handle the problems in administrative public interest litigation.

It must be admitted that under the background of China's current legal system, citizens, legal persons or other organizations in litigation may indeed be at a disadvantage compared to administrative agencies, but this negates the possibility of these relatively weak subjects filing administrative EPIL. The reason for the possibility is not sufficient; otherwise, according to this logical approach, ordinary administrative litigation will also be unable to proceed due to the disparity of power between the plaintiff and the defendant. Chinese legislators have formulated some special rules for such issues in administrative litigation, such as the defendant's proof system,[48] to balance the litigation power of both parties. In addition, the judiciary, as a neutral judge, also has an important role in balancing interests during the trial. Therefore, it is not necessary to completely deny the opportunity to litigate because the subject's ability to litigate may be lacking before the lawsuit is filed. Moreover, using the prosecutor's office as the sole subject of prosecution will inevitably put it in a situation where it is weak or even overwhelmed. After all, as a special agency for national legal supervision, the prosecutor's office must be responsible for countless supervision matters. If other subjects can be included in the scope of plaintiffs, at least part of the work burden of the Prosecutor's Office in the field of administrative public interest litigation can be relieved.

To think that the prosecutor's office can play a more sufficient role in supervising administrative power than other prosecuting subjects obviously ignores the close

[45] Chengdong (2008), p. 183.

[46] Fei and Yonghe (2020), p. 61.

[47] Yi (2017), p. 16.

[48] In China, in ordinary civil or criminal proceedings, the rule of proof in litigation is "who claims, who gives evidence", that is, the plaintiff proves the cause of his claim. However, in administrative litigation, considering the defendant's superior evidentiary strength over the plaintiff, the defendant's evidentiary rule is adopted.

relationship between the public and the environment. Those who live in the environment are always the most sensitive to environmental issues. Since their own rights and interests may be violated, the public is also particularly concerned about environmental issues. Let the public supervise environmental law enforcement agencies, and its enthusiasm and efficiency will not be lower than the Prosecutor's Office, despite the latter having statutory supervision functions. Although the public may have subjective emotions because of their interests in the case or undertake vexatious and abusive litigation, it is only necessary to make reasonable restrictions on the qualifications of plaintiffs. Article 58 of the Law on the conditions of the plaintiffs of civil EPIL can provide experience for reference.

From an international point of view, in Australia's administrative public interest litigation, the Attorney General, private individuals or interest groups can bring lawsuits to the court against acts of administrative agencies that damage the public interest (or public power).[49] The *Gouriet v Union of Post Office Work* [1978] AC 435 case in the United Kingdom in 1977 established the private prosecutor system. When the public interest of society is actually infringed upon, but the attorney general who is qualified to prosecute refuses to prosecute or delays the prosecution without reason, individual citizens can sue as private prosecutors. Neither the United States nor Germany has denied that other types of subjects other than prosecutors have the right to file environmental public interest lawsuits against government actions.

The author believes that it is of practical significance and feasibility to further expand the scope of plaintiffs in administrative EPIL. First, other subjects can share part of the administrative EPIL work for the prosecutor's office, reducing the pressure of their litigation work. When the prosecutor's office cannot supervise the administrative power in a timely and effective manner, other bodies can be used to supervise the environmental law enforcement behavior of the administrative organ, to carry out environmental administrative supervision work in a legal and timely manner, as well as to supervise the ecological and environmental law enforcers more comprehensively and fully. It is conducive to the effective protection of environmental public interests. Second, the legislative provisions of other countries and the provisions on the qualifications of plaintiffs in civil EPIL in China can provide useful experience and reference for broadening the scope of plaintiffs in administrative EPIL. Some environmental protection organizations have shown a high degree of enthusiasm for litigation and outstanding professional capabilities in safeguarding environmental public interests. They should participate in administrative EPIL, and promptly and effectively urge the government to fully perform its duties. Comparing environmental civil public interest lawsuits, which can only be brought against polluters or destroyers after the results of ecological damage have occurred, the administrative EPIL is more efficient, less costly, and more in line with the precautionary principle and public participation principle. Judging from the practical experience of civil EPIL, before the system was incorporated into legislation, it

[49] Yingping (2012), p. 91.

was feared that there would be a risk of abuse of litigation, but it did not actually occur.

If it would be too aggressive to include ordinary citizens in the scope of plaintiffs in environmental administrative public interest lawsuits, it can be first referred to the existing limitations on the qualifications of plaintiffs in environmental civil public interest lawsuits, and grant some environmental organizations that meet the statutory conditions the right to sue in environmental administrative public interest lawsuit rights.

16.4.2 Remove the Existing Civil EPIL System and Build a Public Interest Litigation System for Ecological and Environmental Damage

16.4.2.1 Rationality Analysis of Removing the Existing Civil EPIL System

The previous section has made a comparative analysis of civil EPIL and EEDCL. It can be seen that many important factors in the two litigation systems have commonalities. It is better to eliminate the possible conflicts between the two and design a connection mechanism. It should be integrated to build a new litigation paradigm for environmental public interest relief with environmental pollution or ecological damage actors as defendants, and then together with administrative EPIL constitute a reconciled environmental public interest relief litigation system.

Some scholars have suggested that administrative weakness is the reason for the emergence of civil EPIL in China. Prioritizing administrative EPIL is tantamount to returning the problem to the administrative department with weak enforcement, which will lead to difficulty and inefficiency in public welfare maintenance. If the civil EPIL is the main method, this situation will be changed. By targeting environmental polluters or ecological destroyers, the damaged environment t can be restored as soon as possible.[50] This kind of thinking obviously places too much expectation on judicial power, and even at the cost of crossing the boundary of power, the work that should be done by the main body of environmental administrative supervision is transferred to the court to complete. Environmental law enforcement agencies have a higher understanding of environmental issues, richer professional knowledge, and stronger technical capabilities than judicial agencies, and are originally responsible for environmental supervision. The new "Environmental Protection Law" has made environmental protection law enforcement more powerful, and it is obviously more reasonable and efficient for the environmental law enforcement department to shoulder the responsibility of maintaining environmental public welfare.

[50] Weijian (2015), p. 3.

If the environmental law enforcement department has administrative omissions or illegal acts, resulting in damage or expansion of environmental public interest, the administrative EPIL system can effectively deal with it. The system design in which the public focuses on law enforcement can not only urge law enforcement to enforce the law actively and correctly, but also have a broader radiation effect in the future. It is more effective than suing polluters or ecological destroyers, with lower costs and better results.

In addition, as mentioned above, at the beginning of designing the civil EPIL system in China, the goal was mainly to prevent the occurrence of the "Tragedy of the Commons". Since the EEDCL system has made the government (and its designated departments and agencies) the plaintiff in such actions, it allows them to seek compensation for damage to the ecological environment. Therefore, the hidden worry of the "Tragedy of the Commons" has been eliminated. It is the government's right and responsibility to safeguard environmental public welfare on behalf of the public. Today's government is trying to reverse the original one-way administrative management concept, changing the power-based administrative control method based on administrative punishment, weakening administrative control in the process of performing environmental supervision functions, and encouraging enterprises and the public to actively participate in environmental protection work.[51] The traditional administrative regulation method has limitations. Substances and activities with potential risks cannot become the object of power regulation until their harmfulness is proven.[52] In essence, acts that are not illegal will not become the object of administrative punishment—yet the damage to the ecological environment cannot be ignored and the problem must be solved through diversified methods. The limitations of traditional rigid law enforcement have led to the demand for flexible law enforcement in modern society. The problems that the government cannot solve with mandatory law enforcement can be solved flexibly through information exchange, administrative contracts, consultations, etc. This has led to the commencement of the ecological environment damage compensation system.

In environmental supervision, the government requires offenders to correct their illegal behavior, and can use administrative power to issue instructions to those under supervision. When the instructions are not complied with, they can implement administrative coercive measures or administrative penalties; however, administrative measures often fail to make up for environmental public welfare. In terms of the payment of damages, the other party can be asked to compensate through negotiation, and if the negotiation fails, the judicial authority can be asked for relief. Now that the government has been granted the litigation qualification to maintain environmental public interest through litigation, other subjects only need to supervise the government to effectively perform its environmental supervision duties, and there is no need to file civil EPIL.

[51] Shouqiu (2013), p. 7.

[52] Jun (2008), p. 96.

16.4.2.2 Construction of the Public Interest Litigation System for Ecological and Environmental Damage

(1) Litigation Location and Name

Article 8 of China's Legislation Law stipulates that the "Litigation and Arbitration System" is a "matter that can only be enacted by law". Therefore, it is only an expedient measure to stipulate the EEDCL system in the form of a policy document, and ultimately, the litigation system must be stipulated in legal form.

The purpose of the establishment of the EEDCL system is to request relief for the damage to the ecological environment itself, mainly for the reduction or destruction of environmental elements. The essence of claiming compensation for such damage is the process of internalizing the external cost of the actor who damages the ecological environment, and does not involve personal or property damage.[53] Tort law, as an important part of the civil law department, remedies private interests. If the difference between ecological environment damage and private rights damage is ignored, and the former must be classified as the adjustment field of civil law, it will confuse "public" and "private", which will not only break the original logic of "private law" in the civil legal system, but may also lead to damage to the public welfare of the ecological environment.[54] Therefore, it is inappropriate to stipulate the liability for ecological and environmental damage in the "Tort Liability Section" of the Civil Code, and it is more reasonable to stipulate this issue in a special legislative system.

Furthermore, the title of the lawsuit must also be changed. The "compensation" in the existing name comes from the "compensation for damages" in tort liability, which obviously cannot cover all the functions and objectives contained in such a lawsuit. To be repaired, there are various ways of assuming responsibility, such as stopping infringement, removing obstacles, eliminating danger, and making apologies. In view of this, it is more accurate and appropriate to change the word "compensation" in the title of the lawsuit to "public interest", which is convenient to demonstrate the purpose of this kind of litigation to relieve environmental public interest. "Eco-environmental damage"can point to the lawsuit against the perpetrator of ecological and environmental damage, so as to clarify the defendant and the reason for the lawsuit.

(2) Plaintiff

As mentioned above, the government, as the trustee in the environmental public welfare trust, is the natural subject to represent and protect the environmental public interest. Therefore, when environmental public welfare is damaged by others, it is more appropriate for the government to file a public interest lawsuit as the plaintiff. Environmental protection organizations and the Prosecutor's Office can focus their

---

[53] Haisong (2018), p. 24.

[54] Yixiang and Xingyu (2018), p. 89.

supervision on government environmental law enforcement. If they find problems in their law enforcement, they can seek solutions through administrative EPIL.

(3) Pre-litigation Procedure

According to the existing ecological and environmental damage compensation system, consultation has been stipulated as a pre-procedure. This model can prompt the problem to be resolved through consultation between the two parties before the lawsuit, which is helpful to save judicial resources and should be retained.

(4) The Burden of Proof of Causality

Article 6 of the "Several Provisions" requires the plaintiff to prove that "there is a correlation between the defendant's behavior of polluting the environment, destroying the ecology and damage to the ecological environment",[55] which is consistent with the statement in the "*Interpretation of Several Issues Concerning the Application of Law in Environmental Tort Liability Dispute Cases*".[56] It is generally believed that in environmental infringement, the infringed is often in a weaker position than the infringer in terms of economic and technological strength, information mastery and evidence collection. Based on the consideration of balancing the probative ability of both parties, the plaintiff only needs to meet the low probability standard of causality before the declaration of consummation. Public interest litigation for ecological and environmental damage is a lawsuit of "official suing the people", and the ability of both parties to produce evidence is completely different from that of the "people suing the people" environmental infringement lawsuit. Moreover, in China's "people suing the official" administrative litigation, the defendant is required to bear the burden of proof of the legality of its administrative actions. If the defendant is instead required to provide evidence in this "official suing the people" lawsuit, it will not be persuasive either from a logical

[55] Article 6 of the Several Provisions of the Supreme People's Court on the Trial of Cases Involving Compensation for Ecological Environment Damage (Trial Implementation) provides: "If the plaintiff claims that the defendant bears the liability for compensation for ecological environmental damage, it shall bear the burden of proof on the following facts:

(1) The defendant has committed acts of polluting the environment or damaging the ecology, or has other circumstances that should bear responsibility in accordance with law;
(2) The damage to the ecological environment, as well as the specific amount of restoration costs needed, damage compensation, and so forth;
(3) There is a correlation between the defendant's behavior of polluting the environment, destroying the ecology and damage to the ecological environment."

[56] Article 6 of the Interpretation of the Supreme People's Court on Several Issues Concerning the Application of Law in the Trial of Environmental Tort Liability Dispute Cases stipulates: "Where the infringed party requests compensation in accordance with Article 65 of the Tort Liability Law, it shall provide evidentiary materials proving the following facts:

(1) The polluter has discharged pollutants;
(2) Damages suffered by the infringed party;
(3) There is a correlation between pollutants or secondary pollutants discharged by polluters and damage."

standpoint or from the perspective of the legitimacy of the rules. Therefore, it is more reasonable for the plaintiff to bear the burden of proof of causation.

(5) The Way of Assuming Responsibility

The "Several Provisions" clearly stipulate the possible liability methods of the defendant in ecological and environmental damage compensation lawsuits.[57] If the defendant's behavior infringing on the ecological environment is still in progress, it must first be ordered to stop the infringement; if the behavior has stopped, there is no room for this liability method. If the defendant's state of obstructing environmental public interests continues, they should be required to remove the obstruction so that the public can continue to enjoy environmental public interests. If the defendant's actions only caused the risk of damage to the ecological environment but did not cause actual damage, the defendant should bear the responsibility for eliminating the danger. If the ecological environment has been damaged, the responsibility for ecological environment restoration should be selected first, which can be repaired by itself, or paid for by other agencies. This method of responsibility is mainly to restore the damaged environment to its original state and to compensate for the damage suffered by various functions of the ecological environment itself, aiming to protect the public interest of the environment. In this restoration process, the public authority is usually involved, and the public interest measurement and protection mechanism is followed.[58] In the event that the damaged ecological environment cannot be repaired, alternative restoration schemes can be selected. The "remedial replanting" adopted by the company can flexibly achieve the value goal of making up for the public welfare of the ecological environment.

When the permanent damage to the ecological environment cannot be repaired and an appropriate alternative restoration scheme cannot be adopted, the defendant shall be liable for compensation. Even if the defendant assumes the responsibility for ecological restoration, the loss of the ecological environment during the period after the ecological environment is damaged until the completion of the ecological restoration work should also be compensated. Some scholars have pointed out that China's ecological and environmental damage compensation has always followed the logic of "filling up damage" in the traditional civil damage compensation, and embodied the characteristics of "pursuing complete compensation, abandoning punitive compensation, and combining ecological restoration with monetary compensation". In turn, it hinders the realization of the institutional function of regulating behaviors that damage the ecological environment.[59] The Civil Code adopts punitive damages for intentional environmental infringement. Taking this as a

[57] Article 11 of the "Several Provisions" stipulates: "Where the defendant violates laws and regulations to pollute the environment and damage the ecology, the People's Court shall, according to the plaintiff's claims and the specific circumstances of the case, reasonably order the defendant to undertake the restoration of the ecological environment, compensate for losses, stop the infringement, remove obstacles, eliminate danger, apologize and other civil liabilities".

[58] Zhongmei (2017), p. 11.

[59] Shouqiu and Yi (2018), p. 61.

reference, in public interest litigation for ecological and environmental damage, if the perpetrator has subjectively or objectively engaged in many instances of behavior that damages the ecological environment, punitive damages may be used. Compensation for damage can be needed, and it can also have a deterrent effect on potential subjects who may cause ecological and environmental damage in the future.

(3) The Relationship Between Administrative EPIL and the Ecological Environment Damage Public Interest Litigation
The maintenance of environmental public welfare should mainly rely on the government to perform its environmental public management responsibilities, including preventive environmental supervision and law enforcement and claims for damage to the ecological environment. It should be noted that government claims are distinct from private law rights. The subject of private rights has the freedom to abandon the right, and this "right" of the government is more similar to a "duty". As the trustee, the government protects environmental public welfare on behalf of all citizens. When such interests are infringed upon, if the damage to environmental public welfare cannot be effectively compensated due to the limitations of administrative law enforcement, the government has the right and the obligation to request judicial relief. Due diligence is an inevitable extension of the orderly exercise of environmental supervision rights for the interests of the client (the public). In other words, "claims" are a derivative of the government's regulatory responsibilities when administrative power is insufficient to compensate for the damage to the environmental public interest. When the government is lax in making a claim, the prosecutor's offices and environmental protection organizations can file an environmental administrative public interest lawsuit to require the government to perform its duties.

16.5 Conclusion

The ternary litigation model of civil EPIL, administrative EPIL and EEDCL that has been constructed in the field of environmental public welfare protection in China has an obvious policy goal orientation and reflects the characteristics of "movement-style" legislation. The existing civil EPIL and the EEDCL system have the defects of overlapping content and redundant system design, which may result in unnecessary costs for the plaintiff and the judiciary. There are defects in content cross-conflict and redundant system design between the civil EPIL and EEDCL systems.

Plaintiffs and judicial organs may incur unnecessary costs, so it is necessary to optimize and integrate the following issues, which include removing the existing civil EPIL system, retaining administrative EPIL, including environmental protection organizations in the scope of plaintiffs in such litigation, and supervising the government's environmental law enforcement. The government can use the power of law enforcement to supervise the ecological environment, and when administrative power is insufficient, it can request that judicial organs provide relief for the

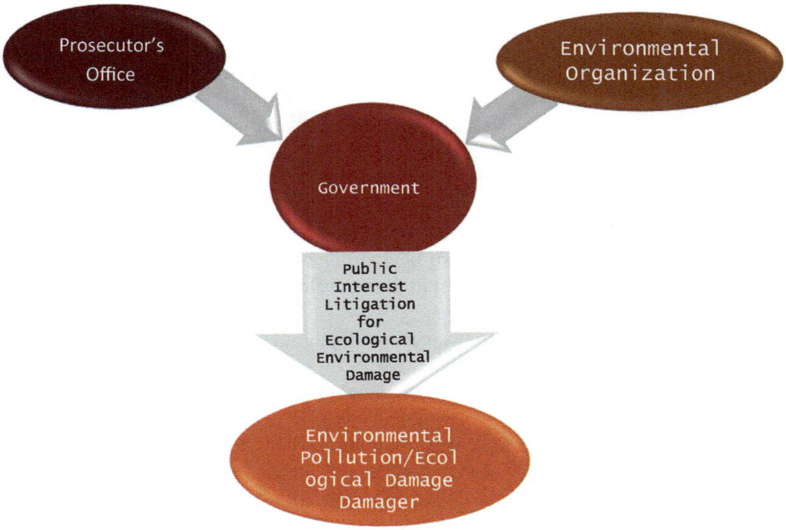

Fig. 16.2 Environmental Public Interest Relief Litigation Dual Model (Source: by author)

damaged environmental public interest through public interest litigation for ecological and environmental damage. This helps to clarify the relationship between the parties, which is clearer than the existing litigation system as shown in Fig. 16.2.

Different litigation models designed by the Chinese government focus on the core goal of environmental public interest protection, although the litigation elements are different, but the value and the function are the same—all pointing to the filling, correction and punishment of environmental damage.

References

Baker U, Wilms J (2001) Freedom and Capitalism: Discuss With the Famous Sociologist Ulrich Baker [M] (trans: Guolin L)

Bao Z (2017) An analysis of the applicable relationship between government civil claim and supervision power in ecological environment damage relief. Legal Forum 3:14–21

Bixin J (2019) The practical development and system improvement of environmental public interest litigation in China. National People's Congress 11:50–54

Chengdong W (2008) Government responsibility theory [M]. China University of Political Science and Law Press, p 65

Dong C (2004) Research on the environmental public interest litigation - focusing OD the environmental citizen suit in the United States of America [D]. Ocean University of China

Fei L, Yonghe X (2020) Public prosecutor's status and institutional construction of procuratorial organs in administrative public interest litigation. Zhejiang Soc Sci 1:59

Ganjie L (2017) Let the reform of the ecological environment damage compensation system become a solid guarantee for people's environmental rights and interests. China Environ News:12–18

Gu G (2014) A review of the "non-public right theory" of state ownership of natural resources. China Law Rev 4:141–158
Haisong C (2018) Reflection and reconstruction of ecological environment damage compensation system - the perspective of constitutional interpretation. Orient Law 6:20–27
Haiyong X (2015) The triple nature of basic functions of claim rights. People's Just 11:53–59
Hao L (2019) Research on the nature and related issues of ecological environment damage compensation litigation - analysis from the perspective of civil environmental public interest litigation. Adm Law Rev 4:55–66
Houfu Y (2006) Sierra Club v. Morton, secretary of the interior, et al. World Environ 6:28–33
Jiang H (2020) Research on the role of court in environmental regulation–thinking from the mode selection of environmental public interest litigation. J Beijing Inst Technol (Soc Sci Edn) 1:141–149
Jin W (2018) On ecological damage compensation litigation and construction of rules for convergence among connected litigations—using case of Dystar Nanjing Colors Co., Ltd. and related cases in ecological and environmental damage compensation litigation for reference. Environ Prot 5:35–40
Jing C (2015) Solving the dilemma of "enterprise pollution and government paying"—the relevant person in charge of the Ministry of Environmental Protection interprets the "Pilot Plan for the Reform of the Compensation System for Ecological Environmental Damage": http://www.xinhuanet.com/fortune/2015-12/03/c_1117349597. Accessed 6 July 2022
Jinzhao C (2011) "Active justice" and the anxiety of the rule of law theorists. Tsinghua Law J 3: 107–122
Jun X (2008) Legal principles and methods of environmental administration [M]. China Legal Publishing House
Li Y (2011) The definition of "active justice" in the context of contemporary China. Stud Law Bus 1:129–135
Lihong L (2000) The relationship between administrative powers and judicial powers–analysis from administrative litigation. Mod Law Sci 2:55–58
Lirong S (2018) Connection between ecological and environmental damage compensation and public interest litigation prosecuted by NGOs. J China Environ Manage Cadre Coll 5:6–9
Mingyuan W (2016) On the development direction of environmental public interest litigation in China: analysis based on the theory of the relationship between administrative power and judicial power. China Legal Sci 1:49–68
Pézier G (2002) French administrative law [M]. (trans: Kunming L, Jie Z), 19th edn. National Academy of Administration Press, pp 15–17
Shan L (2017) The comparing study on environment public interest litigation in Chinese and American [D]. Xiangtan University
Shijun Z (2007) Introduction and comment on the legal system of German environmental NGOs participating in environmental protection through environmental litigation - focusing on the "plaintiff qualification" in environmental public interest litigation. J Heilongjiang Adm Cadre Inst Polit Law 4:8–13
Shouqiu C (2013) From environmental rights to environmental protection obligations of the state and environmental public interest litigation. Mod Law Sci 6:3–21
Shouqiu C, Yi Z (2018) Principle of compensation for ecological environment damage in China and its improvement. Acad J Zhongzhou 10:56–62
Shuyi W, Huaqi L (2018) On the litigation of ecological environment damage compensation in my country. Learn Pract 11:68–75
Stewart (2002)The reconstruction of U.S. administrative law[M] (trans: Jiao S). The Commercial Press, p 21
Tao B (2007) Environmental public interest litigation. Law Press China, pp 5–15
Tao P (2016) Conflict handling rules between judicial power and administrative power [J]. Legal Sci (J Northwest Univ Polit Sci Law) 6:36–43

Weijian T (2015) Administrative Public Interest Litigation and Civil Public Interest Litigation can be Transformed into Each Other[N] 3
Xi W (2016) On the legislative sequence of environmental public interest litigation system. Tsinghua Law J 6:101–114
Xiaobo B (2019) Clarification of the object of civil environmental public interest litigation. J China Univ Geosci (Soc Sci Edn) 3:29–41
Xiaoxia S (1998) The essence of judicial power is the right to judge—the ten major differences between judicial power and administrative power. Law Sci 8:34–36
Yi L (2017) The judicial practice and theoretical exploration of public interest litigation brought by the Procuratorate. J Natl Prosecut Coll 2:4–18
Yingfu H, Xisheng H (2018) Connection dilemma and outlet between administrative consultation and judicial relief on ecological damage. J China Univ Geosci (Soc Sci Edn) 1:30–39
Yingping Z (2012) On plaintiff qualification for administrative public interest litigation in Australia. J Adm Law Res 3:88–97
Yixiang X, Xingyu L (2018) Embodiment and limitation of environmental interests in specific provisions of the civil code. J China Univ Geosci (Soc Sci Edn) 6:81–90
Yucheng S (2019) Theoretical reflection and legal construction of compensation system for eco environment damage. Acad J Zhongzhou 10:85–92
Zejian W (2002) Legal thinking and examples of civil law—the basic theoretical system of the right to claim. University of Political Science and Law Press, pp 30–45
Zhejiang People's Publishing House 119
Zhongmei L (2000) A new vision of environmental law. China University of Political Science and Law Press, pp 89–136
Zhongmei L (2017) Legal analysis of the "ecological environmental damage compensation". Legal Forum 3:5–13

Chapter 17
Drawbacks in the System Design of the Environmental Public Interest Litigation and Ways for Improvement

Xi Wang

Abstract With the complexity and risk of modern society, professional social public affairs need to be managed by professional administrative organs. The functions and duties of the People's Procuratorates and social organizations are obviously different from those of the Administration. The functions and duties of administrative organs shall be fully respected. The law should guarantee the effective exercise of executive power. Article 58 of the Civil Procedure Law and the related judicial Interpretation on public interest litigation have some serious flaws, which lead to arrogating executive power, weakening administrative power, tolerating idle administration, and improper expansion and dysfunction of judicial power. Viewed as a whole, the political and economic costs for implementing this system are too high and will do more harm than good. This chapter analyzes the disadvantages of the current environmental public interest litigation system and puts forward some suggestions for improvement.

Keywords Environmental public interest litigation · Civil EPIL · Administrative EPIL · Civil Procedure Law · Administrative Litigation Law · National Governance System

17.1 Introduction

A litigation system is a legal institutional arrangement for litigation actions. In China, the litigation system includes a civil litigation system, an administrative litigation system and a criminal litigation system. These three systems of litigation embody three types of legal relationships, namely civil, administrative and criminal

The author would like to thank Zhang Haoyu (graduate student) and Guo Xiang (PhD student) of Law School of Kunming University of Science and Technology for their assistance.

X. Wang (✉)
School of Law, Kunming University of Science and Technology, Kunming, China

X. Wang et al. (eds.), *Environmental Public Interest Litigation in China*,
https://doi.org/10.1007/978-3-031-26526-6_17

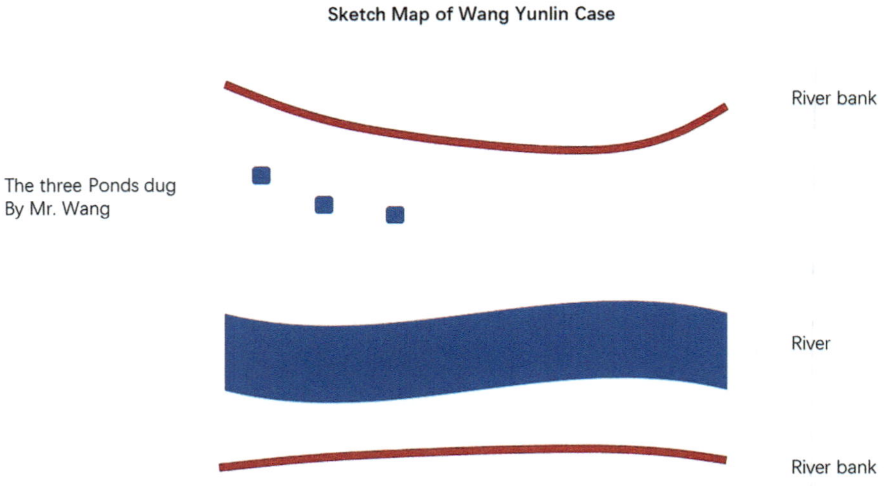

Fig. 17.1 The problems revealed by the Wang Yunlin Case (by author)

relationships. As each of the three legal relationships has its own unique internal logic and has its own purpose and function, they should not be confused by the system design of a legislature. The arrangement of the litigation system is a major legal issue that involves both the state governance system and the social welfare and people's vital interests. In this sense, the litigation system has its constitutional dimension and is a constitutional issue. Since it is a constitutional issue, the design and adjustment of the litigation system must be careful and prudent, and strive to meet the fundamental purpose of the litigation system. Otherwise, any slight adjustment of the litigation system may cause serious wrongs in the practice of law and undermine the overall efficiency of the operation of the state machinery and social justice.

Article 58 of the current Civil Procedure Law and related legal documents provide for a system of civil environmental public interest litigation (civil EPIL). From the perspective of theory and judicial practice, there are some drawbacks that cannot be ignored in the institutional design of the civil EPIL system. This chapter analyzes the current civil EPIL system. The analysis is based on the Wang Yunlin Case (Fig. 17.1), which was decided by the Intermediate People's Court of Yingkou City, Liaoning Province, in 2019.[1] This chapter reveals the serious drawbacks of the civil EPIL system and puts forward suggestions to improve the public interest litigation system of China.

On the merits, the case of Wang Yunlin is a simple case. According to the description in the judgement, the defendant farmer Wang Yunlin dug three large ponds on the benchland within the confines of the Biliu River in Gaizhou City,

[1]Supreme People's Court's Network on Chinese Judicial Documents, Civil Judgment of First Instance, Wang Yunlin Case (2019) (Liao 08 Min Chu No. 5).

Liaoning Province, for fish and crab farming. The local water conservancy department, Gaizhou City Water Conservancy Bureau, believed that these three large pounds endangered the safety of embankments, hindered river flooding, and damaged the water ecology of rivers. The Water Affairs Inspection Authority of Gaizhou City, which is under the Gaizhou City Water Conservancy Bureau, visited the site three times and was obstructed by the families of the defendant. Because of the obstruction, the People's Procuratorate of Yingkou City ("Yingkou Procuratorate") prosecuted Mr. Wang in the Yingkou Intermediate People's Court ("Yingkou Intermediate Court"). Mr. Wang was sentenced 1 year of imprisonment by the court. In addition to the criminal case, the Yingkou Procuratorate filed a civil EPIL against Mr. Wang Yunlin for allegedly endangering the safety of the embankment and the flooding of the river. The Yingkou Intermediate Court ruled against Mr. Wang, ordering him to restore the river to its original state or to pay the costs of a third party who restored the site on his behalf, and to bear the costs of appraisal and litigation costs associated with the proceedings. During the trial, the defendant Wang argued that in this case "the procuratorate should give preference to environmental administrative public interests litigation (administrative EPIL) rather than a civil EPIL." What Mr. Wang meant here is that the procuratorate should sue the local Water Conservancy Bureau rather than sue him. A case against the Water Conservancy Bureau would be a case against a governmental agency, which belongs to another category of litigation—administrative EPIL.

However, from the perspective of improving the governance system and governance capacity of the state, the Wang Yunlin case is not a simple one. This case, like all the other civil EPIL cases, reveals that the civil EPIL system is a system which is in contrary to the fundamental principles of public interest litigation and public affairs governance. In this case, the defendant Mr. Wang's claim that "the procuratorate should give preference to administrative EPIL rather than a civil EPIL" was ignored by the Yingkou Intermediate Court. However, the ignorance of the claim of Mr. Wang by the Yingkou Intermediate Court does mean that there is not a problem. Compared with the focal issue that the Yingkou Intermediate Court held, i.e., whether Wang's behavior caused ecological damage and whether he should be liable in tort and how he should make recompense for the damage caused, the issue raised by Mr. Wang is more fundamental. This is because if an administrative EPIL could be pursued to resolve the issues in this case, then there would be no need for the People's Procuratorate to opt for filing a civil EPIL. Consequently, if the choice of civil EPIL is unnecessary, then the institutional design of Article 58 of the Civil Procedure Law regarding civil EPIL is also unnecessary.

17.2 Civil EPIL and Administrative EPIL: China's Current Public Interest Litigation System

In China, public interest litigation (PIL) is a new form of litigation that has been flourishing in recent years. So far, the development of the PIL system has not been finalized. Currently, there are two forms of PIL: civil EPIL and administrative EPIL.

17.2.1 The Civil EPIL System

With respect to civil EPIL, Articles 58(1) and (2) of the Civil Procedure Law authorize "the organs and relevant organizations prescribed by law" and the People's Procuratorates to initiate civil EPIL.[2] According to the Civil Procedure Law and relevant laws and judicial interpretations,[3] in judicial practice, only the People's Procuratorates and certain qualified social organizations have the right to file environmental civil public interest lawsuits.

[2]Article 58 (1) of the Civil Procedure Law (2021) provides that "For conduct that pollutes environment, infringes upon the lawful rights and interests of vast consumers or otherwise damages the public interest, an authority or relevant organization as prescribed by law may institute an action in a people's court". Article 58(2) of the Civil Procedure Law (2021) provides that "Where the people's procuratorate finds in the performance of functions any conduct that undermines the protection of the ecological environment and resources, infringes upon consumers' lawful rights and interests in the field of food and drug safety or any other conduct that damages social interest, it may file a lawsuit with the people's court if there is no authority or organization prescribed in the preceding paragraph or the authority or organization prescribed in the preceding paragraph does not file a lawsuit. If the authority or organization prescribed in the preceding paragraph files a lawsuit, the people's procuratorate may support the filing of a lawsuit."

[3]Relevant legal and judicial interpretations are mainly as follows: Article 58, Environmental Protection Law (2014); Article 1, Interpretation of the Supreme People's Court on Several Issues Concerning the Application of Law in the Trial of Environmental Civil public interest Litigation Cases (2015) (Fa Shi(2015) No. 1); Article 284, Interpretation of the Supreme People's Court on the Application of the Civil Procedure Law of the People's Republic of China (2015) (Fa Shi (2015) No. 5); Article 25(4), Administrative Litigation Law (2014); Article 13, 15, 21, Interpretation of the Supreme People's Court and the Supreme People's Procuratorate on several Issues Concerning the Application of Law to Procuratorial Public Interest Litigation Cases (2020) (Fa Shi (2018)) No. 6.

17.2.2 *The Administrative EPIL System*

With regard to administrative EPIL, the current law only authorizes the People's Procuratorate to have standing to prosecute.[4] The current Chinese law does not authorize social organizations to file administrative EPILs.

The People's Procuratorates and certain qualified social organizations are granted 'plaintiff status' in civil EPIL. In terms of administrative EPIL, only the People's Procuratorate enjoys plaintiff status. It is worth noting here that the People's Procuratorate is entitled to be a plaintiff in both civil EPIL and administrative EPIL. This is precisely why the defendant Mr. Wang argued that the Yingkou Procuratorate should file an administrative EPIL.

17.3 Deviation of Means from Ends: Shortcomings of the Civil EPIL System

As mentioned above, Article 58 of the Civil Procedure Law authorizes both people's procuratorates and certain qualified social organizations to file a civil EPIL in court. In the context of the Wang Yunlin case and a large number of similar civil EPIL cases, both authorizations seem to be made to protect the public interest in the environment. However, they have actually caused a situation which has a detrimental effect on the proper functioning of the state governance system, for the reasons that will be discussed below.

17.3.1 *The Legitimacy of Authorizing the People's Procuratorate to Initiate Civil EPIL Is in Doubt*

In reality, a fact that cannot be ignored in China is that a somewhat competent administrative authority must be available to take action against the alleged illegal acts of the defendant in a civil EPIL, given the extensive and full administration of the of the country within its territory. For example, in the Wang Yunlin case, the Gaizhou Water Conservancy Bureau was such an administrative authority. It enjoys

[4] Article 25, paragraph 4, of the Administrative Litigation Law provides that "where the people's procuratorate finds in the performance of functions that any administrative authority assuming supervision and administration functions in such fields as the protection of the ecological environment and resources, food and drug safety, protection of state-owned property, and the assignment of the right to use state-owned land exercises functions in violation of any law or conducts nonfeasance, which infringes upon national interest or public interest, it shall offer procuratorial recommendations to the administrative authority, and urge it to perform functions in accordance with the law. If the administrative authority fails to perform functions in accordance with the law, the people's procuratorate shall file a lawsuit with the people's court in accordance with the law".

pre-emption in administrative jurisdiction for the alleged illegal acts committed by Mr. Wang. This is not only a jurisprudential, but also a constitutional requirement. The Constitution of the PRC provides that: "Local people's governments at or above the county level shall, according to law, administer economic, educational, scientific, cultural, public health, sports, urban and rural development, finance, civil affairs, public security, ethnic affairs, judicial administration, family planning and other administrative work within their respective administrative areas".[5] This constitutional provision establishes the priority position of executive power in the management of public affairs in society. The public affairs such as embankment safety, river flooding control and river ecological environmental protection involved in this case are undoubtedly within the jurisdiction of the local government department concerned (such as the Gaizhou City Water Conservancy Bureau), and not the responsibility of the local people's procuratorate.

In the field of environment and resource protection, as in all other fields of public goods, the basic, first and universal relationship among the major actors or players is the regulatory relationship between administrative authorities and regulatees. This relationship involves the control of negative externalities caused by industrial and business operations and the provision and guarantee of environmental public goods. In the Wang Yunlin case, the relationship between Mr. Wang and the Gaizhou City Water Conservancy Bureau is such a relationship. Wang's excavation of the square pond in the river should have been, and was in fact, under the full jurisdiction of the Gaizhou Water Conservancy Bureau in the first place. The relationship between the plaintiff (Yingkou People's Procuratorate) and the defendant (Mr. Wang) is a judicial relationship which is very different in nature and function. It is a relationship subsidiary to the regulatory relationship. As a judicial relationship, it exists for safeguarding the successes of the regulatory relationship in a way of case-by-case judication. It is supposed to exist as a safeguard for the proper functioning of the regulatory relationship between Mr. Wang and the Water Conservancy Authority. In political life, the regulatory relationship and the judicial relationship have different statuses and functions and must not be confused with each other; otherwise, the functioning of the state apparatus would be in disarray. However, as authorized by Article 58 of the Civil Procedure Law, the People's Procuratorates and certain qualified social organizations can file civil EPILs against alleged violators of environmental laws, without involving the relevant administrative authorities. Once such proceedings are instituted, the judicial relationship, as the second "backstop"/"last line of defence", improperly replaces the regulatory relationship. One scholar has pointed out that "the dramatic shift in the role of judicial power in environmental public interest issues is likely to lead to the intervention of the judicial power into the scope of the executive power, resulting in confusion between the roles of judicial power and administrative power, and causing tension and even conflict between the two in terms of the division of competence and functional positioning.[6]

[5] Article 107, The Constitution of the PRC.

[6] Wang (2016a), pp. 49–68.

In this case, for the purpose of restoring the ecological environment of the damaged river, the Yingkou Procuratorate chose to file a civil EPIL against the farmer Mr. Wang under Article 58 of the Civil Procedure Law, instead of filing an administrative EPIL against the Yingkou Water Conservancy Bureau as Mr. Wang had argued for. The Yingkou Intermediate Court accepted the case and ruled against the defendant Mr. Wang. Assuming that Mr. Wang fully implemented the judgment of the Yingkou Intermediate Court, backfilled the three ponds, removed the dams surrounding them and restored the ecological environment of the river, the purpose of this civil EPIL would have been achieved perfectly. This seems to be an outcome that is beneficial to all parties (including the community at large). However, the plausibility of this outcome is questionable from the point of view of the soundness of the state governance system. This positive outcome cannot conceal the following four absurdities:

17.3.1.1 Arrogating Executive Power

In this case, the Yingkou Procuratorate "rushed to the front line of environmental supervision and enforcement" (in the words of the defendant Mr. Wang) by filing a civil EPIL in the court. By filing this case, the Yingkou Procuratorate in fact arrogated the administrative power of the Gaizhou Water Conservancy Bureau. It caused the replacement of the regulatory relationship, which is primarily focused on the management of social and public affairs, with a judicial relationship which is supposed to be the secondary, backup relationship behind the regulatory relationship. According to the judgment in this case, the Water Conservancy Bureau's Water Supervision Branch in Gaizhou City, which is under the jurisdiction of Yingkou City, had conducted three on-site inspections on Mr. Wang's square ponds over a period of more than 2 years from 2017 to 2019. Among the three on-site inspections, the local police station even took part in the third inspection. Knowing that the Gaizhou City Water Conservancy Bureau had already taken administrative measures against Mr. Wang, the Yingkou Procuratorate chose to file a civil EPIL, bypassing the Water Conservancy Bureau, and confronted the defendant Mr. Wang directly in court. Although the Yingkou Procuratorate had won the case and achieved its aims of forcing Mr. Wang to restore the area to its original state and to safeguard the security of the river flow, which sounds perfect in terms of morality, it actually constituted a negation and arrogation of the local government's administrative power over water conservancy. The original duty of the People's Procuratorate is to supervise the trials of the courts and to prosecute suspected criminals. When the People's Procuratorate, in performing its duties, discovers malfunctions of local regulatory authorities, such as the malfunctions of the Gaizhou Water Conservancy Bureau, it should remind and urge the administrative organ to solve any problems itself, instead of overstepping the regulatory authority and doing the job instead.

In this case, Mr. Wang was suspected of endangering the safety of the embankment, obstructing river flooding, and damaging the water ecological environment. Those suspected wrongs undoubtedly fall within the jurisdiction of the local administrative organs (the Gaizhou City Water Conservancy Bureau), rather than the jurisdictions of the People's Procuratorate and People's Court. One scholar has pointed out that "public welfare protection is the primary duty of administrative organs. Even in public interest litigation, their primary duty cannot be completely exempted".[7] Another scholar has stated that by granting the Procuratorate the status of a plaintiff in civil EPIL, it will "disrupt China's current legal system", "distort the procedural structure of civil litigation", and "lead to the infinite expansion of Procuratorate".[8] Another scholar notes that "this use of private law norms in civil EPIL to judge the public responsibility of actors for violations of public law norms does not make logical sense at all".[9] A further scholar has pointed out that "administrative power is the power to manage, and judicial power is the power to judge, and they should not be confused with each other".[10] Thus, Article 58 of the Civil Procedure Law confuses the distinction between judicial power (including procuratorial power and judicial power) and administrative power by granting the procuratorial authorities the status of plaintiff in civil EPIL, which is suspected of allowing the procuratorial authorities to arrogate administrative power.

17.3.1.2 Weakening the Executive Power

In reality, matters of pollution and damage to the environment and ecological conservancy are often under the competence of the administrative authority. The government's ecological and environmental departments, forestry departments, land departments, water departments, marine administration departments, etc. are all such administrative authorities. In the Wang Yunlin case, the matter of Mr. Wang's excavation of square ponds is within the competence or jurisdiction of the Gaizhou City Water Conservancy Bureau and its Water Affairs Inspection Authority. But this administrative power was undermined by the intervention of the Yingkou Procuratorate and the Yingkou Intermediate Court through the civil EPIL. The Yingkou Procuratorate bypassed the Gaizhou City Water Conservancy Bureau and its Water Affairs Inspection Authority and went straight to file a civil EPIL against Mr. Wang. The Yingkou Intermediate Court also ignored the existence of the Gaizhou City Water Conservancy Bureau and accepted the case. Under this situation, the competent administrative authority in charge of local water affairs—the Gaizhou Water Conservancy authority—has thus become an onlooker on local public water affairs. The administrative power on public water affairs therefore became null and void. Under such a situation one may wonder what is the point of the existence of the Water Conservancy Authority? Why should the people pay taxes to support such an onlooker?

[7] Gong (2019a), pp. 127–147.
[8] Zhang (2011), pp. 134–140.
[9] Gong (2019b), pp. 106–118.
[10] Bo (2020), pp. 34–47.

17.3.1.3 Tolerating Idle Administration

In the Wang Yunlin case, instead of urging the Gaizhou City Water Conservancy Bureau to take administrative measures against Mr. Wang through an administrative EPIL, the Yingkou Procuratorate turned a blind eye to the inaction of the Water Conservancy Bureau and went straight to file a civil EPIL against Mr. Wang. In doing so, the Yingkou Procuratorate had tolerated the idle administration of the Gaizhou City Water Conservancy Bureau. Similarly, the acceptance and hearing of the civil EPIL brought by the Yingkou Procuratorate in the Yingkou intermediate Court had the same effect of tolerating and indulging the idle administration. There is no doubt that such tolerance and indulgence should not exist, since one of the main duties of the People's Procuratorate is to supervise administrative acts. The courts also have the duty to supervise administration through administrative proceedings.

17.3.1.4 Improper Expansion and Dysfunction of Judicial Power

In the Wang Yunlin case, the Yingkou Intermediate Court had become a "collaborator" of the Yingkou Procuratorate in arrogating administrative power by accepting the case. Without the trial and verdict of the Yingkou Intermediate Court, this "arrogation" by the Yingkou Procuratorate would not have been possible. This "cooperation" of the Yingkou Intermediate Court is also wrong from the perspective of distinguishing the executive power from judicial power and the modesty of judicial power. In this case, although Mr. Wang solemnly raised the issue of the arrogation of the Procuratorial power over the administrative power in his defense, due to the institutional arrangement of Article 58 of the Civil Procedure Law, not only is the choice of the Yingkou Procuratorate for a civil EPIL beyond reproach, but also the Yingkou Intermediate Court cannot be blamed. In addition, Article 15 of the Interpretation of the Supreme People's Court and the Supreme People's Procuratorate on Several Issues Concerning the Application of Law to Procuratorial Public Interest Litigation Cases (2020) (Fa Shi (2018) No. 6) provides that "The People's Procuratorate shall file a civil public interest lawsuit in accordance with the provisions of paragraph 2 of Article 58 of the Civil Procedure Law. ... The people's court shall register the case". This means that the People's courts must accept civil EPILs brought by the People's Procuratorates, regardless of whether the filing of such litigation would give rise to the aforementioned serious problems. This provision binds the courts and procuratorates together, making the courts co-operators in arrogating executive power. It leads to an improper expansion of judicial power. The courts lose the ability to identify improper legislation and uphold legislative justice in such circumstances. It inevitably leads to dysfunction of the judiciary.

17.3.2 The Rationality of Authorizing Social Organizations to File Civil EPIL Is Also in Doubt

Just as the People's Procuratorate is different from the administrative authorities, "it is indisputable that environmental organizations are less representative of the public interest than administrative authorities".[11] In addition to authorizing the People's Procuratorate to file civil EPILs, Article 58(1) of the Civil Procedure Law also authorizes social organizations qualified under the law to file civil EPILs in courts. Like the abovementioned authorization for the People's Procuratorate, this authorization is also inappropriate (although it has been welcomed by environmental social organizations).

Let us assume that in the Wang Yunlin Case, a legally qualified environmental social organization in Yingkou City files a civil EPIL against Mr. Wang. In this case, the social organization would play the same role as the Yingkou Procuratorate as a plaintiff in the case. By filing the lawsuit, the social organization would also arrogate the power of the Gaizhou City Water Conservancy Bureau, and go straight to court against Mr. Wang. This situation and its impacts are not dissimilar to that of the civil EPIL brought by the Yingkou Procuratorate. In this hypothetical scenario, the only difference is that the plaintiff is changed from the People's Procuratorate to a social organization. Therefore, the absurdities caused by social organizations filing civil EPIL is no different from that caused by the abovementioned civil EPIL filed by the People's Procuratorate. Those four absurdities, as mentioned above, are still there.

To summarize, when filing a civil EPIL under Article 58 of the Civil Procedure Law, either the People's Procuratorate or the social organizations qualified by law will necessarily give rise to the four absurdities mentioned above. These absurdities reveal that, as a means, the institutional design of Article 58 of the Civil Procedure law does not achieve the desired outcome or purpose of a civil EPIL.[12] According to the State Legislature (the Standing Committee of the National People's Congress), the purpose of giving the People's Procuratorate the standing to file civil EPILs is to "strengthen the protection of national interests and public goods".[13] In addition, according to the Supreme People's Procuratorate, the purpose of giving the People's procuratorates the standing to file civil EPILs is to "give full play to the role of legal supervision by procuratorial organs".[14] However, Article 58 of the Civil Procedure Law, which grants procuratorial authorities and social organisations standing as plaintiffs in civil EPILs, runs counter either to the purpose of the congressional authorization or to the purpose of public interest litigation itself. This means that a

---

[11] Hu (2016), pp. 168–176.

[12] Wang (2016b), pp. 101–114.

[13] Decision of the Standing Committee of the National People's Congress on Authorizing the Supreme People's Procuratorate to Launch the Pilot Program of Initiating Public Interest Actions in Certain Areas, [EB/OL] (2015-07-01).

[14] Supreme People's Procuratorate Plan for the Pilot Project of Reform of Instituting Public Interest Litigations by the Procuratorial Organs, [EB/OL] (2015-07-02).

civil EPIL, whether it is filed by the People's Procuratorate or by environmental social organizations, is bound to have an effect that is in contrary to the purposes of either the National Congress or the civil EPIL itself. Viewed this way, the civil EPIL system is a pseudo-proposition which violates the scientific laws of a litigation system.

17.4 Effective Interface Between Procuratorial and Executive Powers: The Administrative EPIL System

In the Wang Yunlin case, Mr. Wang argued that "Once the procuratorial organ rushed to the front line of environmental supervision and enforcement, there is a risk of confusing the boundaries of state powers. China's public interest litigation should be mainly administrative EPIL filed by the procuratorial organs ... This is more in line with the supervisory role of the procuratorial organs, and gives the regulatory power to the environmental protection authorities of the government. When environmental regulatory authorities are reluctant to perform their duties and procuratorial organs do not include them as defendants, people will lose their confidence in the regulatory administrations. Therefore, it is necessary to strengthen the accountability of the regulatory administration, in order to make the regulatory administrations to operate in accordance with the law and to enhance their credibility". Such an argument, although not accepted by the court, was not without merit.

Assuming that the Yingkou Procuratorate had filed an administrative EPIL for Mr. Wang's illegal acts, all the absurdities arising from the filing of a civil EPIL would not exist.

First, there would be no arrogation of executive power. As far as the Yingkou Procuratorate is concerned, because what it filed in the court will be an administrative EPIL, instead of a civil EPIL, it is performing its duty in supervision over the regulatory administrations. A Judicial Interpretation issued by the Supreme People's Court and the Supreme People's Procuratorate provides a sound procedure for the People's Procuratorate to file an administrative EPIL.[15] Article 21 of the Judicial Interpretation has sound provisions for the People's Procuratorates to file administrative EPILs. According to the Article, if the Yingkou Procuratorate had chosen to file an administrative EPIL, what it would have had to do first was to issue a Procuratorial Recommendation to the Gaizhou City Water Conservancy Bureau, urging it to take administrative measures against Mr. Wang. If the Gaizhou City Water Conservancy Bureau failed to take the administrative measures in accordance with the law after receiving the Procuratorial Recommendation from the Yingkou

[15] Article 21, Interpretation of the Supreme People's Court and the Supreme People's Procuratorate on several Issues Concerning the Application of Law to Procuratorial Public Interest Litigation Cases (2020) (Fa Shi (2018) No. 6).

Procuratorate, the Yingkou Procuratorate may file an administrative EPIL against it in the court. In the former case, the Yingkou Procuratorate urges the Gaizhou City Water Conservancy Bureau to perform its duty of regulation through a Procuratorial Recommendation. In the latter case, by the administrative EPIL, the Yingkou Procuratorate compels the Gaizhou City Water Conservancy Bureau to carry out its regulatory duty by filing an administrative EPIL. Therefore, in both cases, the correction of Mr. Wang's violations was achieved through the direct intervention of the regulatory authority—the Gaizhou Water Conservancy Bureau. In this way, i.e., the way of administrative EPIL, the Yingkou Procuratorate was not suspected of overstepping the regulatory power of the Gaizhou Water Conservancy Bureau, and a smooth and effective interface between the procuratorial power and the executive powers was achieved. What the Yingkou Intermediate Court trialed is a typical lawsuit against a governmental authority (the Gaizhou Water Conservancy Bureau), instead of a somewhat questionable civil EPIL, and there will be no question of arrogation of executive power.

Secondly, there would be no weakening of executive power. In this case, the Yingkou Procuratorate forced the Gaizhou City Water Conservancy Bureau to perform its regulatory duty either through a Procuratorial Recommendation or through an administrative EPIL, instead of arrogating its regulatory power by filing a civil EPIL. The regulatory power of the Gaizhou City Water Conservancy was fully respected. In this way, Mr. Wang corrected his violations under direct regulation of the Gaizhou City Water Conservancy Bureau and its inspection authority. Thus, the Gaizhou Water Conservancy Bureau willfully exercised its regulatory power and achieved the desired results of the regulation. The regulatory power of the Gaizhou Water Conservancy Bureau was maintained.

Thirdly, there will be no toleration to idle administration. As the Yingkou Procuratorate files an administrative EPIL, the purpose of the lawsuit will be to urge the Gaizhou City Water Conservancy Bureau and its subordinate inspection authority to take enforcement measures against Mr. Wang's illegal acts. In this case, neither the Yingkou Procuratorate nor the Yingkou Intermediate Court tolerates the idle administration of the Gaizhou City Water Conservancy Bureau, if there is any such idle administration.

Fourth, there will be no improper expansion and dysfunction of judicial power. As far as the Yingkou Intermediate Court is concerned, since what it will try is an administrative EPIL, it will not confront the issues of constitutional legitimacy and reasonableness contained in civil EPILs. Therefore, it will not run the risk of improper expansion and dysfunction of judicial power.

Similarly, if the law authorizes social organizations to file administrative EPILs, the legal consequences of such lawsuits will be as flawless as if they were filed by the People's Procuratorate, and there will be no absurdities such as arrogating regulatory power. Unfortunately, the current law does not authorize social organizations to file administrative EPILs. This leaves legally qualified social organizations with no choice but to devote their valuable resources to file civil EPILs, which are constitutionally questionable.

In summary, this and other civil EPILs all reveal the serious shortcomings of the top-level design of the civil EPIL system. These shortcomings are not conducive to the unified and effective organization of the cause of environmental protection by the state authorities. In terms of legislation, the system design of civil EPIL is not only superfluous, but also constitutionally questionable. The four absurdities revealed in the Wang Yunlin case show that although civil EPILs may result in a beneficial outcome (as in this case, Mr. Wang restored the river benchland to its original state), the result was achieved through incorrect means. This incorrect means is the civil EPIL system created and authorized by Article 58 of the Civil Procedure Law. From the perspective of improving the state governance system, the political, economic and social costs of this incorrect means are too high. It not only sacrifices the harmonious links among the basic components of the state apparatus, namely, the procuratorial, judicial and executive powers, but also, as argued by Mr. Wang, devalues the credibility of these state powers. Therefore, the system of civil EPILs is undesirable.

In contrast, the administrative EPIL system is a correct litigation system and is consistent with the scientific rules of the public interest litigation system. The administrative EPIL system has a sound constitutional basis. The current laws and judicial interpretations not only fully authorize the People's Procuratorate to file administrative EPILs, but also set forth a sound and operable procedure for the administrative EPILs.

17.5 Conclusions and Recommendations

17.5.1 Conclusions

The modernization of the governance system and governance capacity of the state requires a holistic hard look at the public interest litigation system. "This holistic examination inevitably leads to concern and reflection on the rationality of the top-level design of the environmental governance system and the social costs of environmental governance."[16] The complexity and risks inherent in modern society dictate that professional social and public affairs should be left to the professional regulatory authorities. The duties of the People's Procuratorate and social organizations are clearly different from those of the regulatory authorities. The powers and duties of the regulatory authorities should be fully respected. The law should guarantee the effective exercise of regulatory power in accordance with the law. The provisions of Article 58 of the Civil Procedure Law and related legal documents on civil EPIL have caused the four absurdities as mentioned above in judicial practice. Although the civil EPILs may provide some specific remedies to the public

[16] Wang (2020), pp. 12–18.

environmental good, the overall political and economic costs are too high and often do more harm than good.

17.5.2 Suggestions for Improvement

Whereas the system design of the current law and judicial interpretation for the People's Procuratorate to file administrative EPILs is scientific and reasonable as mentioned above,

Whereas the administrative EPIL system can completely replace the system of civil EPIL and achieve the purposes of the civil EPIL system with lower social costs and higher political benefits,

Whereas the system of administrative EPIL is in line with the goal of building a government under the rule of law and the goal of modernization of the state governance system and governance capacity,

By filing an administrative EPIL, legally qualified social organizations can do the same good as that of the People's Procuratorate for the public environment and ecological system.

The suggestions for improving the current public interest litigation system are as follows:

Firstly, amending Article 58 of the Civil Procedure Law and related legal documents so as to abolish the civil EPIL system and withdraw the authorization for People's Procuratorates and social organizations to file civil EPILs.

Secondly, amending the Administrative Litigation Law and the Environmental Protection Law and other relevant legal documents, so as to, on the one hand, clearly grant legally qualified social organizations the standing to file administrative EPILs, and, on the other hand, to require such social organizations to formally notify the competent regulatory authority about the subject matter of the administrative EPIL before filing the administrative EPIL, so as to give the regulatory authority a period of time to perform its administrative duties against the alleged violations of law involved in the case. Only when the period of notification has expired should the social organization be permitted to file the administrative EPIL in court.

References

Bo X (2020) Form the ternary mode to the binary mode - on the reconstruction of litigation system for environmental public interest. J China Univ Geosci (Soc Sci Edn) 4:34–47

Decision of the Standing Committee of the National People's Congress on Authorizing the Supreme People's Procuratorate to Launch the Pilot Program of Initiating Public Interest Actions in Certain Areas. http://www.law-lib.com/law/law_view.asp?id=506525. Accessed 25 Aug 2022

Gong G (2019a) Reflections on the positioning of the nature of civil EPIL. Leg Stud 3:127–147

Gong Xuede (2019b) On the repeated liability in civil EPIL after the sanction of public law. Adm Law Stud 5:106–118. http://www.law-lib.com/law/law_view.asp?id=511934. Accessed 25 Aug 2022
Hu Jin (2016) On functional orientation of public welfare lawsuits by environmental protection organizations. Law Rev 4:168–176
Supreme People's Procuratorate Plan for the Pilot Project of Reform of Instituting Public Interest Litigations by the Procuratorial Organs
Wang M (2016a) On the development direction of environmental public interest litigation: analysis on administrative right and jurisdiction. China Law 1:49–68
Wang X (2016b) On the legislative order of the environmental public interest litigation system. Tsinghua Jurisprud 6:101–114
Wang X (2020) On conceptual model for China's environmental governance: a new paradigm tool. Environ Protect 03-04:12–18
Wang Yunlin Case (2019) (Liao 08 Min Chu No. 5) https://wenshu.court.gov.cn/website/wenshu/181217BMTKHNT2W0/index.html?pageId=354421a6024db2b333f1d17d2a4d60d6&s8=03. Accessed 21 May 2022
Zhang L (2011) Procuratorates are not suitable as plaintiffs in public interest litigation. Law 1:134–140

Index

X. Wang et al. (eds.), *Environmental Public Interest Litigation in China*,
https://doi.org/10.1007/978-3-031-26526-6

GPSR Compliance

The European Union's (EU) General Product Safety Regulation (GPSR) is a set of rules that requires consumer products to be safe and our obligations to ensure this.

If you have any concerns about our products, you can contact us on ProductSafety@springernature.com

In case Publisher is established outside the EU, the EU authorized representative is:

Springer Nature Customer Service Center GmbH
Europaplatz 3
69115 Heidelberg, Germany

Batch number: 10755214

Printed by Printforce, the Netherlands